Computer-aided Analysis of Electric Machines

Computer-aided Analysis of Electric Machines

A *Mathematica* approach

Vlado Ostovic

Prentice Hall
New York   London   Toronto   Sydney   Tokyo   Singapore

First published 1994 by
Prentice Hall International (UK) Limited
Campus 400, Maylands Avenue
Hemel Hempstead
Hertfordshire, HP2 7EZ
A division of
Simon & Schuster International Group

© Prentice Hall International (UK) Limited (1994)

All rights reserved. No part of this publication may be reproduced,
stored in a retrieval system, or transmitted, in any form, or by any
means, electronic, mechanical, photocopying, recording or otherwise,
without prior permission, in writing, from the publisher.
For permission within the United States of America
contact Prentice Hall Inc., Englewood Cliffs, NJ 07632

Printed and bound in Great Britain
at the University Press, Cambridge

Library of Congress Cataloging-in-Publication Data

Ostovic, Vlado
  Computer-aided analysis of electric machines : a Mathematica
approach / Vlado Ostovic.
     p.   cm.
  Includes bibliographical references and index.
  ISBN 0-13-068859-2
  1. Electric machinery—Mathematical models.   2. Mathematica
(Computer program)   I. Title.
TK2211.078   1994
621.31′042′015118–dc20   93–23664
                                                              CIP

British Library Cataloguing in Publication Data

A catalogue record for this book is available from the British Library

ISBN 0-13-068859-2 (pbk)

1 2 3 4 5   98 97 96 95 94

*To Tea, Marko and Thomas
for their love and support*

Contents

Foreword xi

Preface xiii

1. Introduction 1
1.1 Application of Maxwell's Equations in Electric Machine Analysis 1
1.2 Magnetic Circuit of an Electric Machine 18
1.3 Parameters of Conductors in Electric Machines 51

2. Windings, Currents and Air Gap Magnetomotive Force 76
2.1 Classification and Basic Terms 76
2.2 Linear Current Density – Current Sheet 79
2.3 Magnetomotive Force of a Coil and Winding 86
2.4 Winding Factor 96
2.5 Matrix Representation of the Winding Magnetomotive Force 103
2.6 Time Varying Excitation 111
2.7 Generation of the Rotating Field 117
2.8 Representation of the Air Gap Magnetomotive Force in a Rotating Reference Frame 128
2.9 Commutator Windings 131
2.10 Squirrel Cage Winding 137

3. Air Gap Magnetomotive Force and Flux Density 143
3.1 d–q Representation of Spatial Quantities 145
3.2 Influence of Slotting on the Air Gap Magnetomotive Force and Flux Density – Carter Factor 147
3.3 The Total Air Gap Magnetomotive Force and the Air Gap Flux Density in a Commutator Machine 156
3.4 The Total Air Gap Magnetomotive Force and the Air Gap Flux Density in a Synchronous Machine 167

3.5 The Total Air Gap Magnetomotive Force in an Induction Machine ... 176

4. An Electric Machine as a Circuit Element ... 185
4.1 Main and Leakage Inductance Concept ... 185
4.2 The Main Inductance of a Coil and a Winding in a Slotless, Cylindrical, Unsaturated Machine ... 189
4.3 The Main Inductance of a Winding in a Slotted, Cylindrical, Saturated Machine ... 191
4.4 The Main Inductance of a Winding in an Unsaturated Machine with Variable Air Gap Geometry ... 198
4.5 Mutual Inductance between the Windings in a Slotless, Unsaturated Machine ... 204
4.6 Influence of Slotting on Both Sides of the Air Gap on the Main and Mutual Inductances ... 211
4.7 Equivalent Circuit Representation of Windings in Electric Machines ... 220
4.8 Induced Voltages in the Windings of Electric Machines ... 230

5. Force and Torque in Electric Machines ... 235
5.1 The Role of Magnetic Energy in Electromechanical Energy Conversion ... 235
5.2 Force on the Conductors in the Slots of an Electric Machine ... 249
5.3 Torque Produced by the Currents in the Windings – the Torque Function ... 254
5.4 Electromagnetic Torque as a Function of the Air Gap Quantities ... 277

6. Steady State Performance of Induction Machines ... 280
6.1 Construction and Principles of Operation ... 280
6.2 Effects of the Fundamental Spatial and Time Air Gap Flux Harmonics in an Induction Machine ... 282
    6.2.1 Equivalent Circuit for the Fundamental Spatial Harmonic ... 282
    6.2.2 Distribution of Power and Torque–Slip Characteristic ... 285
6.3 Effects of Higher Spatial Harmonics of Flux Density in the Air Gap of an Induction Machine ... 304
6.4 Self Excitation of an Induction Machine ... 307
6.5 Single Phase Induction Machine ... 318
6.6 Capacitor Braking of Single Phase Induction Machines ... 331
6.7 Speed Control of Induction Machines ... 337
    6.7.1 Wound Rotor Machines ... 337
    6.7.2 Squirrel Cage Machines ... 342

7. Steady State Performance of Commutator Machines ... 346
7.1 Principles of Operation of a DC Commutator Machine ... 346
7.2 Induced Voltage and Electromagnetic Torque ... 349
7.3 Armature Reaction ... 354
7.4 Commutation ... 356

Contents   ix

    7.5 D.C. Generators   358
    7.6 D.C. Motors   365
    7.7 A.C. Commutator Motors   368
    7.8 Speed Control of Commutator Motors   369

8. Steady State Performance of Synchronous Machines   374
    8.1 Constructional Features   375
    8.2 Armature Reaction and the Synchronous Reactance Concept   377
    8.3 Performance of a Cylindrical Rotor Synchronous Machine
        Operating on an Infinite Bus   379
    8.4 Salient Pole Machine   392
    8.5 Permanent Magnet Synchronous Motor   395

9. Fundamentals of Electric Machine Dynamics   397
    9.1 A Machine's Differential Equations   400
    9.2 Transient Phenomena in Commutator Machines   402
        9.2.1 Separately Excited Motors   402
        9.2.2 Shunt Motors   406
        9.2.3 Series Motors   409
    9.3 Transient Phenomena in Induction Machines   411
    9.4 Transient Phenomena in Synchronous Machines   424
        9.4.1 Physics of Transients in a Synchronous Machine   424
        9.4.2 Park's Equations   429

Appendix   434
    A 1.1   434
    A 1.2   436
    A 1.3   442
    A 2.3   445
    A 2.4   446
    A 2.5   446
    A 2.6   450
    A 2.7   451
    A 2.8   455
    A 2.9   458
    A 2.10   460
    A 3.2   461
    A 3.3   466
    A 3.4   473
    A 3.5   482
    A 4.3   487
    A 5.3   492
    A 6.2.2   496
    A 6.5   499

A 8.3	504
A 9.2	505
A 9.3	509
References	513
Index	515

Foreword

It may come as a surprise to many that electrical machines and the evolution of computers and computing tools are inextricably linked. Although the first digital computer was proposed by Babbage, the first true working computer was designed and built by James Thompson, Lord Kelvin's brother, in 1876. This machine, which used a ball and disc integrator, was capable of solving a second order non-linear differential equation. When this machine was finally perfected by Vannevar Bush in 1925 the very first application was for the analysis of pulsating torques of a synchronous motor-compressor set [1926]. By the early 1940s when digital computers appeared, these *mechanical differential analyzers* were already being used to simulate application problems involving the transient behaviour of electrical machines. Shortly after World War II, the digital computer made its entry and was almost immediately used to study the load flow problem by Jennings and Quinlan in 1946. As digital computational tools evolved, so did the tools for the analysis of electrical machines. Researchers in electrical machines have contributed greatly to today's powerful finite element, finite difference and boundary element methods for solving partial differential equations.

With this book Vlado Ostovic continues this symbiosis by helping pioneer a new direction in the use of computers, that of using Mathematica® as a teaching tool. The subject of electrical machines is perhaps unique in electrical engineering in that an understanding of its behaviour requires a deep appreciation of both its spatial and temporal attributes and how these two phenomena are linked. Certainly one of the perennial challenges to any teacher of electrical machines is the need to illustrate clearly the differences between time and space harmonics: a task that Dr Ostovic illuminates in exemplary fashion.

Our usual form of teaching electrical machines requires a degree of abstraction which challenges even the most talented. It is, perhaps, a major contributor to the decrease of interest in this subject over the past several decades. Consider the fact that we study a three dimensional, cylindrical, rotating device which is severely non-linear having numerous oddities such as teeth, slots, projecting poles, winding coils, etc., and attempt to represent, to the student's satisfaction, both in transient and steady state behaviour by the use of a few arrows (vectors) on a piece of a paper. What a leap of faith!

By actively using this book, perhaps with other teaching material, the student is led gently, layer by layer, into the mysteries of electrical machinery. Furthermore, Dr Ostovic makes the road one of discovery rather than an exercise in memory retention. Even the hardened expert should take great delight in seeing, for the first time, the MMF waveform of an AC machine dance across her/his computer screen! Enjoy.

T.A. Lipo
University of Wisconsin
Madison, WI
4 July 1993

Preface

Among the courses in the electrical engineering curriculum, the one which deals with electric machines seems to be one of the most demanding for the students. This is probably the main reason for a permanent decrease in the number of electric machines students all over the world. Simply following one of the most general principles of nature, articulated in electrical engineering as Ohm's law, the majority of students choose low resistance paths through the curriculum, leaving courses about machines to the most persistent among them.

As in no other course, the content of an electric machine course demands a profound understanding of the physical principles of electromechanical energy conversion, along with skills in various mathematical disciplines: trigonometry, numerical analysis, ordinary and partial differential equations, etc. The physical behaviour of the machine is often hidden behind the complex mathematical apparatus which describes it, and it is sometimes difficult to avoid the *deus ex machina* explanations of a machine's performance. There are numerous examples which illustrate this approach, including harmonic torque components, air gap flux waves, self excitation of saturated machines, dynamic behaviour, and many others. An experience in teaching machines shows that students very often miss the physical point while fighting with the mathematical description. Therefore, a kind of *big brother*, capable of answering all kinds of maths questions, has been needed for a long time in the area of machine analysis.

The advent of affordable personal computers on which highly sophisticated mathematical software such as Mathematica® from Wolfram Research can be executed, promises a change in the teaching of electric machines. Besides its ability to answer practically every analytical or numerical question which can arise in the analysis of electric machines, this software has enormous graphic features, starting from drawing 2D and 3D diagrams to animation. These properties of Mathematica® software were an inspiration to write a book about electric machines in an unusual way, explaining the physics of a machine's behaviour with the idea in mind that every mathematical formulation which results from physical analysis can be solved with a minimum of effort. Mathematica® notebooks, in which mathematical support of physical explanation is given have been provided, enabling one to repeat all the results obtained in the book. In accordance with the structure of the Mathematica® language, the enclosed software can be easily changed to meet students' needs, without the necessity to recompile the program. Besides various modes of three-dimensional graphical representation, this software enables animation of curves and diagrams. This type of representation, possible only on the computer, allows the air gap wave quantities to appear in front of the reader's eyes in their full guise.

The book is divided into nine chapters and the Appendix, which includes Mathematica® software. The basic physical effects in electric machines are analyzed in the first chapter. The second chapter concentrates on the most important part of the machine – its windings, and the effects of the currents flowing through them. In the third chapter the air gap mmfs and flux densities in various types of machines are analyzed. The fourth chapter deals with the electric machine as seen by the power system, or by the source connected to its terminals. The torque expressions as functions of the machine's type and mmf, flux and inductance harmonics are derived in the fifth chapter. Steady state performances of induction, commutator, and synchronous machines are analyzed in Chapters 6, 7, and 8, respectively. In the ninth chapter, the fundamentals of the dynamics of all types of machines are analyzed. The Appendix is organized to follow the content of the chapters. The Mathematica® software used to obtain the results in the first chapter is given in the first section, A1, of the Appendix; the software for the second chapter is listed in the second section, A2, of the Appendix, etc. An IBM PC as well as Apple Macintosh disk containing the notebooks listed in the Appendix are available upon request from the publisher

I wish to express deep gratitude to my dear friends Victor and Cynthia Stefanovic, who were involved in creating this book from the very beginning. Victor had many useful suggestions regarding electric drives aspects of application of machines. Cynthia spent many hours of invaluable work translating the raw manuscript into readable English. Last, but not least, Tom Lipo's enthusiasm helped me believe that the book will find its way to the reader – a diligent student in the *Electric Machines Wonderland*.

1 Introduction

1.1 Application of Maxwell's Equations in Electric Machine Analysis

Electromagnetic phenomena in electric machines are described by Maxwell's equations. These general relations are valid at any time instant and at every point in the machine. However, depending on the medium in which the electromechanical energy conversion takes place, some electromagnetic phenomena in electric machines are more pronounced than others. Electric machines are traditionally built to utilize the magnetic field as a catalyzer in the electromechanical energy conversion. Nevertheless, micro machines which utilize the electrostatic field energy have been developed recently. Their application is extremely limited, and they will not be treated in this book. The term electric machine will be here used to denote only those machines in which the magnetic field energy is an intermediary in the energy conversion from electric to mechanical and vice versa.

The magnetic field effects dominate in the physics of the machine's operation, defining its two most important parameters: the induced voltage and the mechanical torque. The electrostatic field neither creates useful torques, nor induces voltages in the machine's windings. It must only be considered when dimensioning the winding insulation of machines.

The two field strength vectors, **E** in the electric, and **H** in the magnetic field, produce the field density vectors **J**, **D** and **B**, when applied to a medium. The relationships between the field strength and field density vectors are defined by the medium electromagnetic parameters: its *electrical conductivity* σ, *dielectric constant* ε, and *magnetic permeability* μ in the following manner:

$$\mathbf{J} = \sigma \mathbf{E} \qquad (1.1)$$

$$\mathbf{D} = \varepsilon \mathbf{E} \qquad (1.2)$$

$$B = \mu H \tag{1.3}$$

The electromagnetic parameters of the medium are tensors. The tensor nature of σ, ε, and μ causes spatial shift between the field strength and the field density vectors. The electromagnetic parameters of the medium are considered tensors in the microscopic analysis of electromagnetic fields, when they are calculated at various points in space. However, the spatial relationship between the field strength and field density vectors is seldom important in the macroscopic analysis, since the material properties of a complete medium are usually represented by only one parameter. As a rule, the parameters σ, ε, and μ are considered scalars in the macroscopic analysis of electric machines.

Equation (1.1) states that the current density magnitude J is proportional to the magnitude of the applied electrical field E. This equation is known as the basic form of Ohm's law. When the conductivity is independent of the value of the electrical field strength or the current density, the medium is said to be electrically *linear*. If, however, the conductivity varies as a function of the applied electrical field and current density, the medium is said to be electrically *nonlinear*.

When applied to an electrical insulator, the electric field creates the electrical field intensity vector **D**. The proportionality between vectors **D** and **E** is expressed in Eq. (1.2) with the dielectric constant ε. The dielectric constant is equal to the product of the relative dielectric constant ε_r, and the dielectric constant of free space ε_0:

$$\varepsilon = \varepsilon_r \varepsilon_0 \tag{1.4}$$

The dielectric constant of free space is equal to

$$\varepsilon_0 = 8.854 \times 10^{-12} \frac{As}{Vm} \text{ or } \frac{F}{m} \tag{1.5}$$

Electric current produces a magnetic field characterized by the magnetic field strength vector **H**, and magnetic field density vector **B**. Vector **H** is related to vector **B** as expressed in equation (1.3). The magnetic permeability μ is defined as the product of the relative magnetic permeability μ_r and the magnetic permeability of free space μ_0:

$$\mu = \mu_r \mu_0 \tag{1.6}$$

The magnetic permeability of free space is equal to

$$\mu_0 = 4\pi \times 10^{-7} \frac{Vs}{Am} \text{ or } \frac{H}{m} \tag{1.7}$$

Introduction

The relationship between the magnetic permeability and the dielectric constant of free space is expressed in the following equation:

$$c = \frac{1}{\sqrt{\mu_0 \varepsilon_0}} = 2.998 \times 10^8 \ \frac{m}{s} \tag{1.8}$$

with c denoting the speed of light in free space.

Relative permeability defines physically the magnetic properties of matter. *Diamagnetic* materials have relative permeability slightly below one, *paramagnetic* materials have relative permeability slightly above one, whereas the relative permeability of *ferromagnetic* materials is always a function of the applied magnetic field strength H, reaching in some cases 100,000. The categorization of materials in electrical engineering after their magnetic properties is, however, not so subtle: they are simply grouped into *magnetic* materials (iron, rare earth alloys for permanent magnets), and *nonmagnetic* materials (air, copper, aluminum, some steel alloys).

The relationship between the magnitude of the magnetic field strength vector **H** and the flux density vector **B** is usually represented in the magnetization, or B–H curve. Magnetization curves for various media are shown in Fig. 1.1.

Magnetic field strength magnitude H_0 produces in the air the flux density that can be found from the B–H curve of air in Fig. 1.1. The flux density magnitude B_{air} is represented in this case by the y-coordinate of point 1 in Fig. 1.1.

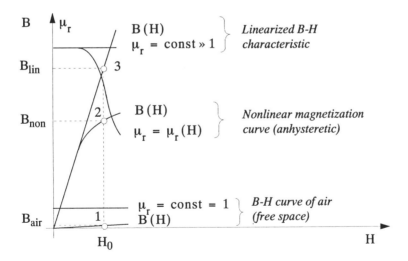

Fig. 1.1 *Magnetization, or B–H curves for various materials and corresponding relative permeability curves*

The field strength H_0 in Fig. 1.1 produces a flux density B_{non} in a magnetic material that is much higher than the flux density in the air B_{air}. The difference between B_{air} and B_{non} in reality is much greater than the one shown in Fig. 1.1, in which the B-axis scale had to be compressed for the sake of clarity. Because of the same reason the curvature of the B–H characteristic near the origin is neglected in this figure. This curvature, however, can be seen in Fig. 1.3.

Comparing the flux densities B_{air} and B_{non} one can realize why electromagnetic devices are always built using much iron: the current that creates the magnetic field strength H_0 produces incomparably higher flux densities in the iron than in the air. More flux density in a machine results in a higher value of induced voltage and higher torque, without increasing the machine's size and weight.

The nonlinear magnetization curve of a real magnetic material is very often linearized, especially when a machine is analyzed as a part of a power system, or control circuit. The linearization is usually carried out by drawing a tangent line to the nonlinear B–H curve at the origin of the coordinate system, as shown in Fig. 1.1. One can see in this figure that the flux density values B_{lin} for a linearized magnetization curve are always higher than those for a nonlinear curve, B_{non}.

Along with the B–H curves, the dependence of the relative permeability on the value of the field strength in various magnetic materials is given in Fig. 1.1. The relative permeability of iron is a nonlinear function of the field strength, as a consequence of the nonlinear B–H curve. The relative permeability μ_r of iron is constant and very high at low values of the field strength. The relative permeability starts to decrease as the iron becomes saturated, and falls to single digit values at extremely high magnitudes of the field strength.

Magnetic properties of ferromagnetics are explained in Weiss's *domain theory*, according to which a magnetic material can be visualized on the atomic level as consisting of domains – miniature permanent magnets – created by elementary currents. The elementary currents within the domains are fictitious: they produce only magnetic effects, but do not dissipate losses. The domains can be oriented in all directions in space when there is no external field. If an external magnetic field is applied to the magnetic material, a force on the domains is created, which tends to align them with the external field. This way, the domains in the magnetic material always support the external field. Such an action of domains can be expressed in terms of the flux density induced by an external field in the following manner: for a given value of the applied field strength a material with domain structure creates a flux density which is several orders of magnitude greater than in a material without domains.

The stronger the external field, the more domains align parallel with it. At extremely high values of the applied field, all domains within the magnetic material are oriented in the direction of the external field. Every further increase of the external field produces in the magnetic material the same effects as in free space, since there are no more domains which could contribute to the external field. A measure of such action of domains is the intensity of magnetization M, defined in scalar form as:

Introduction

$$B = \mu_0(H + M) = \mu_0\mu_r H \qquad (1.9)$$

The intensity of magnetization M represents the additional response of iron to an external magnetic field. The relationship between the magnitude of the vector of external magnetic field strength H and the magnitude of the vector of intensity of magnetization M is described by the modified Langevin equation [10]:

$$M_{an} \equiv M = M_s \left(\coth \frac{H + \alpha M}{a} - \frac{a}{H + \alpha M} \right) \qquad (1.10)$$

where M_s is the saturation magnetization [A/m], i.e. the maximum possible magnitude of M, α is a dimensionless parameter representing the intensity of inter-domain coupling, and a [A/m] is a material parameter which controls the shape of the magnetization curve. The curve described in Eq. (1.10) is *anhysteretic*, because it does not allow for hysteretic losses. In an anhysteretic curve only one value of H is related to a given value of M, and vice versa. The relationship between the quantities H and M in Eq. (1.10) is given implicitly. The solution procedure of this equation for typical values of parameters M, α, and a is given in the Appendix. The results of computation carried out in this section are shown in Fig. 1.2.

A finite number of domains which support the external field in the magnetic material causes *saturation*. At extremely high values of the applied field strength H, the intensity of magnetization M becomes constant, since all the domains are already oriented in the direction of the external field.

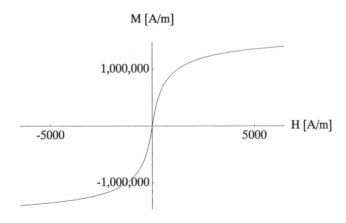

Fig. 1.2 *Langevin's function describing the relationship between the magnitude of external field strength H, and the intensity of magnetization M*

When H increases to infinity, M reaches asymptotically the value of M_s, which in the example illustrated in Fig. 1.2 is equal to 1.6×10^6 A/m.

Besides saturation, the domain structure of magnetic materials causes another effect – *hysteresis*. There exist many processes in nature which have hysteretic character. It is common for all hysteretic processes that the response of the system in which they take place is dependent not only on the parameters of the system and the characteristic of excitation, but also on the conditions in the system at a previous operating point. In our case, the value of magnetization M in a magnetic material at the current operating point due to hysteresis is dependent not only on the applied field strength H, but also on the value of magnetization M at the point from which the magnetic material arrived at the current operating point.

The hysteresis in a magnetic material is caused by friction between the domain walls. Due to friction, not as many domains are oriented in the direction of an alternating external field as in the anhysteretic case. The change of intensity of magnetization M is therefore not simultaneous with the change of field strength H. The relationship between these two quantities is described by the following differential equation [10], the solution of which gives the hysteresis loop:

$$\frac{dM}{dH} = \frac{1}{(1+c)} \frac{M_s \left(\coth \frac{H+\alpha M}{a} - \frac{a}{H+\alpha M} \right) - M}{\frac{\delta k}{\mu_0} - \alpha \left[M_s \left(\coth \frac{H+\alpha M}{a} - \frac{a}{H+\alpha M} \right) - M \right]} + \frac{c}{(1+c)} \frac{dM_{an}}{dH} \quad (1.11)$$

where

$$\frac{dM_{an}}{dH} = M_s \left[\frac{a}{(H+\alpha M)^2} - \frac{1}{a} \operatorname{csch}^2 \frac{H+\alpha M}{a} \right] \quad (1.12)$$

The coefficient c in Eqs. (1.11) and (1.12) is equal to the ratio between the slopes of the hysteretic and anhysteretic magnetization curves at the origin of the (H,B) coordinate system. The parameter δ is defined as

$$\delta = 1 \text{ for } \frac{dH}{dt} > 0 \text{ or for } M_{an} > M$$
$$\delta = -1 \text{ for } \frac{dH}{dt} < 0 \text{ or for } M_{an} < M \quad (1.13)$$

and k is determined by the energy of friction losses between domain walls.

Introduction

The solution procedure for Eqs. (1.11) and (1.12) for typical values of parameters of the magnetic material is given in the Appendix. The hysteresis curve obtained this way is shown in Fig. 1.3.

Besides the vectors of flux and current densities in a magnetic and an electric field, **B**, **D**, and **J**, one can define the scalar quantities – *fluxes*. The flux of a vector through a surface is equal to the dot product of the vector and surface area. Following this definition of flux, one can introduce

– the magnetic flux, Φ

$$\Phi = \int_S \mathbf{B} \cdot d\mathbf{S} \tag{1.14}$$

– the electric flux, Ψ

$$\Psi = \int_S \mathbf{D} \cdot d\mathbf{S} \tag{1.15}$$

– the electric current, I

$$I = \int_S \mathbf{J} \cdot d\mathbf{S} \tag{1.16}$$

S in the equations above denotes the area through which the flux of a particular vector is calculated.

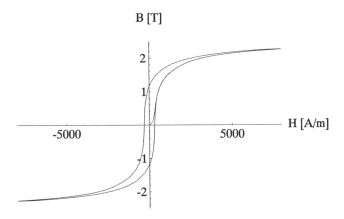

Fig. 1.3 *Hysteresis loop of a ferromagnetic material*

The conductivity of electric conductors (metals) is much higher than the conductivity of air. Therefore, the electric current flows completely through a conductor. The difference between the permeability of iron and air is, however, not as radical as the difference between the conductivity of metals and air. Hence, there always exists a component of magnetic flux passing through the air, parallel with the flux in the iron. This component of flux is denoted as Φ_{air} in Fig. 1.4 (a) and (b).

Since the relative permeability is a function of saturation, the component of flux passing through the air, Φ_{air} in Fig. 1.4 (b), increases relative to the flux Φ_{Fe} through the magnetic material as the iron permeability decreases. The more the magnetic material is saturated, the more flux passes through the air instead of through the iron.

The source of the electric field is a positive charge, and the electric field sink is a negative charge. Therefore, the lines of flux in an electric field connect the charges with opposite signs. The lines of magnetic flux, however, have no source or sink. These lines are closed around the current(s) which produce them, see Fig. 1.5.

The relationship between the magnitude of the magnetic field strength vector **H** and current density **J**, which produces the field, is defined at any point in space by Ampère's circuital law

$$\nabla \times \mathbf{H} = \mathbf{J} \qquad (1.17)$$

which is in electric machines analysis more commonly used in its integral form

$$\oint \mathbf{H} \cdot d\mathbf{l} = \sum I = F \qquad (1.18)$$

The sum of currents ΣI in Eq. (1.18) is called *the magnetomotive force (mmf)* F.

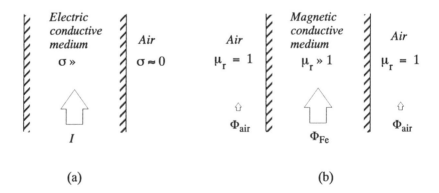

(a)   (b)

Fig. 1.4 *The current lines in an electric field (a) and flux lines in a magnetic field (b)*

Introduction

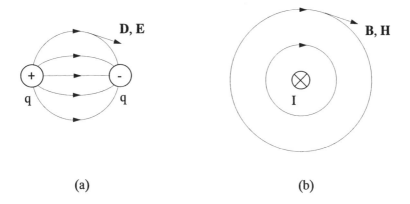

Fig. 1.5 *The flux lines distribution in free space: the electric field created by two charges with opposite signs (a), and the magnetic field created by the current carrying conductor (b)*

The magnetomotive force is expressed in ampereturns. It is named after its analog in electric circuits – the electromotive force. Despite its name, the magnetomotive force *does not have a physical meaning of force*. However, just as an electromotive force, the magnetomotive force acts as a source in magnetic circuits. The line integral on the left hand side of Eq. (1.18) is comprised of infinitesimal products of magnetic field strengths and differential arc lengths along a closed loop. When summed along a closed loop, these infinitesimal products give the mmf drops in the loop, as in Kirchhoff's voltage law. Ampère's circuital law expresses the fact that the sum of the mmf drops along a closed loop in a magnetic field is equal to the total current linked by the loop.

The continuity of magnetic flux, along with Ampère's circuital law, helps one define *the boundary conditions* for the flux lines in an electric machine. The boundary conditions relate the values of magnetic field vectors to the permeabilities of iron and air. When the boundary conditions are known, one can find the flux lines distribution in the two most typical cases: conductor in the air gap, Fig. 1.6 (a), and conductor in a slot, Fig. 1.6 (b). The flux distribution in a machine determines the values of its main and leakage reactances.

The first condition that lines of flux must satisfy when passing from the iron to the air can be derived by utilizing Fig. 1.7, which illustrates the definition of the magnetic flux through a surface. The magnetic flux through an infinitesimal surface dS in Fig. 1.7 is a continuous scalar function, defined in Eq. (1.14). Since the flux density **B** is a vector, a unit vector **n** perpendicular to the differential surface dS has to be defined, in order to satisfy Eq. (1.14).

Denoting by $\mathbf{B_n}$ the component of the flux density perpendicular to the surface, or the normal component, one can define the magnetic flux through the surface S as

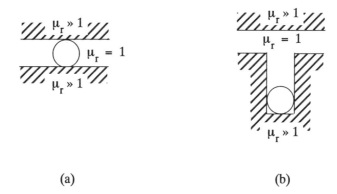

(a)   (b)

Fig. 1.6 *The conductor placed in the air gap (a) and in a slot (b) of an electric machine*

$$\Phi = \int_S \mathbf{B} \cdot \mathbf{n} \, dS = \int_S B_n \, dS \qquad (1.19)$$

or, all the flux which leaves the iron is equal to the flux which enters the air. Equation (1.19) can be further elaborated by employing Fig. 1.8. In this figure, a line of flux crossing the air to iron boundary is sketched, along with the vectors of flux density and field strength. Since the flux through a surface is determined by the normal component of the flux density, one can write for the air–iron boundary

$$B_{Fe, n} = B_{air, n} \qquad (1.20)$$

where $B_{Fe,n}$ is the magnitude of the normal component of the flux density on the iron side of the boundary, and $B_{air,n}$ is the analogous value on the air side.

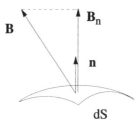

Fig. 1.7 *Illustrating the definition of the magnetic flux through a surface*

Introduction

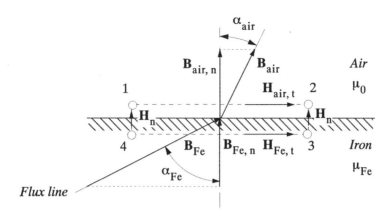

Fig. 1.8 *Boundary conditions for the magnetic field vectors*

Another boundary condition which magnetic field vectors have to satisfy can be derived by applying Ampère's law to the integration loop 1234 in Fig. 1.8:

$$H_{air,t} l_{12} + (-H_n) l_{23} + (-H_{Fe,t}) l_{34} + H_n l_{41} = \sum I \qquad (1.21)$$

where $H_{air,t}$ and $H_{Fe,t}$ are the magnitudes of the tangential components of the magnetic field strength vector in the air and iron, respectively. H_n is the magnitude of the normal component of the field strength, and l_{12}, l_{23}, l_{34}, and l_{41} are the distances between points 1, 2, 3 and 4 in the same figure. The total current linked by the integrating loop, $\sum I$, is equal to zero in this case. Since the boundary surface between the air and iron is infinitesimally thin, the mmf drop along the paths l_{23} and l_{41} is equal to zero. Applying these conditions to Eq. (1.21), one obtains

$$H_{air,t} = H_{Fe,t} \qquad (1.22)$$

or, *the magnitude of the tangential component of the magnetic field strength remains unchanged when crossing the air–iron boundary,* if there is no current on the boundary. Replacing the field strengths in Eq. (1.22) by the flux densities, one obtains

$$\frac{B_{air,t}}{\mu_0} = \frac{B_{Fe,t}}{\mu_{Fe}} \qquad (1.23)$$

Since the tangents of angles α_{air} and α_{Fe} in Fig. 1.8 can be expressed as

$$\tan\alpha_{air} = \frac{B_{air,t}}{B_{air,n}} \qquad (1.24)$$

$$\tan\alpha_{Fe} = \frac{B_{Fe,t}}{B_{Fe,n}} \qquad (1.25)$$

one can rewrite Eq. (1.23) as

$$\frac{\tan\alpha_{air}}{\tan\alpha_{Fe}} = \frac{1}{\mu_r} \qquad (1.26)$$

where μ_r denotes the relative permeability of the iron. *The angles of the flux lines on the iron to air boundary are dependent on the saturation of the iron*, since μ_r varies as a function of the flux density. To illustrate the previous considerations, a typical value of the relative permeability in unsaturated iron, $\mu_r = 5000$, will be inserted in Eq. (1.26). The result is

$$\tan\alpha_{air} = 0.0002 \ \tan\alpha_{Fe} \qquad (1.27)$$

Large variations of the angle α_{Fe} cause very small changes of the angle α_{air}. This means that the flux lines in the air are almost perpendicular to the iron surface, when iron is not too saturated. Following this conclusion, an approximate distribution of the lines of flux for the configurations in Fig. 1.6 (a) and (b) can be sketched as shown in Fig. 1.9 (a) and (b).

Flux line distributions in Fig. 1.9 (a) and (b) are derived by assuming that the largest part of the magnetic flux *crosses* the air gap. However, the conductor(s) that carry the current back can in some cases be placed in a machine so that the resulting flux goes through the iron *parallel* to the air boundary, instead of being perpendicular to it.

An accurate solution of the field equations in an electric machine can be obtained by applying the finite element method procedures. The result of the finite element computation is usually represented in the form of the flux lines distribution in the iron, slots and air gap. Such a detailed approach is, however, outside the scope of this book.

The boundary condition for the tangential component of the magnetic field strength, expressed in Eq. (1.22), is illustrated in the following example. The metallic bar in Fig. 1.10 is placed in the close vicinity of a laminated iron core, through which passes an externally driven alternating magnetic flux $\Phi_\sim$.

Introduction 13

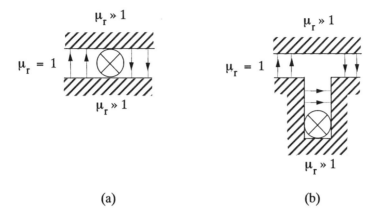

Fig. 1.9 *The lines of flux produced by the current carrying conductor placed in the air gap (a) and in a slot (b) of an electric machine*

If the amplitude of the flux density in the laminated iron core in Fig. 1.10 is so high that the iron core is deep in saturation, the magnetic field strength in the core becomes extremely large, due to the nonlinear character of its B–H curve. Following Eq. (1.22), the magnetic field strength in the metallic bar close to the contact surface iron core–metallic bar has the same amplitude as the magnetic field strength in the iron core on the other side of the contact surface. Since the magnetic field is alternating, it induces currents in the metallic bar. The ampereturns of the induced currents in the metallic bar are proportional to the value of the magnetic field strength in it.

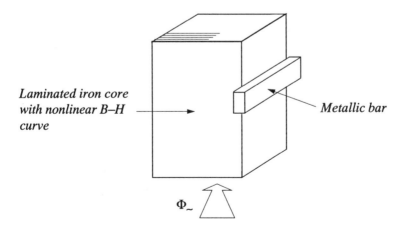

Fig. 1.10 *Illustrating boundary conditions in a magnetic field*

When the flux density in the laminated core in Fig. 1.10 exceeds the value of 1.7–1.8 T, depending on the type of magnetic material, the accompanying field strength becomes so high that the bar currents, induced through the previously described mechanism, melt the bar. Therefore, it is a rule when designing electric machines and transformers that no metallic parts may be attached to the parts of their magnetic circuit through which passes a high alternating magnetic flux.

The lines of magnetic flux are closed curves without source(s) and sink(s). This is expressed by Maxwell's equation

$$\nabla \mathbf{B} = 0 \qquad (1.28)$$

or, in integral form,

$$\oint_S \mathbf{B} \cdot d\mathbf{S} = 0 \qquad (1.29)$$

where S is a closed surface. The magnetic flux through a closed surface is equal to zero, or, the flux that enters a closed surface must be equal to the flux that leaves it. A closed surface in a rotating electric machine is the rotor cylinder. A rotating electric machine is usually designed to conduct the magnetic flux through the rotor surface only *radially*. Such a rotating machine is called *heteropolar*, since the flux periodically enters and exits its rotor cylinder. As illustration, the flux distribution in a four pole heteropolar machine is shown in Fig. 1.11 (a). The magnetic poles around the rotor periphery of a heteropolar machine alter periodically, i.e. N–S–N–S, etc. None of the lines of flux in a heteropolar machine goes in the axial direction. A heteropolar electric machine can have any even number of poles.

In *homopolar* machines, on the contrary, all lines of flux enter (exit) the rotor *axially* and exit (enter) *radially*. A homopolar rotating machine can have only one pole, since the flux lines always go through its rotor periphery in the same direction. The flux distribution in a homopolar machine is shown in Fig. 1.11 (b).

Ampère's circuital law defines the magnitude of the magnetic field strength vector produced by the current in the conductor. It is irrelevant whether the current is constant or time varying, since the *instantaneous* value of the current produces the *instantaneous* value of the magnetic field strength.

An inverse phenomenon, in which a magnetic field drives an electric current, is, however, possible only if certain conditions in a time varying field are satisfied.

A time varying magnetic field which links a conductor induces a time varying electric field along the conductor. The relationship between the vector of the induced electric field $\mathbf{E}$ and the vector of the magnetic flux density $\mathbf{B}$ is given in Maxwell's equation

$$\nabla \times \mathbf{E} = -\frac{d\mathbf{B}}{dt} \qquad (1.30)$$

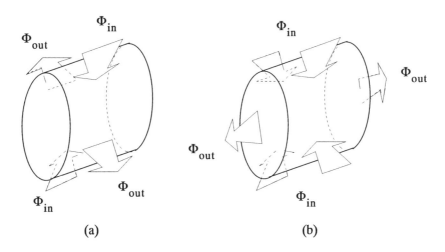

Fig. 1.11 *The magnetic flux through the rotor of a heteropolar (a) and homopolar (b) machine*

which is also known as Faraday's law. The physical meaning of Eq. (1.30) can be expressed as a property of the induced electric field **E** to drive current(s), the ampereturns of which tend to compensate the ampereturns creating the flux characterized by the flux density **B**.

Equation (1.30) is illustrated in Fig. 1.12, in which the electric field induced in the portion of the conductor limited by the two dashed lines is shown. These lines denote the part of the conductor which links the flux Φ, and in which the electric field **E** is therefore induced. There is no voltage induced by the flux variation in the rest of the conductor.

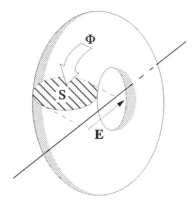

Fig. 1.12 *The time varying flux density Φ and the induced electrical field E*

The torus that limits the magnetic flux Φ in Fig. 1.12 defines a *flux tube*. A flux tube is a geometrical figure in which all the lines of flux are perpendicular to its bases and no lines of flux cut its sides. The form of a flux tube is in most cases defined by the geometry of the magnetic circuit. The induced voltage in the portion of the conductor which links flux Φ in Fig. 1.12 is equal to

$$e = -\frac{d\Phi}{dt} \qquad (1.31)$$

According to equation (1.31), *the voltage is induced in a portion of the conductor which links a time varying flux*. To obtain the induced voltage in a conductor, it is *not necessary that the conductor forms a loop* through which passes a time varying flux. This is of particular interest when the induced voltage per conductor in a machine is evaluated, in which case the effects of the rest of the flux linking conductor loop is not important.

When a conductor links the same flux w times, as shown in Fig. 1.13 for the number of turns w equal to 2, the total induced voltage is equal to the sum of the induced voltages in each turn. Therefore, one can write for the induced voltage in a coil with w turns

$$e = -w\frac{d\Phi}{dt} \qquad (1.32)$$

It is assumed in Eq. (1.32) that the same flux Φ links each turn of the coil. In reality, however, the amount of linked flux varies from turn to turn of a coil, due to a change of the fringing flux.

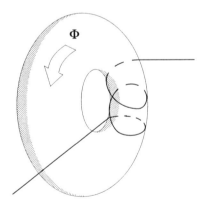

Fig. 1.13 *A flux tube defined and linked by two turns of a coil*

Introduction

Therefore, instead of a fictitious value wΦ, which is equal to the total flux linked by w turns, the *flux linkage* Ψ is introduced. The induced voltage expressed in terms of the flux linkage is

$$e = -\frac{d\Psi}{dt} \tag{1.33}$$

The flux linkage Ψ is equal to the product of the inductance L and current i

$$\Psi = Li \tag{1.34}$$

Substituting Eq. (1.34) into (1.33), one obtains

$$e = -\frac{d\Psi}{dt} = -L\frac{di}{dt} - i\frac{dL}{dt} = -L\frac{di}{dt} - i\omega\frac{dL}{d\gamma} \tag{1.35}$$

where γ is the coordinate along which moves the coil with inductance L and

$$\omega = \frac{d\gamma}{dt} \tag{1.36}$$

The voltage can be induced by varying the flux linkage in two ways: either by changing the current in the coil, or by changing the coil's inductance. The first mechanism for inducing the voltage is characteristic for transformers; hence, it is called the *transformer* voltage

$$e_t = -L\frac{di}{dt} \tag{1.37}$$

The second mechanism for inducing the voltage is characteristic for a coil which moves relative to a magnetic field. Therefore, it is called the *motional* voltage

$$e_m = -i\omega\frac{dL}{d\gamma} \tag{1.38}$$

Although the mechanisms for inducing the transformer and motional voltage are different, their effects in electric machines are equal.

According to Faraday's law, every change of magnetic flux induces currents in the surrounding conductive media. These currents tend to keep the flux constant. An *increase* of flux induces the currents, the flux of which acts *against* the increased flux; a *decrease* of flux, on the contrary, induces the currents, the flux of which *supports* the decreased flux. The induced currents tend to conserve the state in space in which the flux is changing, acting against every flux change.

1.2 Magnetic Circuit of an Electric Machine

Magnetic flux in an electric machine is produced by the current carrying coils and/or permanent magnets. The lines of flux pass through the machine linking either only the stator windings, or only the rotor windings, or all of them. The paths through which the lines of flux pass form *the magnetic circuit of the machine*. The physical parts of the magnetic circuit are the iron core, slots, and air gap. For illustration, a portion of the magnetic circuit of a machine that has teeth on both stator and rotor is shown in Fig. 1.14. The lines of common stator and rotor flux, two of which are shown in this figure, pass through the stator yoke, stator tooth, air gap, rotor tooth, rotor yoke, another rotor tooth, air gap, another stator tooth and then back to the stator yoke. In a machine there also exist the lines of flux that pass only through the stator or the rotor part of the magnetic circuit. The flux that links only the stator, or the rotor windings, is called *the leakage flux*.

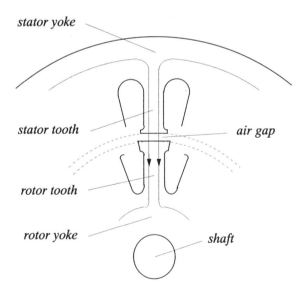

Fig. 1.14 *A portion of the magnetic circuit of an electric machine*

Schematic elements of the magnetic circuit are magnetic sources and magnetic resistances, or *reluctances*. Magnetic sources drive the fluxes in magnetic circuits, just as electric sources drive currents in electric circuits.

The reluctances, however, have a different role from the resistors in electric circuits. The resistor in an electric circuit *converts* the electric into thermal energy (heat). The reluctance, on the other hand, *stores* the magnetic energy. The higher the reluctance, the more current is necessary to drive the same amount of flux in a machine. The reluctance of the iron path of a magnetic circuit is nonlinear, as a consequence of the nonlinear iron B–H curve. The air gap reluctance is linear, and it usually dominates in the machine's magnetic circuit.

To solve the magnetic circuit of an electric machine means to determine the magnetic state in it, as a function of the applied mmfs, fluxes, and machine reluctances. Once the link between the fluxes and mmfs (currents) is established, one can define all other electromagnetic quantities including the induced voltages and torque. Magnetic circuits are often solved by applying *the magnetic equivalent circuit method* [6]. The magnetic circuit of a machine in the magnetic equivalent circuit method is copied into an electric circuit, which is solved for the currents, voltages and resistances. These quantities are then transformed back to their magnetic analogs. The transformation of the machine's magnetic circuit into an equivalent electric circuit is based upon the analogies between the magnetic and electric field quantities.

The relationship between the magnetic field quantities **H** and **B** is defined in Eq. (1.3). This equation is formally analogous to the basic form of Ohm's law, given in Eq. (1.1). According to this analogy, the current density **J** in electric circuits corresponds to the flux density **B** in magnetic circuits. The electric field strength **E** is in the same manner analogous to the magnetic field strength **H**. A further step in expanding the similarity between Eqs. (1.1) and (1.3) is the introduction of the magnetic flux Φ as an analog to the electric current I. In addition, the magnetomotive force F depends on the magnetic field strength **H** in exactly the same manner as the electromotive force V (voltage) depends on the electric field strength **E**.

The electrical resistance R_{el}, defined as the ratio between the voltage V and current I, is a parameter of the electrical flux tube which conducts the current I, and between the ends of which the voltage drop V can be measured. If the cross-sectional area S of the flux tube is constant, its resistance R_{el} is

$$R_{el} = \frac{1}{\sigma}\frac{l}{S} \qquad (1.39)$$

with l denoting the length of the flux tube and σ the electrical conductivity of the medium in the flux tube. The electrical resistance R_{el} of the flux tube in a medium that has constant conductivity and a longitudinally variable cross-sectional area is equal to

$$R_{el} = \int_0^1 \frac{1}{\sigma S(x)} dx = \frac{1}{\sigma}\int_0^1 \frac{dx}{S(x)} \qquad (1.40)$$

The reluctance in a magnetic circuit is defined in the same manner as the electrical resistance in Eq. (1.40). However, one should distinguish two cases which can occur in the magnetic fields: the flux tube in the air, and the flux tube in a magnetic material.

The relative permeability of the air is constant and equal to one. Therefore, the reluctance of a flux tube in the air varies only as a function of its dimensions

$$R_m = \frac{1}{\mu_0}\int_0^1 \frac{dx}{S(x)} \qquad (1.41)$$

with the flux tube dimensions given in Fig. 1.15.

The reluctance of magnetic material is not constant. It begins to demonstrate its nonlinear character at approximately 0.7–1 T (absolute), depending on the type of material. The relative permeability starts to decrease above this value of the flux density, reaching single digit numbers at very high flux densities. If the cross-sectional area of the flux tube is constant, its reluctance is equal to

$$R_m(\Phi) = \frac{1}{\mu(\frac{\Phi}{S})} \frac{1}{S} \qquad (1.42)$$

where the permeability μ is expressed as a function of the flux density $B = \Phi/S$.

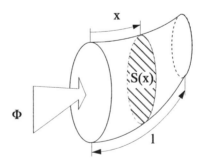

Fig. 1.15 *Illustrating calculation of the reluctance*

Introduction

The mmf drop on the element which has the reluctance given in Eq. (1.42) is

$$F = \Phi R_m(\Phi) \qquad (1.43)$$

When the flux tube's cross-sectional area is variable, i.e. $S = S(x)$ as illustrated in Fig. 1.15, a constant flux Φ in the flux tube results in the variable flux density $B(x) = \Phi/S(x)$ along it. Examples of a flux tube with variable cross-sectional area are the teeth in an electric machine. Being a function of the flux density, the permeability in this case becomes dependent on the linear coordinate x along which the reluctance is calculated, since the flux density is also a function of x.

The reluctance of the flux tube in a magnetic medium with a nonlinear B-H curve and a variable cross-sectional area varies as a function of both the flux tube geometry and material permeability

$$R_m(\Phi) = \int_0^l \frac{1}{\mu\left(\frac{\Phi}{S(x)}\right)} \frac{dx}{S(x)} \qquad (1.44)$$

If the mmf drop F on the reluctance in Eq. (1.44) has to be evaluated for a given flux Φ through it, the starting point is Ampère's circuital law

$$F = \int_0^l H(x)\,dx \qquad (1.45)$$

$$F = \int_0^l \frac{B(x)}{\mu\left(\frac{\Phi}{S(x)}\right)} dx = \Phi \int_0^l \frac{dx}{S(x)\mu\left(\frac{\Phi}{S(x)}\right)} = \Phi R_m(\Phi) \qquad (1.46)$$

The dependence of the permeability on the flux density is seldom known in analytical form. Therefore, the mmf drop on a portion of the magnetic circuit with a variable cross-sectional area is usually evaluated by applying Simpson's rule to the integrals in Eq. (1.46). Geometrically, this mmf drop is equal to the area beneath the curve $H(x)$. The procedure for calculating the mmf drop in this case is illustrated in Fig. 1.16.

A tooth with a variable cross-sectional area is shown in Fig. 1.16 (a), whereas the dependence of a tooth cross-sectional area on a tooth longitudinal coordinate x is given in Fig. 1.16 (b).

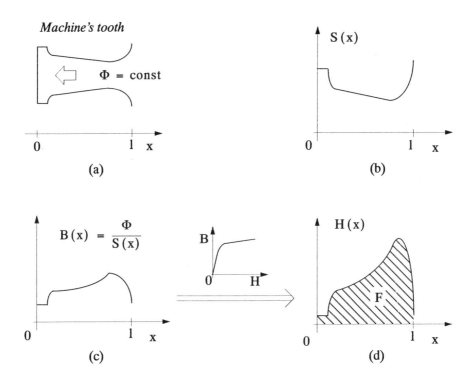

Fig. 1.16 *Illustrating calculation of the mmf drop on a tooth*

The flux density distribution along the coordinate x, shown in Fig. 1.16 (c), is obtained from the curve S(x) in Fig. 1.16 (b) by applying the relationship B(x)=Φ/S(x). In order to find the field strength distribution H(x), shown in Fig. 1.16 (d), one has to utilize the B–H curve of the magnetic material. The mmf drop F is then evaluated by numerical integration of curve H(x). Therefore, the mmf drop on the tooth can be represented by the shaded area in Fig. 1.16 (d). In terms of electric circuit theory, the reluctance of a flux tube in which the medium has constant permeability is *linear* – its value is independent of the value and/or direction of the flux through it. The reluctance of the iron part of a magnetic circuit is, on the contrary, *nonlinear*, since the iron permeability depends on the amount of flux passing through it. The symbols for linear and nonlinear reluctance, which are used in the analysis of magnetic circuits, are shown in Fig. 1.17 (a) and (b). A current carrying coil with w turns is represented in the machine's magnetic equivalent circuit as an mmf source. The magnetomotive force of this source is equal to F=iw, where i is the current in the coil. An mmf source is analogous to the voltage source in electric circuits.

Introduction

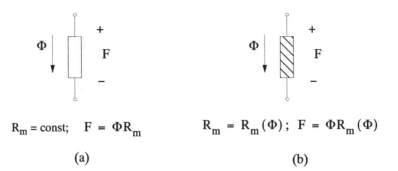

Fig. 1.17 *Symbols for linear (a) and nonlinear (b) reluctance*

A flux source in the magnetic circuit is represented by a current source in the equivalent electric circuit. The symbols for mmf and flux source are shown in Fig. 1.18.

The energy stored in a magnetic circuit can be derived by following the simple considerations of the energy distribution in an R–L circuit, represented in Fig. 1.19.

The voltage equation for the circuit in Fig. 1.19 yields

$$v = iR + \frac{d\Psi}{dt} = iR + w\frac{d\Phi}{dt} \qquad (1.47)$$

After multiplying this equation by idt, one obtains

$$vidt = i^2Rdt + id\Psi = i^2Rdt + wid\Phi \qquad (1.48)$$

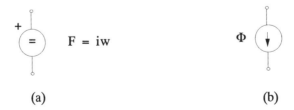

Fig. 1.18 *Symbols for an mmf (a) and flux source (b)*

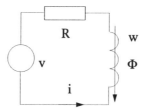

Fig. 1.19 *Illustrating energy distribution in an R–L circuit*

The term vidt in Eq. (1.48) is equal to the differential of electrical energy supplied from the source. i^2Rdt is the differential of electrical energy converted into thermal energy (heat) in the resistance. $id\Psi$, or $wid\Phi$, is the differential of energy supplied to the coil. This term converts into various forms of energy, depending on the characteristics of the medium in which the coil is placed. The following cases can occur:

(a) A coil in the air, no motion in the magnetic circuit. The term $id\Psi$, or $wid\Phi$, is completely converted into the differential of energy stored in the coil's magnetic field, or the differential of *the magnetic energy* dW_{mg}:

$$dW_{mg} = wid\Phi = id\Psi \qquad (1.49)$$

There are two substantial differences between the terms i^2Rdt and dW_{mg}. The electrical energy converted to heat is *dissipated* on the resistor and cannot be returned to the source in a simple manner. The magnetic energy is, however, always completely *returned* back to the source, when the source is disconnected from the inductance where the energy was stored. The electrical energy dissipated into heat increases in time, independent of the current waveform. The stored magnetic energy, on the contrary, is constant as long as the current in the circuit is constant.

Substituting the term wi in Eq. (1.49) with the ampereturns F, one can further write

$$dW_{mg} = Fd\Phi \qquad (1.50)$$

Introducing Hl instead of F, and BS instead of Φ, Eq. (1.50) becomes

$$dW_{mg} = lSHdB = V\frac{B}{\mu_0}dB \qquad (1.51)$$

Introduction

where V=lS denotes the volume in which the magnetic energy is stored.

The magnetic energy can be represented graphically as an area in the H–B coordinate system, as shown in Fig. 1.20 (a).

The accumulated magnetic energy at point $B=B_1$ is equal to

$$W_{mg,1} = \frac{V}{\mu_0} \int_0^{B_1} B dB = V \frac{B_1^2}{2\mu_0} \qquad (1.52)$$

Besides the magnetic energy, an auxiliary quantity, *the magnetic coenergy* W'_{mg}, is sometimes used. As opposed to the magnetic energy, the magnetic coenergy has no physical meaning. It is defined as

$$dW'_{mg} = VBdH \qquad (1.53)$$

and represented as the shaded area in Fig. 1.20 (b). The following relation between the magnetic energy and coenergy is valid for both linear and nonlinear magnetizing curves:

$$W_{mg} + W'_{mg} = BH \qquad (1.54)$$

(b) A coil with the iron core and air gap, no motion in the magnetic circuit. If there is an iron core in a magnetic circuit, the term $iwd\Phi$ is separated into two parts: the differential of stored magnetic energy dW_{mg}, and the differential of the iron core losses dW_{Fe}:

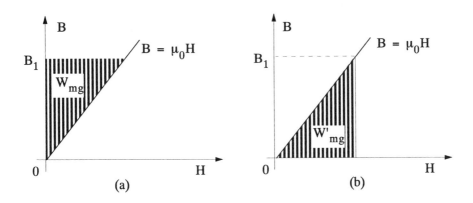

Fig. 1.20 *Illustrating the definition of the magnetic energy (a) and coenergy (b)*

$$iwd\Phi = dW_{mg} + dW_{Fe} \qquad (1.55)$$

The magnetic energy in the circuit is stored in the iron core, $dW_{mg,Fe}$, and in the air gap, $dW_{mg,air}$. Therefore

$$dW_{mg} = dW_{mg,Fe} + dW_{mg,air} \qquad (1.56)$$

In a very rough, but, from the point of view of the magnetic energy distribution, accurate approximation, an electric machine can be reduced to a coil on the iron core and the air gap. The machine in this model is represented by its three basic parts: winding(s), iron core, and air gap, as shown in Fig. 1.21 (a).

The F–Φ characteristic of the magnetic circuit in Fig. 1.21 (a) is shown in Fig. 1.21 (b). The F–Φ characteristic of a magnetic circuit is analogous to the V–I characteristic of an electric circuit. The F–Φ characteristics of the parts of the magnetic circuit in Fig. 1.21 (b) are obtained from their B–H curves, which are similar to the ones shown in Fig. 1.1. The flux density B in this conversion has to be multiplied by the cross-sectional area S of the element of the circuit, and the field strength H has to be multiplied by the element's length l. This gives the mmf drop F

$$H = \frac{B}{\mu} \Rightarrow F = lH = \frac{1}{\mu}\frac{l}{S}\Phi = R_m\Phi \qquad (1.57)$$

Since the gap reluctance is constant, the F–Φ characteristic of the air gap is a line. This characteristic is denoted as *Air gap* in Fig. 1.21 (b).

The F–Φ characteristic of the iron part of the magnetic circuit is nonlinear, as a consequence of the nonlinear iron B–H curve. This curve is denoted as *Iron core* in Fig. 1.21 (b). The F–Φ characteristic of the series connection of the air gap and iron core reluctances is denoted as *Common* in the same figure. This characteristic gives the flux in the core as a function of the coil ampereturns $F = iw$. The mmf drops on the iron and air gap for the same value of flux Φ_c are comparable to each other, as opposed to the field strength values at the same flux densities. The field strength in the air gap for the same value of the flux density is incomparably higher than the field strength in the iron, as shown in Fig. 1.1.

There is one more detail which can be seen by observing the F–Φ characteristics in Fig. 1.21 (b). The resultant curve, denoted as *Common*, is linear in a much wider interval of the mmfs than the F–Φ characteristic of the iron, denoted as *Iron core*. The resultant curve includes the influence of the air gap which does not exist in the *Iron core* characteristic.

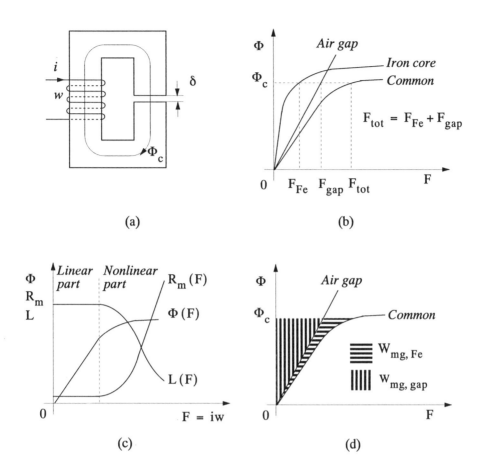

Fig. 1.21 *The magnetic circuit with the iron core and air gap (a), its F–Φ characteristic (b), the dependence of the circuit parameters on the coil ampereturns (c) and the magnetic energy distribution in the circuit (d)*

The F–Φ characteristic of a magnetic circuit with an air gap is linear in a significantly wider interval of the coil current than the F–Φ characteristic of a circuit without the air gap. However, the drawback of introducing the air gap is that there will be a higher value of current for a given flux than in a circuit without the air gap.

The magnetic circuit of an electric machine is incomparably more complex than the one shown in Fig. 1.21 (a). A machine usually has several windings, as opposed to only one coil in this example. The machine's windings consist of coils which often overlap each other (the winding is said to be *distributed*), whereas the coil in Fig. 1.21 (a) is *concentrated* at one part of the magnetic circuit. An equivalent to the two-dimensional F–Φ characteristic of the magnetic circuit in Fig. 1.21 (b) would be a family of multidimensional surfaces, the graphical representation of which is meaningless. However, such surfaces are useless, since the state in the magnetic circuit of a machine is always calculated at a particular operating point. This point can be defined in different manners. If all the *fluxes* in the machine are known, the *ampereturns* of mmf sources as functions of the fluxes are calculated. If the *ampereturns* of all the mmf sources are known, the value of flux in one of the branches in the circuit is evaluated at an observed operating point. Very often, the parameters which define the state in the magnetic circuit are represented as a combination of the mmfs, fluxes and additional conditions defined by the type of the machine's windings and other properties of the machines. One can find more about this topic in [6].

The F–Φ characteristic of the complete magnetic circuit, denoted by *Common* in Fig. 1.21 (b), determines another important parameter of the coil – its *inductance*. The inductance L of the coil is equal to the ratio between the flux linked by the coil, wΦ, and the current i in the coil:

$$L = \frac{w\Phi}{i} = \frac{w^2\Phi}{F} = \frac{w^2}{R_m} \qquad (1.58)$$

The inductance in the linear part of the F–Φ characteristic is constant, since the reluctance of the circuit in this region is constant, too. As the iron becomes saturated, the reluctance increases and the inductance decreases, as shown in the *Nonlinear part* of the characteristics in Fig. 1.21 (c).

The magnetic energy stored in the magnetic circuit for a given flux Φ_c is shown in Fig. 1.21 (d). The total magnetic energy is equal to the sum of the magnetic energy stored in the air gap, $W_{mg,gap}$, and the magnetic energy stored in the iron, $W_{mg,Fe}$.

The differential of energy dW_{Fe} in Eq. (1.55) represents the energy converted into *iron core losses*. These losses are dissipated in the iron in two modes: as *the hysteresis loss* and *the eddy currents loss*

$$dW_{Fe} = dW_{hys} + dW_{eddy} \qquad (1.59)$$

The friction between the domain walls causes a time delay of the flux density vectors in domains relative to the external field strength. A motion accompanied by friction is always a source of loss. When the B–H curve of a magnetic material has hysteretic character, the phase shift between the applied field strength and flux density causes hysteretic loss in the magnetic material. With sinusoidal magnetization, the flux density in a coil is 90° lagging the induced voltage, and the field strength is in phase with the current in the coil.

When the magnetization curve of the magnetic material has no hysteresis, the vectors H and B are in phase, preserving the angle of 90° between the flux and the induced voltage. There are no hysteresis losses in such a magnetic material, since there is no component of the current which is in phase with the induced voltage.

When the magnetic material B–H curve has hysteresis, the fundamental components of vectors H and B are out of phase, as shown in Fig. 1.22. Since the induced voltage is 90° ahead of the flux, a component of the current *in phase* with the induced voltage is generated. This component of the current carries from the source the power which covers the hysteresis loss in the magnetic material.

The mechanism of the hysteresis loss generation is illustrated in Fig. 1.23. Assume that at the beginning of the magnetization cycle both the flux density and field strength are equal to zero. The field strength increases from 0 to H_{max}, and then goes back to zero. In the time interval when H increases, the flux density varies following curve 1 in Fig. 1.23 (a). When H decreases, the flux density is defined by curve 2 in the same figure. The operating point on the hysteretic B–H curve is not only a function of the current values of the field vectors, but also of the previous state in the magnetic material.

The magnetic energy accumulated in the magnetic material in the period when H increases is proportional to the area defined by curve 1, the B-axis, and the line which connects points (0, B_{max}) and (H_{max}, B_{max}) in Fig. 1.23 (a). Analogously, the magnetic energy returned to the source in the period when H decreases from H_{max} to 0, when dB<0, is proportional to the area defined by curve 2, the B-axis, and the line connecting the points (0, B_{max}) and (H_{max}, B_{max}). *The magnetic energy supplied to the core is greater than the energy returned to the source*. The difference between the two is equal to the hysteresis loss in the magnetic material. The hysteresis loss during the complete cycle is proportional to the area of the hysteresis curve, as shown in Fig. 1.23 (b).

The *power* P_{hys} dissipated through the mechanism of the hysteresis loss is proportional to the derivative of the energy of the hysteresis loss with respect to time. This is equivalent to the multiplication of the energy of the hysteresis loss per cycle by frequency.

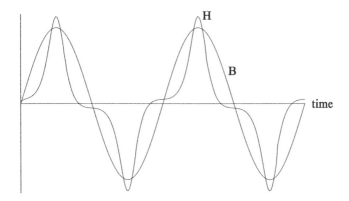

Fig. 1.22 *The sinusoidal applied flux and distorted field strength in a magnetic material having hysteretic B–H curve*

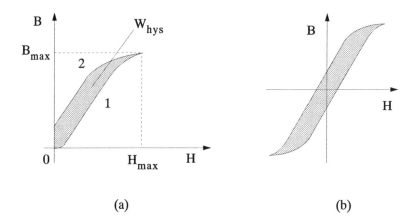

Fig. 1.23 *Graphical representation of the hysteresis loss in the magnetic material*

The power of hysteresis loss is also proportional to the value of the maximum flux density in one cycle, B_{max}, elevated to a power between 1.6 and 2.3, depending on the type of magnetic material:

$$P_{hys} = c_{hys} \, fB_{max}^{1.6 \div 2.3} \qquad (1.60)$$

with c_{hys} being dependent on the technological characteristics of the material.

Another mechanism for dissipating losses in the magnetic material is eddy current. A time varying magnetic flux induces the electric field in the magnetic material, according to Eq. (1.30). Since iron is an electrical conductor, the induced electric field drives the currents in the core itself, as shown in Fig. 1.24 (a). The magnitude of the eddy currents is proportional to the cross-sectional area of the surface through which the eddy currents flow, and the conductivity of the iron. Therefore, either the cross-sectional area or the iron's conductivity have to be minimized if the eddy currents are to be decreased.

Laminating the core decreases the iron cross-sectional area (faced by the eddy currents) through which eddy currents must flow. The sheets in Fig. 1.24 (b) are electrically insulated to each other, so that each eddy current is limited within the sheet through which it passes. The flux in each sheet is n times lower than the total flux, if there are n sheets. Therefore, the induced voltage which drives the eddy currents in each sheet is also n times lower. In addition, the cross-sectional area of the surface through which the eddy currents flow decreases n times, making the resistance of each sheet n times higher.

If E denotes the induced voltage which drives eddy currents in the solid iron core in Fig. 1.24 (a), and R stands for the corresponding iron resistance, the eddy current loss

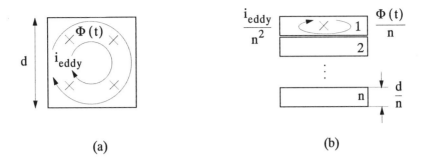

Fig. 1.24 *Illustrating the eddy currents loss minimization*

in the solid iron can be expressed as

$$P_{eddy,s} = \frac{E^2}{R} \tag{1.61}$$

If, however, the iron core consists of n sheets, which occupy the same area as the solid iron, and which are insulated one to another, the eddy current loss per sheet is equal to

$$P'_{eddy,1} = \frac{\frac{E^2}{n^2}}{nR} = \frac{1}{n^3}\frac{E^2}{R} \tag{1.62}$$

The total eddy current loss in n sheets is n times higher

$$P_{eddy,1} = \frac{1}{n^2}\frac{E^2}{R} \tag{1.63}$$

The eddy current loss reduces n^2 times if the iron core is manufactured of n sheets instead of solid iron. Therefore, the iron core of electromagnetic devices which carry a time varying flux is always laminated.

The electrical conductivity of the iron core can be substantially decreased by the addition of silicon, which is a usual means of reducing the eddy current loss in low frequency devices. The percentage of silicon varies within single digit numbers. A higher percentage of silicon would deteriorate the mechanical properties of the lamination and cause fast wear of the punching tools which cut iron plates into the stator and rotor lamination.

(c) A coil with the iron core and the air gap, a part of the magnetic circuit moves relative to the others. There is a wide variety of forms of magnetic circuit which belong to this category. One of them is shown in Fig. 1.25 (a).

The F–Φ characteristic of the magnetic circuit in Fig. 1.25 (a) is dependent on the position of the plunger relative to the iron core. When the plunger is far from the core and none of the lines of the core flux links it, the F–Φ characteristic is identical to the one denoted as *Common* in Fig. 1.21 (b). This characteristic is denoted as *Plunger far from the core* in Fig. 1.25 (b). Another extreme position of the plunger is denoted as *Plunger completely in the gap* in Fig. 1.25 (b). The plunger in this position is placed so that it fills the gap completely, making the F–Φ characteristic of the magnetic circuit very similar to the *Iron core* characteristic in Fig. 1.21 (b). The only difference between the two characteristics is caused by tiny air gaps between the plunger and core. The accumulated magnetic energy in the magnetic circuit in Fig. 1.25 (a) is not only a function of the current in the coil, but also of the plunger's *position*. This is illustrated in Fig. 1.25 (b), in which the two shaded areas denote accumulated energies in the circuit at two extreme plunger positions and for the coil ampereturns F_1.

If there is no external force, the work for the plunger's motion can come only from the energy $iwd\Phi$, which the source supplies to the coil. Therefore, one can write

$$iwd\Phi = dW_{mg} + dW_{Fe} + dW_{mech} \qquad (1.64)$$

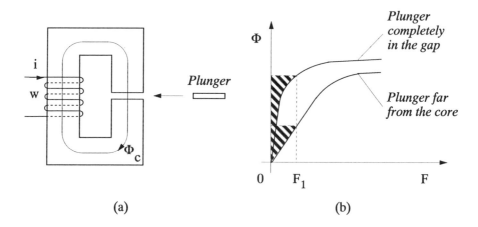

Fig. 1.25 *Magnetic circuit with a moving part (a) and its F–Φ characteristic (b)*

Introduction

where dW_{mech} denotes the differential of mechanical energy (work):

$$dW_{mech} = Fds \qquad (1.65)$$

F in equation above denotes the force on the plunger, and ds is the differential length of the path along which the work is performed.

The analysis of the magnetic circuit elements will be concluded with permanent magnets. Permanent magnets are made of alloys having a wide hysteresis loop, as shown in Fig. 1.26 (a). B_{res} in this figure denotes the *residual flux density*, whereas H_{coer} stands for the *coercive force*. The residual flux density is an ideal quantity which would exist in the permanent magnet if it were magnetically short-circuited with a material having a reluctance of zero. The coercive force does not have a physical meaning of force. It is equal to the amount of ampereturns necessary to demagnetize the permanent magnet.

A simple magnetic circuit with a permanent magnet is shown in Fig. 1.26 (b). Ampère's circuital law applied to the circuit in this figure gives

$$H_{PM}l_{PM} + H_{Fe}l_{Fe} + H_\delta \delta = 0 \qquad (1.66)$$

The zero on the right hand side of Eq. (1.66) stands because there is no current linked by the line of flux along which the integration is carried out. H_{PM} in Eq. (1.66) denotes the field strength in the permanent magnet, l_{PM} is the permanent magnet's height, H_δ is the field strength in the air gap, δ is the length of the air gap, H_{Fe} is the field strength in the iron and l_{Fe} is the total length of the line of flux in the iron. The field strength in the permanent magnet can be expressed from Eq. (1.66) as

$$H_{PM} = -\frac{H_{Fe}l_{Fe} + H_\delta \delta}{l_{PM}} \qquad (1.67)$$

The operating point of a magnetic circuit which contains a permanent magnet is in the *second* quadrant of the (H,B) coordinate system, where the field strength H is negative, and where the B–H curve of the permanent magnet is usually considered linear.

A typical operating point in the second quadrant is illustrated in Fig. 1.26 (c). In this figure the operating point P is positioned in the intersection of the permanent magnet B–H line and the B–H curve of the rest of the magnetic circuit.

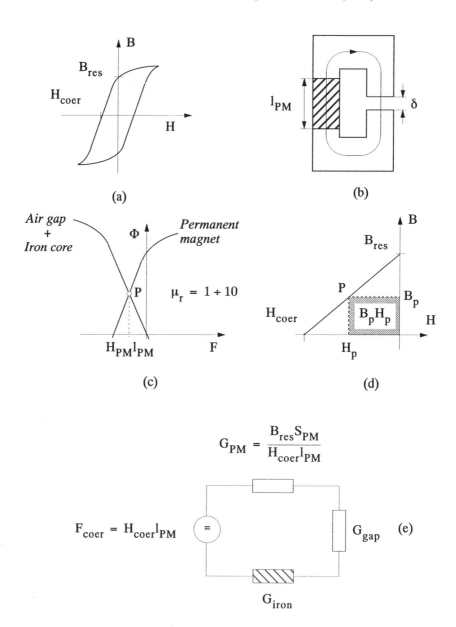

Fig. 1.26 *B–H characteristic of permanent magnet (a), magnetic circuit with permanent magnet (b), its F–Φ characteristic (c), magnetic energy stored in the air gap (d), and its magnetic equivalent circuit (e)*

Introduction

The relative permeability of permanent magnets at the operating point P is usually a single digit number, mostly as a consequence of the air gap's high reluctance. The higher the air gap reluctance, the lower the slope of the curve denoted as *Air gap + Iron core* and, therefore, the lower the relative permeability.

The magnetic equivalent circuit of the iron core with permanent magnets from Fig. 1.26 (a) is shown in Fig. 1.26 (e). F_{coer} in this figure denotes the mmf of the magnet which is proportional to its coercive force, G_{PM} is the permanent magnet's internal permeance, and G_{gap} and G_{iron} are the air gap and iron permeances, respectively.

The magnetic energy stored in the air gap of the magnetic circuit in Fig. 1.26 is proportional to the product of the flux density B and field strength H, because the relationship between the two is linear. Therefore, the magnetic energy can be represented as the area in the (H,B) coordinate system. The magnetic energy stored in the air gap at operating point P is proportional to the shaded area in Fig. 1.26 (d). Considering the B–H curve of the permanent magnet in the second quadrant to be linear, the maximum of the magnetic energy is delivered to the air gap at the operating point at which the flux density is equal to $B_{res}/2$, and the field strength is $H_{coer}/2$. This is analogous to an electric circuit with a voltage or current source delivering maximum power when the resistance of the load is equal to the internal resistance of the source.

To *solve a magnetic circuit* means to find the fluxes in all of its branches and the ampereturns of all the mmf sources in it. When the fluxes in the branches are known, one can find the reluctances of all the passive elements of the magnetic circuit. The kind of applied electrical sources defines the solution strategy for the magnetic circuit of a device connected to the sources. If an electromagnetic device is connected to the *current source(s)*, the ampereturns of the mmf sources are known, and the branch fluxes have to be calculated. When the device is connected to the *voltage sources*, two cases can be distinguished: D.C. applied voltages, and A.C. applied voltages. In both cases the *steady state* is considered, which commences after the R.M.S. values of the currents become constant.

The steady state value of the currents in an RL circuit fed from D.C. voltage sources is defined only by the applied voltage(s) and circuit resistance(s). The magnetic circuit parameters in this case participate in the determination of the currents only during the transient state. Since the steady state values of the currents are proportional to the applied voltages, the ampereturns of the mmf sources in the magnetic circuit of a device fed from D.C. voltage sources are proportional to the amounts of the applied voltages, independent of parameters of the magnetic circuit.

In the case of A.C. applied voltages, the voltage drops on the coil resistances can be neglected if the magnetic circuit is not too saturated. The applied voltage in each coil is almost equal to the induced voltage, which is proportional to the amplitude of the flux linkage. Therefore, the *fluxes* in some branches of the magnetic circuit are proportional to the amplitudes of the applied sinusoidal voltages. These fluxes determine the values of permeances in the magnetic circuit, which then define the ampereturns of the mmf sources.

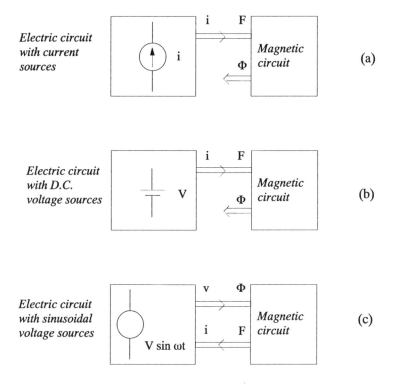

Fig. 1.27 *Illustrating the principles of solutions of magnetic circuits at steady states*

The current in an unsaturated and moderately saturated RL circuit fed from a sinusoidal voltage source is defined solely by the parameters of the magnetic circuit. Previous considerations can be visualized by means of Fig. 1.27.

In the following numerical example, the application of the magnetic equivalent circuit method in solving a simple magnetic circuit will be illustrated. The circuit consist of a coil on the iron core with the air gap, as shown in Fig. 1.28.

Let us first assume that the nonlinear magnetization curve of the iron core is anhysteretic. The ampereturns of the coil in Fig. 1.28 (a), F = i w, drive the total flux Φ_{tot}. A part of the total flux, denoted by Φ_σ, passes through the surrounding air, instead of through the air gap. The line of flux denoted by Φ_σ in Fig. 1.28 (a) represents the flux which in reality does not uniformly link all turns of the coil. The reluctance of the air through which the leakage flux is closed is denoted by R_σ in Fig. 1.28 (b). In the same figure the R_{gap} denotes the reluctance of the air gap.

Introduction

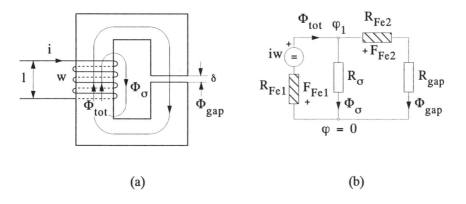

Fig. 1.28 *A coil on an iron core with the air gap (a) and its magnetic equivalent circuit (b)*

The exact values of both reluctances, R_σ and R_{gap}, can be found for a given geometry of the iron core from the flux distribution obtained by applying the finite element procedures. However, satisfactorily accurate values of R_σ can be calculated by assuming that the leakage flux Φ_σ is closed within a rectangular that has sides not greater than three times the sides of the iron core rectangle, as shown in Fig. 1.29.

The length of the flux tube through which the leakage flux passes can be taken in this analysis to be equal to the coil height l. Without significant deterioration of the accuracy of computation, one can express the reluctance of the leakage flux path as

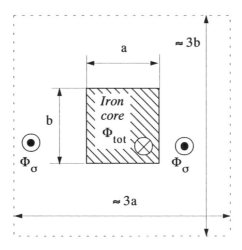

Fig. 1.29 *Illustrating the definition of the iron core leakage flux area*

$$R_\sigma \approx \frac{1}{\mu_0} \frac{1}{10ab} \qquad (1.68)$$

The air gap reluctance R_{gap} is calculated as

$$R_{gap} \approx \frac{1}{\mu_0} \frac{\delta}{1.1ab} \qquad (1.69)$$

with δ denoting the air gap length, as shown in Fig. 1.28 a). The area of the air gap in Eq. (1.69) is taken to be 10% larger than the area ab of the iron that faces the air gap, which is a usual assumption in the analysis of magnetic circuits with short gaps. In this way the property of the flux lines to repel one another, when there is room for it, is allowed for.

The element denoted by R_{Fe1} in Fig. 1.28 (b) is the reluctance of the part of the core on which the coil with w turns is placed. Another nonlinear element in this circuit, R_{Fe2}, represents magnetically the rest of the iron core. The dependence of the reluctance of these two elements on the value of flux through them is determined by the iron core B–H curve and by the core dimensions. Data for the iron core B–H curve are usually supplied in tabular form, as a result of measurements on the specimen. The magnetization curve defined this way is discrete, which is inconvenient for further analysis. Therefore, the input data are usually converted into a continuous function by using the interpolating curves for the values of the arguments between the measured points. The interpolating curve for the measured B–H curve data used in this example is shown in Fig. 1.30 (a). This curve, as well as results of all other computations in this example, is obtained by executing the statements listed in the Appendix.

The data of an anhysteretic B–H curve are usually supplied for the first quadrant only. However, in computations, both signs of the flux density and field intensity are utilized. Therefore, practical definitions for the mmf drop as a function of flux, as well as for the reluctance, are

$$F(\Phi) = \text{Sign}(\Phi) \, lH\left(\frac{|\Phi|}{S}\right) \qquad (1.70)$$

$$R_{Fe}(\Phi) = \frac{lH\left(\frac{|\Phi|}{S}\right)}{|\Phi| + 10^{-9}} \qquad (1.71)$$

Introduction

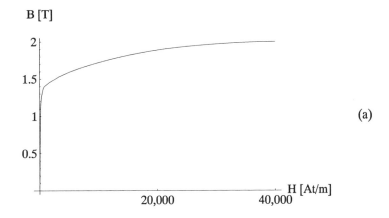

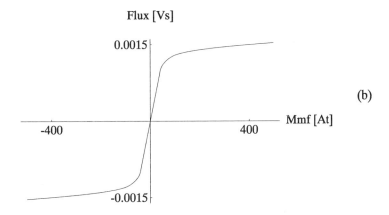

Fig. 1.30 *The interpolation of the B–H curve (a) and the F–Φ characteristic of the magnetic circuit in Fig. 1.28 (b)*

with the term 10^{-9} in the denominator of Eq. (1.71) being added to prevent division by zero when the flux is zero.

One should recognize in Eq. (1.71) the definition of the reluctance of an element with a constant cross-sectional area, introduced in Eq. (1.42). This equation may be used here, since the cross-sectional area of the iron core in this example is constant. Using Eqs. (1.70) and (1.71), one can define the mmf drops F_{Fe1} and F_{Fe2}, as well as the reluctances R_{Fe1} and R_{Fe2} as

$$F_{Fe1}(\Phi_{tot}) = \text{Sign}(\Phi_{tot}) \, \text{lH}\left(\frac{|\Phi_{tot}|}{S}\right)$$

$$F_{Fe2}(\Phi_{gap}) = \text{Sign}(\Phi_{gap}) \, \text{lH}\left(\frac{|\Phi_{gap}|}{S}\right)$$

$$R_{Fe1}(\Phi_{tot}) = \frac{\text{lH}\left(\frac{|\Phi_{tot}|}{S}\right)}{|\Phi_{tot}| + 10^{-9}} \qquad R_{Fe2}(\Phi_{gap}) = \frac{\text{lH}\left(\frac{|\Phi_{gap}|}{S}\right)}{|\Phi_{gap}| + 10^{-9}}$$

According to Ampère's circuital law, the source that is driving the flux in the circuit in Fig. 1.28 (b) has ampereturns $F = iw$. The ampereturns F drive all the fluxes in this circuit.

The magnetic circuit in Fig. 1.28 (b) can be solved by utilizing any of the methods for electric circuits analysis. If the node potential method is applied, one obtains

$$\varphi_1\left[\frac{1}{R_\sigma} + \frac{1}{R_{gap} + R_{Fe2}(\Phi_{gap})}\right] = \Phi_{tot} \qquad (1.72)$$

The node potential method in magnetic circuits relates the fluxes which enter/exit a node in a circuit to the magnetic scalar potentials (φ) of the nodes. The magnetic scalar potential φ is defined in the same manner as the electric scalar potential, i.e.

$$H = -\nabla \varphi \qquad (1.73)$$

The voltage drop across electric resistance is equal to the difference between the electric potentials at the resistance's ends. Analogously, the mmf drop across a reluctance is equal to the difference between the magnetic scalar potentials at the ends of the reluctance. Additional information about the mmf distribution within each branch of a magnetic circuit is obtained by applying Kirchhoff's voltage law. In this case, the potential φ_1 can be expressed in terms of the coil ampereturns $F = iw$ as

$$\varphi_1 = iw - F_{Fe1}(\Phi_{tot}) \qquad (1.74)$$

Introduction

The air gap flux Φ_{gap} is equal to

$$\Phi_{gap} = \Phi_{tot} - \Phi_\sigma = \Phi_{tot} - \frac{\varphi_1}{R_\sigma} = \Phi_{tot} - \frac{iw}{R_\sigma} + \frac{F_{Fe1}(\Phi_{tot})}{R_\sigma} \qquad (1.75)$$

Substituting Eqs. (1.74) and (1.75) into (1.72), one obtains for the circuit in Fig. 1.28 (a) the nonlinear relationship between the coil ampereturns iw and the total flux linked by the coil turns Φ_{tot} as

$$[iw - F_{Fe1}(\Phi_{tot})]\left[\frac{1}{R_\sigma} + \frac{1}{R_{gap} + R_{Fe2}\left(\Phi_{tot} - \frac{iw - F_{Fe1}(\Phi_{tot})}{R_\sigma}\right)}\right] = \Phi_{tot}$$

$$(1.76)$$

The nonlinear relationship between the coil ampereturns and the flux, expressed in Eq. (1.76), is a consequence of the nonlinear B–H curve of the iron core. Therefore, the curve obtained as a solution of Eq. (1.76) has to reflect the character of the iron core magnetization characteristic, shown in Fig. 1.30 (a).

As in any other nonlinear circuit, the strategy of the solution of Eq. (1.76) is strongly determined by the choice of the circuit parameters and the operating point. For the set of parameters defined in the Appendix, and the B–H curve of magnetic material shown in Fig. 1.30 (a), the solution of Eq. (1.76) for the various values of flux is given in Fig. 1.30 (b).

The reluctance R_m, defined as

$$R_m = \frac{iw}{\Phi_{tot}} \qquad (1.77)$$

is the Thevenin equivalent reluctance as seen from the terminals of the mmf source iw in Fig. 1.28 (b). The equivalent reluctance R_m of a nonlinear magnetic circuit is a function of the applied mmfs. For the circuit parameters chosen in this example the reluctance depends on the values of the mmf as shown in Fig. 1.31 (a).

When the applied mmf in a magnetic circuit is large enough, the iron core becomes saturated. In the saturated region, the same amount of increase of the mmf results in a smaller increase of flux than in an unsaturated interval. Therefore, the ratio between the flux and current decreases as the magnetic material becomes more saturated.

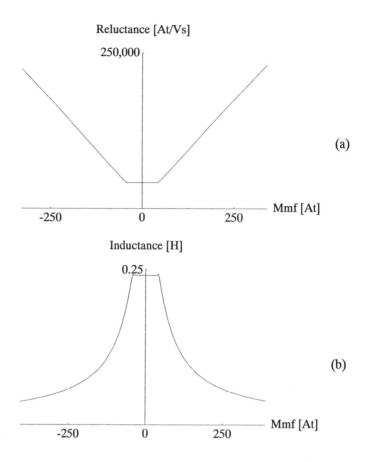

Fig. 1.31 *The equivalent reluctance of the magnetic circuit (a) and the coil inductance (b)*

Since the coil inductance is defined as the ratio between the coil flux linkage and the current through it, it will decrease along with the current increase. The inductance of the coil as a function of the coil ampereturns is shown in Fig. 1.31 (b).

The response of the magnetic circuit analyzed in this example to sinusoidal applied current and sinusoidal applied flux is shown in Fig. 1.32. When a *sinusoidal current* is applied to the coil, the flux in the iron core looks as shown in Fig. 1.32 (a). One can notice in this figure the characteristic flattening of the flux, which appears when the magnetic circuit is saturated. The flux waveform in a linear magnetic circuit, however, always follows the current waveform. When the total flux is *sinusoidal*, the current in the coil is distorted, as shown in Fig. 1.32 (b). For the positive and negative flux values that drive the magnetic circuit into saturation, the current has to increase and decrease faster than flux, due to the nonlinear B–H curve of the iron.

Introduction

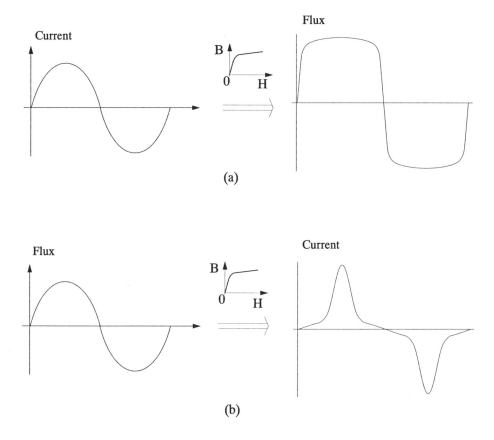

Fig. 1.32 *The total flux as a function of the applied sinusoidal current in the coil (a), and the coil current as a function of the applied sinusoidal flux (b)*

One can see in Fig. 1.32 that a nonlinear magnetic circuit introduces *higher time harmonics* into the electric circuit. Neglecting hysteresis, the higher time harmonics are only odd, because an anhysteretic B–H curve of a magnetic material is always symmetric with respect to the origin, i.e. B(–H) = – B(H). The response of a nonlinear magnetic circuit is always nonlinear, independent of the type of applied electric source. This is a very important conclusion, which has far-reaching effects on the behaviour of electric machines and power systems theory.

The time response of the magnetic circuit analyzed in this example to the various kinds of applied voltages can be found by solving Kirchhoff's voltage equation for the electric circuit. Since the voltage drop across the coil is expressed as a time derivative of the flux linkage, the voltage equation is not algebraic, but differential. In this case, the voltage differential equation yields

$$v = iR + w\frac{d\Phi_{tot}}{dt} = iR + e \qquad (1.78)$$

with i denoting the current, R the coil resistance, w the number of turns, Φ_{tot} the average flux linked by one turn in the coil and e the induced voltage in the coil. The differential equation (1.78) is nonlinear, since the relationship between the unknowns i and Φ_{tot} are determined by a nonlinear B–H curve. However, if the magnetization curve is linearized, as shown in Fig. 1.1, the coil inductance becomes constant. Equation (1.78) becomes linear and can be solved analytically. In this case one can write

$$v = iR + L\frac{di}{dt} \qquad (1.79)$$

The analytical solution of Eq. (1.79) is

$$i = \frac{V}{R}(1 - e^{-t/\tau}) \qquad (1.80)$$

with V standing for the amplitude of the applied D.C. voltage, and τ denoting the electrical time constant of an RL circuit

$$\tau = \frac{L}{R} \qquad (1.81)$$

The value of the inductance L in this example can be read from Fig. 1.31 (b). Since the circuit is linearized, the value of L must be taken for a very small argument of the coil ampereturns. The inductance of the coil in this example is 0.25 H. The solution of Eq. (1.79) is graphically represented in Fig. 1.33 (a) for the coil current. The core flux is shown in Fig. 1.33 (b). The dependence of the flux Φ_{tot} on time in this example is derived from the current dependence on time following the relation

$$\Phi_{tot} = i\frac{L}{w} \qquad (1.82)$$

The current reaches its steady state value V/R after approximately five time constants. The same time is necessary for the flux to reach its steady state value LV/(wR). The tangential line drawn at the origin of the coordinate system to the current curve intercepts the steady state value of the current in the circuit at the point (τ, V/R).

Introduction

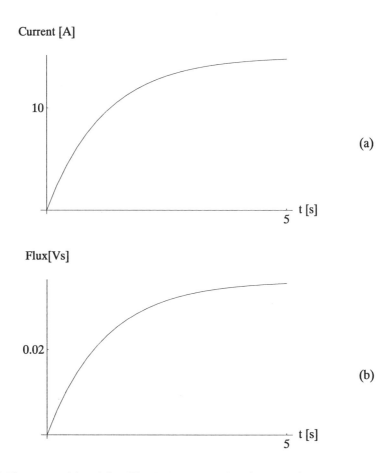

Fig. 1.33 *The current (a) and flux (b) response to a D.C. voltage in a linear RL circuit*

The electrical time constant can be defined only in a magnetic circuit which has a linear B–H curve. In a nonlinear magnetic circuit the inductance is not constant, therefore τ defined as a time *constant* is meaningless. However, the ratio between the coil inductance and resistance is important in the analysis of a nonlinear RL circuit, too. This is illustrated in the next example, which shows the response of a nonlinear RL circuit to a D.C. voltage. For the current and flux responses shown in Fig. 1.34, as well as for all other numerical calculations in this section, the circuit parameters and solution procedures are given in the Appendix.

The first conclusion that one can draw from comparing the current responses in a linear, Fig. 1.33 (a), and a nonlinear, Fig. 1.34 (a), RL circuit, is that the current reaches its steady state incomparably faster in a nonlinear circuit. The same is valid for the flux. A simple explanation for this difference follows.

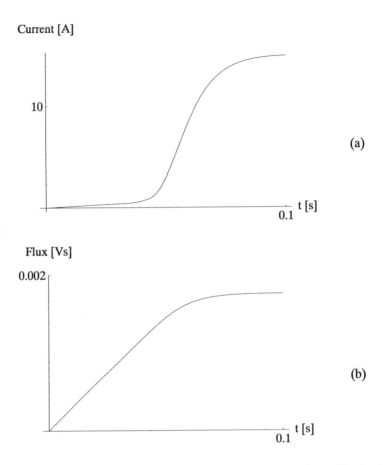

Fig. 1.34 *The current (a) and flux (b) response to a D.C. voltage in a nonlinear RL circuit*

Assuming that the initial values of all variables are equal to zero, the flux in a nonlinear magnetic circuit increases in the same manner as the flux in the linear circuit, as long as the iron is not saturated. When the iron becomes saturated, the current in the coil becomes much higher for the same values of flux in a nonlinear than in a linear circuit. The higher current causes a higher voltage drop across the resistance, leaving less source voltage to cover the lower induced voltage. The lower induced voltage means a lower time rate of change (increase) of the flux, and, consequently, a lower amount of flux. Once the iron is saturated with the flux still increasing, the coil current very quickly reaches its maximum value of V/R.

This maximum value of current is equal in a nonlinear and linear circuit, because the current is D.C. and its maximum value is determined only by the applied voltage and resistance in the circuit. However, the same current corresponds to less flux in the nonlinear than in the linear case, as seen by comparing the responses in Fig. 1.34 (b) and Fig. 1.33 (b).

Transient states in nonlinear magnetic circuits are much faster than in linear ones, because the steady state values of the fluxes are lower and they are reached faster in nonlinear than in linear circuits. Formally, the time constant of a nonlinear magnetic circuit is much shorter than the time constant of a linear circuit.

When the applied voltage is sinusoidal, the current and flux responses contain the D.C. components and the components which are characteristic for the sinusoidal applied voltage. The D.C. components exist in the circuit response to any kind of applied voltage. When the applied voltage is sinusoidal, the voltage differential equation (1.78) becomes

$$iR + w\frac{d\Phi}{dt} = V_{max}\sin(\omega t + \varphi) \qquad (1.83)$$

where φ is the initial phase shift of the voltage. In a linear circuit, where $w\Phi = Li$, the solution of the equation above is

$$i = \frac{V}{\omega^2 L^2 + R^2}\left[(\omega L\cos\varphi - R\sin\varphi)e^{-t/\tau} + \omega L\cos(\omega t + \varphi) + R\sin(\omega t + \varphi)\right]$$

$$(1.84)$$

with the D.C. component of the current being equal to

$$i_{DC} = \frac{V}{\omega^2 L^2 + R^2}(\omega L\cos\varphi - R\sin\varphi)e^{-t/\tau} \qquad (1.85)$$

The D.C. component of the current response is, among others, a function of the initial voltage phase shift, φ. Since the reactance ωL of the coil is usually much greater than its resistance R, the D.C. component of the current response will be greatest if the RL circuit is energized at those time intervals when the voltage crosses the zero value, i.e. when $\varphi \sim 0$ and, therefore, $\cos\varphi$ is around 1.

The character of the solution of the differential equation (1.83) does not change if the circuit is nonlinear. However, the amplitudes of the currents and the fluxes may vary substantially from the linear case. The solution of Eq. (1.83) for the nonlinear magnetic circuit, with the parameters given in the Appendix, is represented graphically in Fig. 1.35. At the beginning of the transient, the *inrush current* flows through the coil. The values of the peak amplitudes of the inrush current are dependent on the maximum value of the applied voltage, V_{max}, as well as on the angle φ. The inrush current effect can be explained by Fig. 1.35 (b).

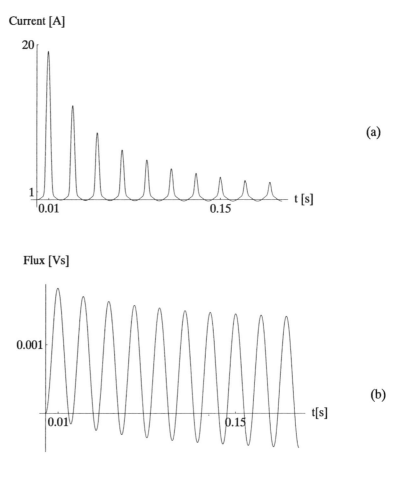

Fig. 1.35 *The current (a) and flux (b) response of a nonlinear magnetic circuit to a sinusoidal applied voltage, according to Eq. (1.83). The initial voltage phase shift, φ, is equal to zero.*

The D.C. decaying component of the transient flux is superimposed on the sinusoidal flux response, causing relatively high amplitudes of the flux at the beginning of the transient. Due to the nonlinear B–H curve of the iron, this high flux is accompanied by very high values of the current, as shown in Fig. 1.35 (a). One can see in Fig. 1.35 (b) that the A.C. component of the flux is negligibly distorted from the sinusoidal shape. Despite the extremely distorted current at the beginning of the transient, the A.C. component of the flux is almost sinusoidal.

Introduction

The influence of the initial phase shift φ in Eq. (1.83) on the inrush current waveform is shown in Fig. 1.36. By varying the initial phase shift φ (shown on the y-axis of this diagram) from 0 to $\pi/2$, one varies the amplitudes of the inrush current. The minimum amplitudes of the inrush current are obtained when the transient begins in a time interval where the applied voltage is at its maximum, i.e. for $\varphi = \pi/2$. A simple physical explanation of this effect is that the flux has only a quarter of the period of the applied voltage to reach its maximum. After this time interval, the applied voltage, and therefore the time rate of change of the flux, change their signs. The flux itself, accordingly, starts to decrease if it was previously increasing, and vice versa.

If the coil is energized in the time interval when the voltage crosses zero level, the flux has enough time (half the period of the applied voltage) to reach the values which correspond to high coil currents. The dependence of the inrush current waveform on the initial phase shift φ in the complete interval of 2π can be reconstructed from Fig. 1.36. The current response in interval $-\pi/2 \leq \varphi \leq 0$ is the mirror image in the y-direction of the response in Fig. 1.36, with respect to the z-axis. The inrush current for the values of the initial phase shift φ in interval $\pi/2 \leq \varphi \leq 3\pi/2$ is equal to the negative value of the inrush current in interval $-\pi/2 \leq \varphi \leq \pi/2$.

The flux response for the same values of parameters of the RL circuit as the one used in Fig. 1.36 is shown in Fig. 1.37. The flux is sinusoidal, following the sinusoidal applied voltage. Similar to the inrush current, the amplitudes of flux at the beginning of the transient are maximum when the initial phase shift φ is equal to 0, and minimum when $\varphi = \pi/2$.

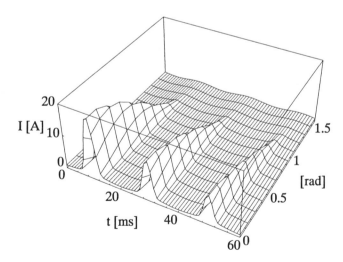

Fig. 1.36 *The inrush current in a nonlinear RL circuit, as a function of time and the initial voltage phase shift φ in Eq. (1.83)*

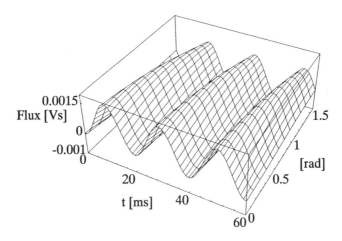

Fig. 1.37 *The flux in a nonlinear RL circuit, as a function of time and the initial voltage phase shift φ in Eq. (1.83)*

The flux response in the complete range of the initial phase shift φ can be reconstructed from the flux waveform in Fig. 1.37 by utilizing the same principles as in the case of the inrush current.

So far, the responses of the magnetic circuit in this example have been calculated by assuming that the magnetic material has an anhysteretic B–H characteristic. When the hysteresis is allowed for, the responses of the circuit differ from the ones shown in the previous figures. If the magnetic material has the magnetization characteristic described in Eq. (1.11) and shown in Fig. 1.3, the response of the magnetic circuit to the applied sinusoidal flux density looks as shown in Fig. 1.38. In this figure the current in the coil during the first two periods of the applied voltage is shown. The parameters of the circuit, along with the solution procedure to obtain the solution in Fig. 1.38, are given in the Appendix.

The applied flux density in this example has zeros for the values of the x-axis variable t/T in Fig. 1.38 equal to k/2, where k = 1,2,3,.... The resultant current shown in this figure, however, has zeros which do not coincide with the zeros of the flux density. This means that the fundamental term of the current has a component which is *in phase* with the applied voltage. This component transfers energy from the source, which covers hysteresis loss in the magnetic material.

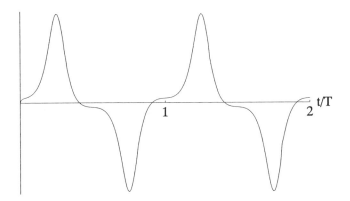

Fig. 1.38 *Current in a coil on magnetic core having hysteretic B–H curve. The applied flux density B is sinusoidal.*

1.3 Parameters of Conductors in Electric Machines

The conductors in an electric machine form the coils, which further build the windings. The conductors are usually placed in the machine's slots, except in some special cases, when the complete winding is in the machine's air gap. A rotor, on which, for the sake of clarity, only two slots are drawn, is shown in Fig. 1.39. The coil which is placed in these two slots has its *active part* – the conductors in the slots, and the *end windings*. When the coil carries current, magnetic flux is generated in the machine. The part of the flux which links other coil(s), besides the one which produces flux, is called *the main flux*. The part of the flux which links only the coil itself, is called *the leakage flux*. The leakage flux has several components, of which the most pronounced are *the end winding leakage flux*, and *the slot leakage flux*. Due to the complex geometry of the end windings, the end winding leakage inductance cannot be found analytically. Instead, empirical formulas, and, lately, a three-dimensional finite element method are used to evaluate it.

The slot leakage flux can be determined accurately by using analytical methods only if the teeth and the part of the yoke between the teeth are not too saturated. In this case the mmf drop in iron can be neglected, when compared to the mmf drop across the slot. The complete slot ampereturns which drive the slot leakage flux cover the mmf drop in the slot.

This is illustrated in Fig. 1.40 (a), in which a flux line of the slot leakage flux created by iz ampereturns is shown. The height of z conductors is negligible as compared to the slot height h. Ampère's circuital law for the flux line in Fig. 1.40 (a) gives

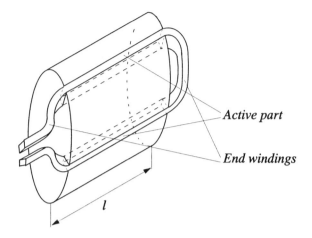

Fig. 1.39 *The winding nomenclature in an electric machine*

$$Hb = iz \qquad (1.86)$$

In this case the field strength H along the slot height is constant. With the substitution $B=\mu_0 H$, one can express the differential of the leakage flux in the slot as

$$d\Phi = \mu_0 \frac{iz}{b} l\, dr \qquad (1.87)$$

where l is the axial length of the machine, as illustrated in Fig. 1.39. Since all the lines of flux are linked with the same number of conductors z, the differential of the flux linkage is equal to

$$d\Psi = z\, d\Phi = \mu_0 \frac{iz^2}{b} l\, dr \qquad (1.88)$$

and the total leakage flux linked by the conductors in one slot

$$\Psi = \int_0^h d\Psi = \mu_0 l z^2 \frac{h}{b} i \qquad (1.89)$$

The value of the slot leakage inductance $L_{\sigma,s}$ is obtained by dividing the slot leakage flux by the current:

Introduction

$$L_{\sigma,s} = \mu_0 l z^2 \frac{h}{b} \tag{1.90}$$

Introducing *the specific slot permeance*, λ_s, defined as

$$\lambda_s = \frac{h}{b} \tag{1.91}$$

one can express the slot leakage inductance as

$$L_{\sigma,s} = \mu_0 l z^2 \lambda_s \tag{1.92}$$

The slot leakage inductance is proportional to the square of the number of conductors in the slot, the machine's length and the specific slot permeance.

The value of the specific slot permeance for various types of slots, shown in Fig. 1.40 (b), (c) and (d), is as follows:

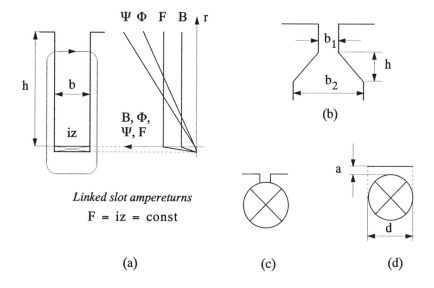

Fig. 1.40 *The slot leakage flux in the part of the slot without current carrying conductors (a) and various shapes of slots (b) – (d)*

- for the trapezoidal part of the slot in Fig. 1.40 (b)

$$\lambda_s = \frac{2h}{b_1 + b_2} \qquad (1.93)$$

where $(b_1+b_2)/2$ is the average length of the flux lines;
- for the round part of the slot in Fig. 1.40 (c) the specific slot permeance does not depend on the radius

$$\lambda_s = \frac{2}{\pi} = 0.637 \qquad (1.94)$$

- for the slot bridge in Fig. 1.40 (d) the specific permeance is a function of the dimensions of the slot wedge, B–H characteristic of the magnetic material of the wedge, and the current in the slot. The dependence of the permeance on the current in the slot in this case is evaluated in a manner similar to that for a tooth, illustrated in Fig. 1.16.

This procedure, however, can sometimes take too long. Therefore, an approximate, experience based computation of the specific slot leakage permeance, which is independent of the type of material of the slot wedge, is performed in the Appendix. The result of this computation for a range of ratio between the slot bridge width a and slot radius d (shown on the x-axis) and magnetic field strength H created by the slot current (y-axis) is shown in Fig. 1.41.

When there are z conductors in the slot, and each of them is carrying current i, as shown in Fig. 1.42, the slot leakage flux is not distributed as the one in a slot without conductors.

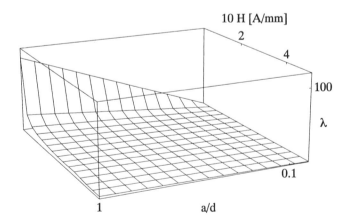

Fig. 1.41 *Specific slot bridge permeance λ of a closed slot shown in Fig. 1.40 (d)*

Introduction

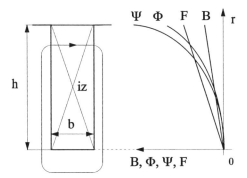

Linked slot ampereturns

$$F(r) = iz\frac{r}{h}$$

Fig. 1.42 *The leakage flux in the slot with z conductors which all carry the same current*

The slot leakage flux in this case does not link all z conductors from the bottom of the slot, as in Fig. 1.40 (a). Instead, the amount of ampereturns linked by the slot leakage flux increases gradually as the coordinate r varies from 0 to h.

Neglecting the skin effect in the conductors, Ampère's circuital law for the loop in Fig. 1.42 gives

$$iz\frac{r}{h} = Hb \tag{1.95}$$

The differential of the slot leakage flux is equal to

$$d\Phi = \mu_0 \frac{izl}{bh} r\, dr \tag{1.96}$$

The number of conductors linked by the slot leakage flux increases proportionally to the radial coordinate r. Therefore, the differential of the leakage flux can be expressed as

$$d\Psi = \frac{r}{h} z\, d\Phi = \mu_0 \frac{iz^2 l}{bh^2} r^2 dr \tag{1.97}$$

After integration from 0 to h, one obtains the slot leakage flux as

$$\Psi = \mu_0 l z^2 \frac{h}{3b} i \qquad (1.98)$$

and the slot leakage inductance

$$L_{\sigma,s} = \mu_0 l z^2 \frac{h}{3b} \qquad (1.99)$$

The specific slot permeance of a slot filled with current carrying conductors can be found from the equation above. It is equal to

$$\lambda_s = \frac{h}{3b} \qquad (1.100)$$

The specific permeance of a slot filled with current carrying conductors is equal to one-third of its value in a slot without conductors. This is caused by the radially distributed excitation which cannot drive as much leakage flux through the slot as the concentrated excitation, represented in Fig. 1.40 (a) by a thin current layer on the bottom of an empty slot.

The slot geometry is often more complex than the single rectangular form analyzed in the previous examples. A slot with complex geometry is usually divided into sections which have simple geometric forms. The total slot leakage flux is then calculated as the sum of the leakage fluxes per each section of the slot.

Two or more adjacent slots in a machine usually carry the same current. The slot leakage flux not only links the conductors in one slot, but all the conductors in n slots, if n adjacent slots carry the same current. Therefore, the total mmf which drives the leakage flux is n times greater, see Fig. 1.43. However, the total reluctance along the path of the leakage flux lines is also increased n times, keeping the slot leakage flux Φ independent of the number of slots n. The flux linkage Ψ is yet n times greater, since the conductors in all n slots link the total slot leakage flux. The total slot leakage inductance of n slots $L_{\sigma,n}$ can therefore be written as

$$L_{\sigma,n} = n L_{\sigma,s} \qquad (1.101)$$

where $L_{\sigma,s}$ is the leakage inductance per slot. Although the number of turns has increased n times, the leakage inductance is only n times, instead of n^2 times, greater than the leakage inductance of only one slot, because the reluctance has also increased n times.

Introduction

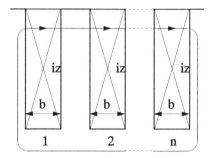

Fig. 1.43 *The slot leakage flux of a group of n slots*

The previous calculations of the leakage slot inductance are derived assuming that the current density along the slot height is constant. This is true when the current in the slot is time invariant. When the current is alternating, the leakage flux in the slot is also alternating. Following Faraday's law, an alternating leakage flux passing through a slot in a tangential direction induces the voltage in the slot conductors (bars). This voltage acts in an axial direction, and drives in the conductors the eddy currents i_e which tend to cancel the slot leakage flux. The total current at any point in the conductor is equal to the sum of the external current I and eddy current i_e at that point. A solid conductor in a slot, carrying an externally driven alternating current I, is shown in Fig. 1.44.

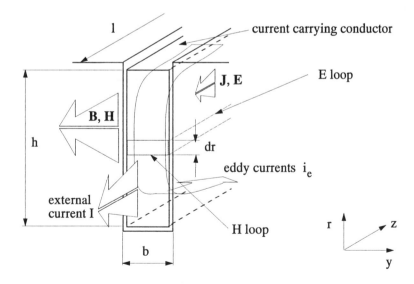

Fig. 1.44 *Illustrating the skin effect quantities in a conductor in the slot of an electric machine*

The total current density, equal to the sum of the current density due to the external current and eddy currents, increases from the bottom to the top of the conductor. The increased current density at the top of the conductor causes more copper loss. Since more lines of flux are concentrated on the top than on the bottom of the slot, the reluctance which defines the slot leakage flux becomes fictitiously greater. The slot leakage inductance therefore decreases. The shift of an A.C. current towards the top of the conductor in a slot caused by time variation of the current is known as *the skin effect*.

The current density and field strength distributions in the conductor along the slot radial coordinate r in Fig. 1.44 can be derived by applying Maxwell's equations. Magnetic field vectors **B** and **H** in the conductor in Fig. 1.44 have only the y-axis (machine tangential) components, and the electric field vectors **E** and **J** have only the z-axis (machine axial) components. Neglecting the difference between the slot width and the conductor width b, Ampère's circuital law for the H loop in Fig. 1.44 gives

$$[H_y(r+dr) - H_y(r)]b = \sum I = \mathbf{J} \cdot d\mathbf{S} = \sigma E_z b\, dr \qquad (1.102)$$

with subscripts y and z denoting the y- and z-components of vectors, respectively. Since

$$\mathbf{H}(r+dr) = \mathbf{H}(r) + \frac{\partial \mathbf{H}}{\partial r} dr \qquad (1.103)$$

one can further write (1.102) as

$$\frac{\partial H_y}{\partial r} = \sigma E_z \qquad (1.104)$$

Applying (1.30) to the E loop in Fig. 1.44, one obtains

$$[E_z(r+dr) - E_z(r)]l = \mu l\, dr \frac{\partial H_y}{\partial t} \qquad (1.105)$$

or

$$\frac{\partial E_z}{\partial r} = \mu \frac{\partial H_y}{\partial t} \qquad (1.106)$$

Solution of the partial differential equations (1.104) and (1.106) gives the distribution of the magnetic and electric field strengths in the bar along the radial coordinate r. The variables of integration in these two equations are time t and the coordinate r.

Assuming a harmonic time variation of the field quantities, one can eliminate time as the variable of integration. This means that the amplitudes of all field vectors are considered constant in time, or that the solution is valid only for steady states. Thus is the system of partial differential equations (1.104) and (1.106) reduced to the system of ordinary differential equations (ODE). According to this reduction, one can write

$$\mathbf{E}(r, t) = \mathbf{E}(r) e^{j\omega t}; \quad \mathbf{H}(r, t) = \mathbf{H}(r) e^{j\omega t}; \quad \mathbf{J}(r, t) = \mathbf{J}(r) e^{j\omega t} \quad (1.107)$$

The solution for each field quantity in Eq. (1.107) is represented as a product of two terms, which are *independent of each other*: the *time solution*, which is an harmonic function since the system is linear, and the *spatial solution*, represented as a function of the spatial coordinate r. Substituting Eq. (1.107) into Eqs. (1.104) and (1.106), and eliminating the magnetic field strength from the resulting system of equations, one obtains the second order ODE for the magnitude of the induced electric field

$$\frac{d^2 E}{dr^2} - j\omega\mu\sigma E = 0 \quad (1.108)$$

Introducing the so called *skin depth*, or *the penetration depth* δ, defined as

$$\delta = \sqrt{\frac{2}{\omega\mu\sigma}} \quad (1.109)$$

one can further write Eq. (1.108) as

$$\frac{d^2 E}{dr^2} - \frac{2j}{\delta^2} E = 0 \quad (1.110)$$

The skin depth is an important quantity, characterized by the electric and magnetic parameters of the material and frequency. At distance δ from the surface the amplitudes of the magnetic and electric field strength are equal to $e^{-1} = 0.368$ of their surface values. The values of the skin depth at 50 and 60 Hz for some materials used in the manufacture of electric machines are given in Table 1.1. As an example, the current density in copper at 50 Hz is equal to 36.8% of its surface value at a distance of approximately 9.6 mm below the surface.

Table 1.1 *The skin depth δ in some materials at low frequencies*

Material	$\mu \left[\dfrac{Vs}{Am} \right]$	$\sigma \left[\dfrac{1}{\Omega m} \right]$	f [Hz]	δ [mm]
Copper	1.256×10^{-6}	50×10^6	50/60	9.6/8.7
Aluminum	1.256×10^{-6}	28×10^6	50/60	13.5/12.3
Regular iron, $\mu_r = 1000$	1.256×10^{-3}	3×10^6	50/60	1.3/1.2
Nonmagnetic iron, $\mu_r = 1$	1.256×10^{-6}	1×10^6	50/60	71/65

The solution of Eq. (1.111) can be expressed as

$$E(r) = C_1 e^{\frac{-1-j}{\delta}r} + C_2 e^{\frac{1+j}{\delta}r} \qquad (1.111)$$

with C_1 and C_2 denoting the constants of integration. Analogously, the solution for the magnetic field strength yields

$$H(r) = \frac{-1-j}{j\omega\mu\delta} \left[C_1 e^{\frac{-1-j}{\delta}r} - C_2 e^{\frac{1+j}{\delta}r} \right] \qquad (1.112)$$

The constants of integration C_1 and C_2 are determined from the boundary conditions. The first boundary condition for the magnetic field strength can be written in the form

$$H(0) = 0 \qquad (1.113)$$

i.e., the magnetic field strength at the bottom of the slot is equal to zero, since the ampereturns which drive the slot leakage flux at the bottom of the slot are also equal to zero. The second boundary condition can also be derived from Ampère's circuital law, which is in this case written for the flux line at the top of the conductor

$$H(h) = \frac{I}{b} \qquad (1.114)$$

Introduction

since this flux line links the complete conductor current I. Applying the boundary conditions from Eqs. (1.113) and (1.114) to the solution in Eq. (1.112), one obtains

$$C_1 = C_2 = \frac{j}{1+j} \frac{\omega\mu\delta}{b} \frac{I}{2\sinh\frac{1+j}{\delta}h} \qquad (1.115)$$

Substitution of the values of constants of integration from Eq. (1.115) into Eqs. (1.111) and (1.112) gives:
- the distribution of the magnetic field strength along the r-axis

$$H(r) = \frac{I}{b} \frac{\sinh\frac{1+j}{\delta}r}{\sinh\frac{1+j}{\delta}h} \qquad (1.116)$$

- the distribution of the induced electric field strength due to the skin effect along the r-axis

$$E(r) = \frac{j}{1+j}(\omega\mu\delta)\frac{I}{b}\frac{\cosh\frac{1+j}{\delta}r}{\sinh\frac{1+j}{\delta}h} \qquad (1.117)$$

- the distribution of the current density due to the skin effect along the r-axis

$$J(r) = \frac{j}{1+j}(\omega\mu\sigma\delta)\frac{I}{b}\frac{\cosh\frac{1+j}{\delta}r}{\sinh\frac{1+j}{\delta}h} = \sigma E(r) \qquad (1.118)$$

The electromagnetic field vectors H, E, and J in Eq. (1.116) – (1.118) are complex numbers. They are shifted in phase to the externally driven current I. The real components of H, E, and J are in phase with I, whereas their imaginary components lead the current I by 90°. The imaginary components of H, E, and J are caused by the component of the induced voltage in the conductor which leads the conductor current by 90°.

Equation (1.118) is graphically represented in Fig. 1.45, in which the trajectory of the current density vector in a three dimensional space defined by the current density's real and imaginary parts and the conductor height is shown. This spiral trajectory is obtained by connecting the points determined by the real and imaginary parts of the current density vector along the conductor height in the slot. The copper conductor in this example is 50 mm high, and 10 mm wide. The externally driven current of 500 A has the frequency of 50 Hz. If the current were uniformly distributed over the whole cross-sectional area of the conductor, as it is in the case of D.C. current, its density would be equal to 500 A/(50 mm × 10 mm) = 1 A/mm^2. At 50 Hz, however, the skin effect is so strong that the maximum value of the current density on the top of the conductor reaches 7 A/mm^2.

One can see in Fig. 1.45 that the amplitude of the current density vector increases as the slot radial coordinate varies from 0 (bottom of the conductor) to 50 mm (top of the conductor). Simultaneously, the phase shift of the current density vector varies in such a manner that the characteristic spiral form in Fig. 1.45 is obtained. Although the current density on the top of the conductor is significantly higher than the one on its bottom, the total current, calculated as an integral of the current density, is always equal to the conductor current I. This can be checked by integrating the expression in Eq. (1.118) from 0 to h. The integral of the imaginary component of current density J in Eq. (1.118) from the bottom to the top of the conductor is always equal to zero. The integral of the real component of the current density J in Eq. (1.118) in the same limits is equal to the applied current I.

Utilizing equations which convert the hyperbolic functions of complex arguments into combinations of hyperbolic and trigonometric functions of real arguments:

$$\sinh(x + jy) = \sinh x \cos y + j \cosh x \sin y \qquad (1.119)$$

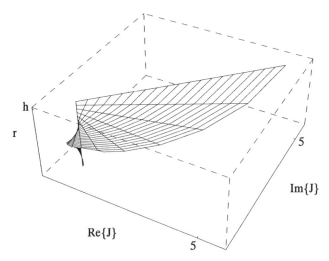

Fig. 1.45 *The trajectory of the current density vector in the conductor in Fig. 1.44*

Introduction

$$\cosh(x + jy) = \cosh x \cos y + j \sinh x \sin y \qquad (1.120)$$

one can express the maximum value of the magnetic field strength from Eq. (1.116) as

$$H(r) = \frac{I}{b} \sqrt{\frac{\cosh(2\frac{r}{\delta}) - \cos(2\frac{r}{\delta})}{\cosh(2\frac{h}{\delta}) - \cos(2\frac{h}{\delta})}} \qquad (1.121)$$

The magnetic field quantities are always expressed in terms of their *maximum values*, because the maximum value of the magnetic field density or field strength define the level of saturation in the material. Sometimes the R.M.S. values of magnetic quantities are used which, however, have no physical meaning. The relationship between the maximum value of the flux density and its artificially defined R.M.S. value depends on the character of the B–H curve and the position of the operating point on it. The R.M.S. value of a magnetic field quantity cannot give any information about the level of saturation; therefore, it is useless.

The distribution of the absolute value of the magnetic field strength H(r) in Eq. (1.121) along the coordinate r in a copper bar which is 50 mm high and has the conductivity $\sigma = 55 \times 10^6 \, \Omega^{-1} m^{-1}$ in the range of angular frequencies between $\omega = 1$ and 1000 rad/s is shown in Fig. 1.46.

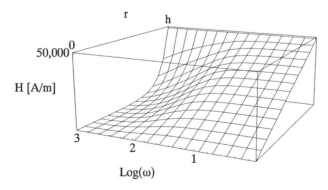

Fig. 1.46 *The distribution of the absolute value of the magnetic field strength in the range of angular frequencies $1 < \omega < 1000$ rad/s, along the coordinate r in a copper bar which is 50 mm high*

The magnetic field strength value at the top of the bar is equal to H=I/b=50000 A/m because it is determined only by the amplitude of the conductor current and the slot width, and is independent of the value of angular frequency ω. At low values of the angular frequency ω, the magnetic field strength distribution along the r-axis is linear, as shown in Fig. 1.42, in which the field strength distribution was analyzed assuming that skin effect was negligible. When the angular frequency begins to increase, the magnetic field strength distribution along the bar height becomes distorted, as a consequence of the skin effect. The majority of the current gets shifted to the bar surface which results in lower values of the field strength in the lower portion of the bar.

The A.C. electric field quantities are always expressed in terms of their R.M.S. values. The R.M.S. value of the voltage or current is a measure of the energy which they can transfer. The R.M.S. value of the current density in Eq. (1.118) is

$$J(r) = \frac{I}{\sqrt{2}} \frac{1}{b\delta} \sqrt{\frac{\cosh(2\frac{r}{\delta}) + \cos(2\frac{r}{\delta})}{\cosh(2\frac{h}{\delta}) - \cos(2\frac{h}{\delta})}} \qquad (1.122)$$

The distribution of the R.M.S. value of the current density along the bar height for the same parameters used in the previous example is shown in Fig. 1.47. As one can see in this figure, at low values of angular frequency ω the current density is distributed uniformly along the bar height.

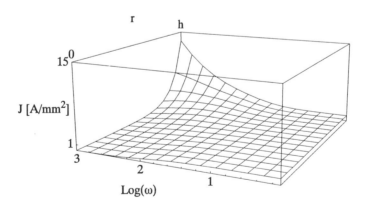

Fig. 1.47 *The distribution of the R.M.S. value of the current density per each ampere of the bar current, in the range of angular frequencies 1 < ω < 1000 rad/s, along the coordinate r in a copper bar which is 50 mm high*

Introduction

As the angular frequency increases, more current is suppressed towards the slot opening. At the same time, the current density on the bottom of the bar decreases, as compared to its D.C. value. The decrease of the current density is caused by an increased amplitude of eddy currents which act against the bar current on the bottom of the bar.

The I^2R losses in the conductor increase as the frequency of the external current I increases, without changing its amplitude. Although the amplitude of the current density J in the lower part of the conductor decreases as the frequency increases, thus decreasing the losses, the increase of the amplitude of J in the upper part of the conductor is disproportionately larger. In the final effect, the I^2R losses in the conductor increase as the frequency increases.

When the *time variation* of the current density is analyzed, one has to allow for the phase shift γ of the current density, which is a function of the radial coordinate, the frequency of the applied current, and the electromagnetic parameters of the conductor. γ is the angle between the external current and the current density vector at a given conductor height. The phase shift γ of the current density can be calculated as the argument of the function J(r) in Eq. (1.118)

$$\gamma = \frac{\pi}{4} + \operatorname{atan}\left(\tanh\left(\frac{r}{\delta}\right)\tan\left(\frac{r}{\delta}\right)\right) - \operatorname{atan}\left(\coth\left(\frac{h}{\delta}\right)\tan\left(\frac{h}{\delta}\right)\right) \qquad (1.123)$$

with π/4 denoting the argument of the term j/(1+j) in Eq. (1.118).

The time variation of the current density in the bar can be obtained by evaluating Eq. (1.107), or by multiplying the spatial current density distribution, expressed in Eq. (1.122), by an harmonic function which represents the current in the conductor. The result of such a multiplication with a cosine function having a frequency of 50 Hz for a bar with a height of 50 mm is shown in Fig. 1.48.

The damped wave character of the current density in the bar is obvious from Fig. 1.48. If the argument γ in Eq. (1.123) is equal to zero, the current densities in all the layers along the bar height are in phase.

When γ≠0, there exist phase shifts between the current densities in adjacent layers, and the current density wave travels along the bar height. The amplitude of the current density increases exponentially in the direction from the bottom (r=0) to the top (r=50 mm) of the slot.

The current and field strength distributions caused by the skin effect change the fundamental electromagnetic parameters of the conductor: its inductance and resistance. The variation of the conductor inductance as a function of frequency can be found by evaluating the magnetic energy stored in the conductor's field. According to Eq. (1.49), the instantaneous value of the magnetic energy, $W_{mg}(t)$, is equal to

$$W_{mg}(t) = \int i\, d\Psi \qquad (1.124)$$

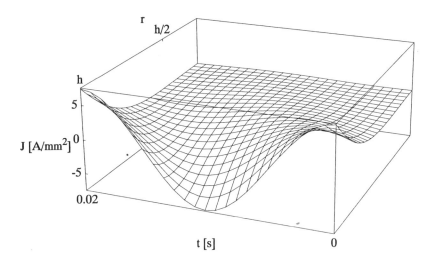

Fig. 1.48 *The time variation of the absolute value of the current density J in the bar along the bar height r during one period (20 ms) of the bar current*

The distribution of the current along the coordinate r can be found either directly, from Eq. (1.116), or by integrating the current density along the coordinate r. In both cases one obtains

$$i(r,t) = i(t) \frac{\sinh \frac{1+j}{\delta} r}{\sinh \frac{1+j}{\delta} h} \qquad (1.125)$$

Since there is only one conductor in the slot, $d\Psi$ is equal to $d\Phi$, and therefore

$$d\Psi = B(r)\, l\, dr = \mu H(r)\, l\, dr \qquad (1.126)$$

The instantaneous value of the magnetic field energy is now

$$W_{mg}(t) = \mu l \frac{i^2(t)}{b} \frac{1}{\sinh^2 \frac{1+j}{\delta} h} \int_0^h \sinh^2 \left(\frac{1+j}{\delta} r\right) dr \qquad (1.127)$$

Introduction

After integrating the expression (1.127) and averaging the calculated instantaneous value of the magnetic energy in one interval of the sinusoidal bar current $i(t) = I_m \sin \omega t$, one obtains

$$\overline{W}_{mg} = \mu l \frac{I_m^2}{2b} \frac{\delta}{2} \frac{\sinh 2\frac{h}{\delta} - \sin 2\frac{h}{\delta}}{\cosh 2\frac{h}{\delta} - \cos 2\frac{h}{\delta}} \qquad (1.128)$$

Introducing a dimensionless *normalized conductor height* ξ which tells how many times the conductor is higher than the skin depth,

$$\xi = \frac{h}{\delta} \qquad (1.129)$$

and applying Eq. (1.99) to substitute the value of the leakage inductance $L_{\sigma,s}$, one can express the conductor inductance due to skin effect L_{AC} as

$$L_{AC} = L_{\sigma,s} \frac{3}{2\xi} \frac{\sinh 2\xi - \sin 2\xi}{\cosh 2\xi - \cos 2\xi} \qquad (1.130)$$

As one can see in Fig. 1.49, in which the curve described in Eq. (1.130) is shown, the inductance of the conductor decreases as the normalized slot height increases.

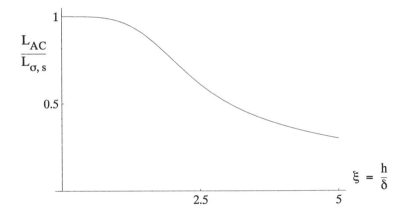

Fig. 1.49 *The dependence of the ratio between the conductor inductance due to skin effect, L_{AC}, and the inductance for the D.C. current, $L_{\sigma,s}$, on the value of the normalized conductor height*

Since the normalized slot height is proportional to the square root of the frequency, the inductance of the conductor decreases as the frequency increases. A simple physical explanation of this effect can be derived from the fact that the flux lines, along with the current, are suppressed at higher frequencies toward the top of the conductor.

Keeping the current constant, the flux linkage decreases as the frequency increases because the slot area through which the flux lines pass decreases. As a final effect, the conductor inductance at higher frequencies is smaller.

The dependence of the conductor resistance due to the skin effect on the electromagnetic characteristics of the material and on the frequency of the sinusoidal current can be derived by utilizing a principle similar to the one applied in the derivation of the dependence of the leakage inductance. Here the so-called *equivalent resistance* R_{AC} is introduced which, multiplied by the square of the conductor current without the skin effect, gives the loss equal to that caused by the skin effect in the conductor.

One can write for the differential of losses

$$dP = VdI = EIJbdr = \frac{lb}{\sigma}J^2 dr \qquad (1.131)$$

Utilizing Eq. (1.122), the total loss is

$$P = \frac{1}{2\sigma b \delta^2} \frac{I^2}{\cosh(2\frac{h}{\delta}) - \cos(2\frac{h}{\delta})} \int_0^h \left[\cosh(2\frac{r}{\delta}) + \cos(2\frac{r}{\delta})\right] dr = I^2 R_{DC} \qquad (1.132)$$

The equivalent resistance due to skin effect R_{AC} can be expressed from the equation above as

$$R_{AC} = R_{DC} \xi \frac{\sinh 2\xi + \sin 2\xi}{\cosh 2\xi - \cos 2\xi} \qquad (1.133)$$

The variation of the ratio between the A.C. and D.C. resistance, as a function of the normalized slot height ξ, is shown in Fig. 1.50. The ratio between the A.C. and D.C. resistance increases as the conductor height increases, provided that all other parameters which define ξ are constant. The penetration depth δ for a given frequency and material parameters is constant. An increase of the conductor height in the range $\xi > 1.4$ does not contribute to the reduction of the A.C. resistance, since the alternating current flows only in a layer in the upper part of the conductor. For higher values of conductor height h, the cross-sectional area through which flows the A.C. current remains constant as the conductor height increases. Therefore, the resistance for the A.C. current does not change, although the conductor height, along with its area, increases.

Introduction

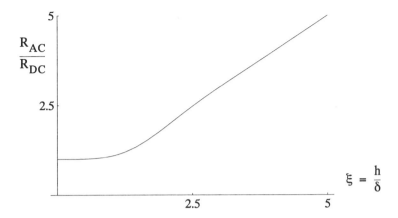

Fig. 1.50 *The increase of the conductor resistance due to skin effect as a function of the normalized conductor height*

The D.C. resistance, on the other hand, decreases as the conductor height increases. The final effect is that the ratio between the A.C. and D.C. resistance increases along with the increase of the relative slot height ξ.

The increase of the rotor bar resistance caused by skin effect is utilized to improve the starting performances of squirrel cage induction machines. Squirrel cage induction machines are usually built with deep bars in the rotor slots. Due to the skin effect, the A.C. resistance of the bars at low rotor speeds is much higher than at the rated point, because the frequency of the rotor currents is higher at low rotor speeds. A higher rotor resistance generates an additional component of the starting torque and decreases the starting current. The physics of the influence of skin effect on the characteristics of the deep bar squirrel cage induction machine is discussed in Chapter 6, devoted to induction machines.

If **several conductors** are placed in a slot of an electric machine one above another, and if they are connected in series, so as to carry the same current I, the current in every conductor except the lowest is exposed to two fluxes: its own slot leakage flux, and the slot leakage flux created by the current I flowing through each of the conductors below the observed. The distribution of the field quantities in the conductors in this case is obtained by solving Eqs. (1.104) and (1.108) with boundary conditions which can be derived by utilizing Fig. 1.51.

The ampereturns linked by the first p–1 conductors in the slot are equal to (p–1)I; the ampereturns linked by the first p conductors are equal to pI. These two data are used to define the field strength on the bottom and on the top of the p-th conductor. Utilizing such obtained values instead of those expressed in Eq. (1.113) and (1.114), one can write the constants of integration C_1 and C_2 in Eqs. (1.111) and (1.112) as

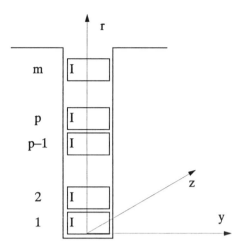

Fig. 1.51 *A slot with m conductors connected in series, each of them carrying the current I. The height of each conductor is h.*

$$C_1 = \frac{j\omega\mu\delta}{(1+j)b} \frac{pI - (p-1)Ie^{\frac{(1+j)h}{\delta}}}{2\sinh\frac{(1+j)h}{\delta}} \qquad (1.134)$$

$$C_2 = \frac{j\omega\mu\delta}{(1+j)b} \frac{pI - (p-1)Ie^{-\frac{(1+j)h}{\delta}}}{2\sinh\frac{(1+j)h}{\delta}} \qquad (1.135)$$

with h denoting the height of each conductor. The current density distribution can be written for the p-th conductor as

$$J = \frac{j\omega\mu\sigma\delta}{(1+j)b} \frac{pI\cosh\frac{1+j}{\delta}y - (p-1)I\cosh\frac{(1+j)(h-y)}{\delta}}{\sinh\frac{(1+j)h}{\delta}} \qquad (1.136)$$

Introduction 71

where the y coordinate in each conductor varies from 0 on the bottom of the conductor to h on its top. Equation (1.136) is also valid for p=1, i.e. when there is only one conductor in the slot. By integrating the current density in this equation for y between 0 and h, and multiplying it by the conductor width b, one obtains as a result the current I in each conductor.

To illustrate previous calculations, the dependences of the real, imaginary and absolute values of the current density on the radial coordinate r are shown in Fig. 1.52. There are five conductors in the slot in this example, each of them carrying the same current I.

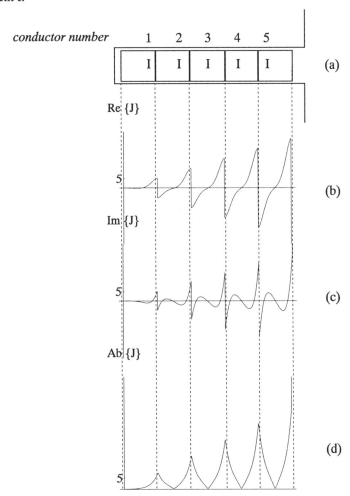

Fig. 1.52 *Five conductors connected in series, one above another, in a slot of an electric machine, carrying the same current I (a), and distributions of the real (b), imaginary (c) and absolute value (d) of the current density [A/mm^2] in the conductors*

The height of each conductor is 50 mm, and its width is 10 mm. The conductor material is copper, the frequency of the external current I of 500 A is 50 Hz. The current density is again expressed in A/mm^2. The insulation between the conductors is assumed very thin in this example. Therefore, the real and imaginary components of the current density change stepwise for the values of the coordinate r which are integer multiples of the conductor height. The dependence of the current density distribution in a conductor on the slot leakage flux created by the current flowing in the conductors below the one being analyzed can be clearly seen in all conductors, except in the lowest. The external leakage fluxes induce in these conductors eddy currents which act against the conductor currents in the conductors' lower halves. In the conductors' upper halves, however, the eddy currents support the conductor currents. This distribution can be seen in both the real (Fig. 1.52 (b)) and imaginary (Fig. 1.52 (c)) component of the current density. Such effect, however, cannot occur in the lowermost conductor, because it is exposed only to its own leakage flux. The current in the lowermost conductor cannot create a leakage flux capable of driving eddy currents in the lower part of the conductor which could cancel the conductor current. In this hypothetical case the leakage flux, as the source of eddy currents, would also have been cancelled.

Although the average value of the current density distributions in Fig. 1.52 (b) and (c) is always equal to I, independently of their extreme values on the bottoms and on the tops of the conductors, the absolute value of the current density increases substantially along the slot radial coordinate r. The absolute value of the current density is an important quantity, because it determines the copper losses in the conductors. The absolute value of the current density in this example is shown in Fig. 1.52 (d).

The trajectory of the current density vector in the third conductor in Fig. 1.52 (a) is shown in Fig. 1.53.

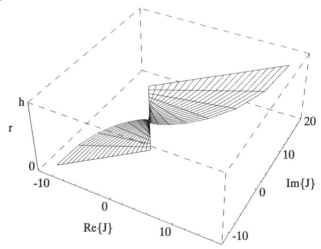

Fig. 1.53 *The trajectory of the current density vector in the third conductor in Fig. 1.52*

Introduction

The unit for the current density is again A/mm^2 in this figure. In every conductor, except the lowest, the current density distribution has qualitatively the same form as the one shown in Fig. 1.53, because the currents in those conductors are exposed both to the external and own slot leakage flux. The high amplitudes of the current density also mean high copper losses in the conductors. Therefore, the height of conductors in the slots of electric machines is always limited to a much lower value than the 50 mm used in this example. When these conductors have to carry large amounts of currents, they are manufactured from a large number of strands, as shown in Fig. 1.54 (c), instead of a solid piece of copper. The strands are insulated from each other, preventing the current from shifting towards the top or the bottom in such a conductor.

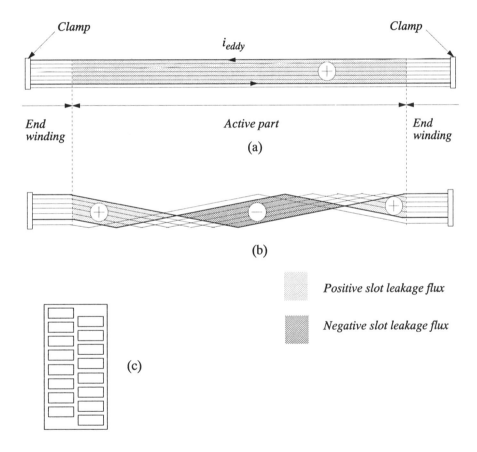

Fig. 1.54 *The eddy currents in untransposed (a) and 360° transposed (b) strands of a conductor, and the cross-sectional view of the conductor with the strands (c)*

In order to perform their function of parallel path for a part of the conductor current, all the strands which form a conductor have to be connected in parallel by the end winding connectors (clamps). However, the parallel connections of the strands in the end windings form loops within the conductor, one of them being drawn bold in Fig. 1.54 (a). The slot leakage flux induces the voltages in these loops, which then drive the so- called *circular currents*. The same mechanism which generates the eddy currents in a solid bar causes the circular currents in the strands. The division of the conductor into strands in itself does not prevent the occurrence of eddy currents in the conductor, because of inevitable end winding connectors which provide the return path for the eddy currents.

However, if the strands are transposed, as illustrated in Fig. 1.54 b), the total leakage flux which drives the eddy currents in the strands can be substantially decreased and, in some cases, completely eliminated. Such a bar is called a *Roebel bar*, after its inventor. When the strands are 360° transposed, as shown in Fig. 1.54 (b), each of them passes along the length of the active part of the machine at various slot heights and always ends up at the same slot height at which it entered the active part. This way, *the slot leakage flux* linked by any two strands which form a loop within the conductor, as illustrated by the bold strands in Fig. 1.54 (b), is always equal to zero. The voltages induced by the slot leakage flux and the circular currents are, therefore, also equal to zero for 360° transposition.

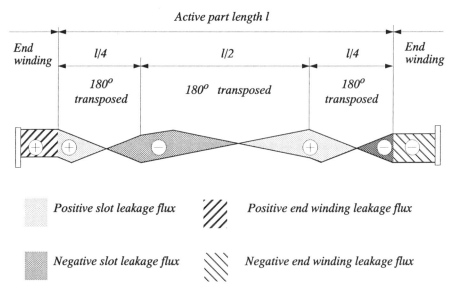

Fig. 1.55 *The 540° transposition of twisted strands in the active part of the conductor*

Introduction

In large electric machines the induced voltage due to leakage flux in the end windings cannot usually be neglected. The induced voltage due to end winding leakage flux within a conductor can be minimized if the so-called 540° transposition in the machine's active part is utilized, as illustrated in Fig. 1.55.

The transposition of strands in slots shown in Fig. 1.55 results in zero slot leakage flux linkage in every current loop defined by the strands and end winding connectors.

The end winding leakage flux on one side of the machine is equal to the negative end winding leakage flux on the machine's opposite side. Therefore, the total leakage flux linked by the strands which are 540° transposed is equal to zero, and the induced voltage which could drive the circular currents in those strands is also equal to zero.

2 Windings, Currents and Air Gap Magnetomotive Force

Although every part of an electric machine plays an important role in electromechanical energy conversion, one must agree that the windings are the heart of the machine. An electric machine communicates with the outer electromagnetic world by means of its electric terminals, to which the ends of the machine's windings are connected. The current carrying conductors in the windings create a magnetic field, which is a catalyst in the energy conversion process from electrical to mechanical form and vice versa. A profound understanding of the processes in a machine is possible only with a clear vision of the action of the windings.

2.1 Classification and Basic Terms

The basic function of a winding in an electric machine is to create a magnetic field which, alone or together with the field(s) created by other winding(s), generates the force between the stator and the rotor.

If all turns of the winding are sited in the same slot or in the interpolar space, the winding is called *concentrated*. The concentrated winding on the stator side, shown in Fig. 2.1 (a), is typical for D.C. machines, switched reluctance and stepper motors, whereas the concentrated winding on the rotor side, shown in Fig. 2.1 (b), is characteristic of low speed synchronous machines, or hydrogenerators.

When the winding in the machine is separated into coils, all of them being sited in different slots, it is said to be *distributed*. If each coil of the distributed winding has a different pitch, it is called *concentric*. The axes of coils in concentric windings always coincide in space, as shown in Fig. 2.2 (a). If all coils have an equal pitch, the winding can be *lap* or *wave*, depending on the way the conductors are connected. An example of a lap winding is shown in Fig. 2.2 (b), and of a wave winding in Fig. 2.2 (c). When placed into the slots, each coil can occupy either the whole slot, or only half. In the first case, the winding is said to be *single layer*, and in the second case it is *double layer*.

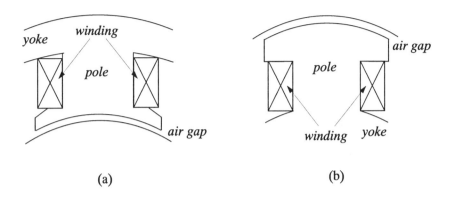

Fig. 2.1 *Concentrated winding on the stator side – a D.C. machine (a), and on the rotor side – a synchronous machine (b)*

The double layer windings allow more freedom in obtaining the desired spatial distributions of the air gap quantities. Therefore, they are used more frequently than the single layer ones.

A special type of winding in electrical machines, derived from a distributed winding, is *the squirrel cage winding*. This winding is utilized in the rotors of squirrel cage induction machines and as a damping winding in the rotors of synchronous machines. The parts of the squirrel cage winding placed in the slots are called *bars,* whereas the conductors which connect the bars from the front and back sides of the rotor are the *rings*. The squirrel cage winding has as many bars as rotor slots and two rings, as shown in Fig. 2.3.

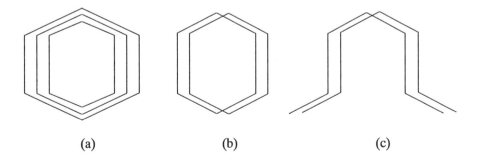

Fig. 2.2 *The coils in an electric machine forming concentric (a), lap (b) and wave winding (c)*

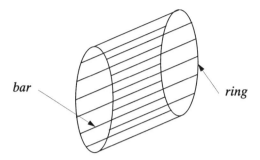

Fig. 2.3 *Squirrel cage winding*

The squirrel cage winding is unique among all types of windings, since the currents in the rings of the squirrel cage winding can flow from or to any bar.

The windings in a polyphase machine are grouped into *phases*. The number of stator phases m determines how many independent currents flow through the stator winding. Since the currents in the bars of a squirrel cage winding are usually all out of phase, the number of phases of a squirrel cage winding is equal to the number of slots N in which the winding is placed. In the special case when the number of slots per pole pair p is even, a squirrel cage has N/p phases.

The peripheral distance between two adjacent zeros of the fundamental components of the air gap quantities is called the *pole pitch* τ_p. Denoting by D the average air gap diameter, and by p the number of pole pairs, the pole pitch τ_p is

$$\tau_p = \frac{D\pi}{2p} \tag{2.1}$$

The *slot pitch* τ_s is defined in a similar manner

$$\tau_s = \frac{D\pi}{N} \tag{2.2}$$

where N is the number of teeth (slots).

The machine's electromagnetic quantities repeat periodically along the air gap periphery. Their fundamental period is $2\tau_p$, which corresponds to the electrical angle 2π. Therefore, the relationship between the mechanical (geometric) angle α and the corresponding electrical angle α_{el} can be written as

$$\alpha_{el} = p\alpha \tag{2.3}$$

The relationship between the machine's tangential, radial and angular coordinates is illustrated in Fig. 2.4.

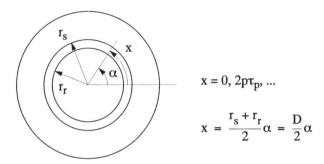

Fig. 2.4 *The relationship between the machine's coordinates*

2.2 Linear Current Density – Current Sheet

The number of conductors which can be placed in the air gap of an electric machine is finite; hence, the spatial distribution of electric and magnetic quantities produced by the windings changes stepwise. A mathematical description of discontinuously (stepwise) changing functions is substantially more complicated than that of continuous functions. Therefore, instead of currents in the conductors, an auxiliary quantity, called *the current sheet,* is often used in the analysis of electric machines.

The current sheet $A(x)$ has the meaning of the linear current density along the rotor periphery. The linear current density is defined as

$$A(x) = \frac{d}{dx} i(x) \qquad (2.4)$$

The current can flow freely in the rotor peripheral cylinder only in some special machines. In these machines it is meaningful to consider the current sheet as being continuously distributed along the peripheral coordinate x, so that it is mathematically sound to evaluate the derivative of the current with respect to the coordinate x.

When the currents are located in the conductors placed around the rotor periphery, a modified definition of the linear current density is used. According to this definition, the current sheet created by the current I which flows through the conductor the width of which is b, can be written as

$$A = \frac{I}{b} \tag{2.5}$$

The relationship between the currents in the conductors and the corresponding values of the current sheet are shown in Fig. 2.5. The spatial coordinate x, introduced in Eq. (2.4) and shown in Fig. 2.5, is called the *tangential*, or *circumferential coordinate*, due to the machine's cylindrical geometry. The relationship between the machine's tangential, radial and angular quantities is shown in Fig. 2.4.

If the dependence of the current sheet A(x) on the coordinate x is known, the total current $i_{a,b}$ in interval [a,b] of the rotor periphery can be expressed as

$$i_{a,b} = \int_a^b A(x)\,dx \tag{2.6}$$

Here the term *total current* is used, which is expressed in amperes. However, the real dimensions of the total current are *ampereturns*, i.e. the product of the current and the number of conductors in the gap through which the current flows. The total current $i_{a,b}$ in interval [a,b] is equal to the area beneath the curve A(x) in the same interval, in accordance with the geometrical meaning of a definite integral. This is illustrated in Fig. 2.6, in which the total current $i_{a,b}$ is represented by the shaded area. The currents shown in Fig. 2.5 create a magnetic field with strength **H**, which can be evaluated by applying Ampère's circular law for currents, Eq. (1.14).

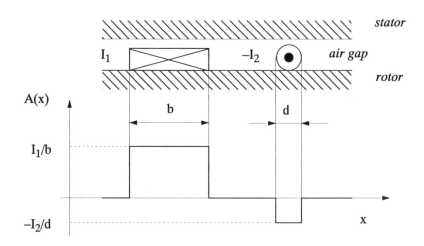

Fig. 2.5 *Current carrying conductors in the air gap and their current sheet*

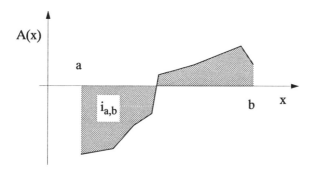

Fig. 2.6 *The current as an integral of the current sheet*

When the current sheet in interval [a,b] is known, one can express the total current $i_{a,b}$ as

$$i_{a,b} = \int_a^b A(x)\,dx = \oint \mathbf{H} \cdot d\mathbf{l} \qquad (2.7)$$

The total current can be expressed either in terms of the integral of the current sheet along the peripheral coordinate x, or in terms of the integral of the magnetic field strength along the integration contour, as illustrated in Fig. 2.7. Every conductor in this figure carries a different current, creating the current sheet distribution denoted by A(x). The longest part of the integration contour in Fig. 2.7 is placed in the stator and rotor irons. In an ideal situation, when the permeability of iron is infinitely high, the field strength in iron is equal to zero.

In reality, the magnetic field strength in iron is much smaller than in the air gap. Nevertheless, the mmf drop in iron must also be taken into account, since the lines of flux in iron are much longer than in the air gap. The length of the flux lines in iron makes the mmf drop in iron influential when the total mmf drop in the magnetic circuit is calculated.

Due to the nonlinear character of the iron core B–H curve, the line integral of the field strength in iron cannot usually be evaluated analytically. Instead, the line integral in Eq. (2.7) is expressed analytically only for a part of the integration contour in the air gap, and then the mmf drop in iron is added. Neglecting for the moment the mmf drop in iron, one can write

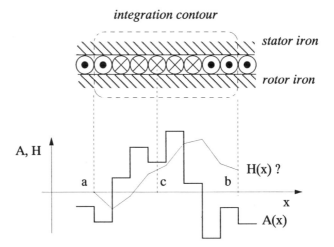

Fig. 2.7 *Conductors in the air gap, each of them carrying a different current. The corresponding current sheet A(x) is changing stepwise along the rotor periphery. The integral of the current sheet within the interval [a, b] is denoted by H(x)?*

$$i_{a,b} = \int_a^b A(x)\,dx = H_b \delta_b + H_a \delta_a + C \tag{2.8}$$

where H_a is the magnetic field strength in the air gap at point a, while H_b is the magnetic field strength in the air gap at point b. The air gap width at point a is denoted by δ_a and at point b by δ_b.

A definite integral is fully defined only if the constant of integration is known. For the current sheet defined in Eq. (2.8), the constant of integration is evaluated from the physical condition that there can be no unipolar flux in a heteropolar machine. The lines of flux in a heteropolar machine can pass through the rotor cylinder only radially, as shown in Fig. 1.8 (a).

Maxwell's divergence equation for the magnetic field in heteropolar machines can be written in the form of a closed surface integral for the rotor as

$$\oint \mathbf{B} \cdot d\mathbf{S} = 0 \tag{2.9}$$

Since there exists no flux passing through the bases of the rotor cylinder and all lines of flux are perpendicular to the rotor lateral surface, vector equation (2.9) can be written in scalar form as

$$\int B dS = \mu_0 \int H dS = 0 \qquad (2.10)$$

because the permeability μ of air, for which the calculation is carried out, is equal to the absolute permeability μ_0 of free space. The integration in the equation above is performed on the rotor lateral surface. Since the magnetic quantities repeat periodically every two poles, one can further write

$$l \int_0^{2p\tau_p} H(x)\,dx = 0 \qquad (2.11)$$

where τ_p denotes the machine's pole pitch, l is its axial length and p is the number of pole pairs. With substitution

$$dx = r\,d\alpha \qquad (2.12)$$

Eq. (2.11) becomes

$$\int_0^{2\pi} H(\alpha)\,d\alpha = 0 \qquad (2.13)$$

The result obtained in Eq. (2.13) can be geometrically expressed as a condition requiring that the area beneath the curve $H(\alpha)$ in interval $[0, 2\pi]$ has to be equal to zero. When this condition is used to define the constant of integration C in Eq. (2.8) a result is obtained, according to which the constant C is equal to zero if the starting point for integration is placed at the maximum or minimum of the current sheet. The physical explanation of this condition lies in the fact that the *magnetic field strength is equal to zero at maxima and minima of the current sheet*. The current sheet and the air gap field strength are 90 degrees out of spatial phase. This important relationship between the flux density, or field strength, on one side, and the current sheet on the other, can be illustrated with an example of a single coil, Fig. 2.8 (a), and a winding in the machine's air gap, Fig. 2.8 (b).

The maxima and minima of the field intensity and magnetic flux density of the coil in Fig. 2.8 (a) lie on the axis A–A. The zeros of the field strength and flux density are placed in the conductors through which the line B–B passes. The centres of the positive and negative current sheets of the coil in Fig. 2.8 (a) are colinear with the line B–B, whereas the zeros of the current sheet distribution coincide with line A–A.

The distributions of the current sheet, field strength and flux density in the machine in Fig. 2.8 (b) are easier to visualize than in the coil in Fig. 2.8 (a), due to the circular distribution of the current carrying conductors around the rotor periphery. However, in a machine one has also to consider that the distributions of the magnetic field quantities in the air gap are different from the corresponding distributions in the stator and rotor yokes. The current sheet in a machine, created by the current carrying conductors, is equal to zero on line A–A. The flux density and field strength *in the air gap* are maximum and minimum on the line A–A, and equal to zero on the line B–B. However, the maxima and minima of the flux density and field strength *in the stator and rotor yoke* lie on the line B–B, whereas their zeros are on the line A–A.

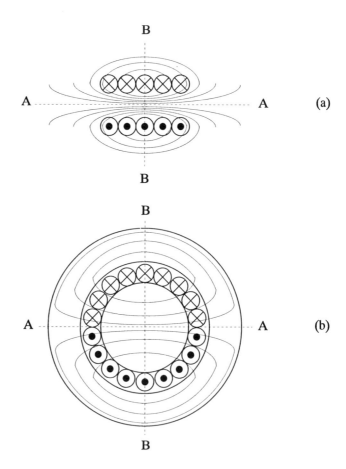

Fig. 2.8 *Approximate distributions of the lines of flux in a coil (a) and in a machine (b)*

Windings, Currents and Air Gap Magnetomotive Force

The yoke quantities in a heteropolar electric machine are shifted by 90 electrical degrees to the corresponding gap quantities. The yoke flux density and field strength are in phase with the air gap current sheet. This conclusion can be useful when the correct position of the winding in rectangular stators has to be chosen, in order to avoid high values of flux at the narrowest parts of the stator yoke.

Since the magnetic field strength in the air gap is equal to zero at maxima and minima of the current sheet, one can define the integration contour in Eq. (2.7) in such a manner that it passes exactly through one of these points. In this case, the value of the field strength H in the air gap at point c in Fig. 2.7 is equal to zero. Therefore, the mmf drop F in the air gap in a radial direction at point c is also equal to zero. This can be clearly visualized in Fig. 2.9, which is obtained by redrawing Fig. 2.7 in such a manner that the integration contour for the field strength is positioned so that one of its sides lies in the centre of the conductors carrying current in the same direction. Ampère's circuital law in this case reduces to

$$F_{x0} = \oint H dl = H_{x0} \delta_{x0} = \int_0^{x0} A(x) \, dx \qquad (2.14)$$

H_{x0} in Eq. (2.14) is the field strength and δ_{x0} is the air gap width at the point x_0. The line integral of the field strength along a closed loop is equal to the integral of the current sheet limited by the loop.

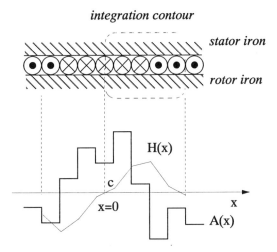

Fig. 2.9 *The current sheet and field strength distribution in the air gap of an electric machine, with correct choice of the zero point for the magnetic quantities*

The relationship between the flux density and the field strength in the air gap is linear, because here the permeability in the air gap is constant and equal to the permeability of free space. The flux density at point x_0 can be expressed from Eq. (2.14) as

$$B_{x0} = \mu_0 H_{x0} = \mu_0 \frac{F_{x0}}{\delta_{x0}} \qquad (2.15)$$

Equation (2.15) is often used because it directly relates the flux density B at point x_0 in the air gap to the value of the mmf F and the corresponding air gap width δ at the same point. Substituting the value from Eq. (1.7) for the absolute permeability of free space, one finds that per each millimeter of the air gap width a magnetomotive force of 800 ampereturns is necessary to produce the flux density of 1 T in the air gap.

2.3 Magnetomotive Force of a Coil and Winding

To find the mmf distribution of a single conductor in the air gap of an electric machine is an undefined task. Only if the return path of the current is known, can one determine the distribution of the flux lines, as illustrated in Fig. 2.10 (a), (b) and (c).

The conductor which carries the current flowing in the direction denoted by an X in Fig. 2.10 is always placed in the air gap. The conductor through which the current flows back is placed at various positions in the machine in Fig. 2.10 (a), (b) and (c). The direction of current in these conductors is denoted by a dot.

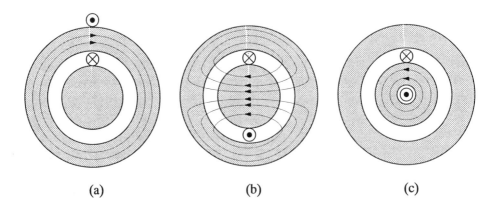

Fig. 2.10 *The approximative distribution of the lines of flux in an electric machine for various placements of the conductors*

Windings, Currents and Air Gap Magnetomotive Force

The two conductors form a coil which produces magnetic flux, the approximate distribution of which is shown in Fig. 2.10 (a), (b) and (c). Only if both the conductors are placed in the air gap, as in Fig. 2.10 (b), do the majority of the lines of flux which they produce pass through the air gap. In the other two cases, when the second conductor is out of the air gap, as shown in Fig. 2.10 (a) and (c), the lines of flux do not pass through the air gap. It should be noted here that the very first electric machines were wound in the manner shown in Fig. 2.10 (a), however, with numerous coils around the air gap periphery. The coils in such machines were connected so that the current in them changed direction periodically in the air gap, in order to obtain a heteropolar machine.

Assuming that all the conductors are placed in the air gap, the expressions for the spatial distributions of the magnetomotive force for different winding configurations and currents will be derived in this section. At the beginning, the mmf distribution of a single coil will be found. The distance y between the sides of the coil is called the *coil pitch*.

Following the definition of the mmf in Eq. (2.14), one can relate the two mmfs in Fig. 2.11 as

$$F_1 = F_2 + wI \tag{2.16}$$

where w denotes the number of turns *per pole*. Applying the condition that in a heteropolar machine there is no unipolar flux

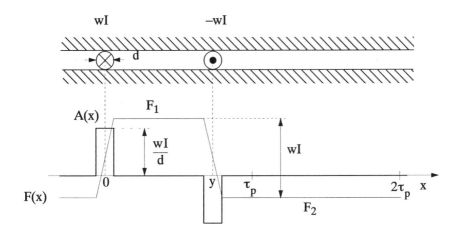

Fig. 2.11 *The current sheet and mmf distribution of a coil with w turns*

$$1\int_0^{2\tau_p} B dx = \frac{\mu_0}{\delta}\int_0^{2\tau_p} F dx = 0 \Rightarrow F_1 y + F_2(2\tau_p - y) = 0 \qquad (2.17)$$

simultaneous solution of equations (2.16) and (2.17) gives

$$F_1 = wI\frac{2\tau_p - y}{2\tau_p} \qquad F_2 = -wI\frac{y}{2\tau_p} \qquad (2.18)$$

The mmf distribution is very often approximated by the Fourier series

$$F(x) = \frac{a_0}{2} + \sum_{n=1}^{\infty}\left(a_n \cos\frac{n\pi x}{\tau_p} + b_n \sin\frac{n\pi x}{\tau_p}\right) \qquad (2.19)$$

where

$$a_n = \frac{1}{\tau_p}\int_0^{2\tau_p} F(x)\cos\frac{n\pi x}{\tau_p} dx \qquad b_n = \frac{1}{\tau_p}\int_0^{2\tau_p} F(x)\sin\frac{n\pi x}{\tau_p} dx \qquad (2.20)$$

After applying the procedures given in the Appendix, one obtains for the rectangular mmf distribution in Fig. 2.11

$$a_n = \frac{wI}{n\pi}\sin\frac{\pi n y}{\tau_p} \qquad (2.21)$$

and

$$b_n = \frac{wI}{n\pi}\left(1 - \cos\frac{\pi n y}{\tau_p}\right) \qquad (2.22)$$

The amplitude of the n-th spatial harmonic of the mmf distribution is n times smaller than the amplitude of the fundamental harmonic. In the special case when the coil has a full pitch, i.e. when $y=\tau_p$, the positive and negative half waves of the mmf distribution have equal absolute values, since

$$F_1 = -F_2 = \frac{wI}{2} \tag{2.23}$$

and the coefficients a_n and b_n are

$$a_n = 0$$
$$b_n = \frac{2}{\pi}\frac{1}{n}wI \quad \textit{for n odd;} \qquad b_n = 0 \quad \textit{for n even} \tag{2.24}$$

Therefore, the Fourier series which approximates the mmf distribution in this case is

$$F(x) = \frac{2}{\pi}wI \sum_{n=1}^{\infty} \frac{1}{n} \sin\frac{\pi n x}{\tau_p} \tag{2.25}$$

The amplitude of the fundamental spatial harmonic is $4/\pi$ times greater than the maximum of the ampereturns F_1 in Eq. (2.23). The period of the n-th spatial harmonic is n times shorter than the period of the fundamental. Therefore, for the n-th spatial harmonic, the machine has np pole pairs instead of p pole pairs.

When there is one chorded coil per pole instead of per two poles of the machine, as shown in Fig. 2.11, the terms of the Fourier series calculated in the Appendix, are

$$a_n = 0$$
$$b_n = \frac{4}{\pi}\frac{1}{n}wI\sin\frac{\pi n}{2}\sin\frac{\pi n y}{2\tau_p} \quad \textit{for n odd;} \qquad b_n = 0 \quad \textit{for n even} \tag{2.26}$$

with w denoting the number of turns per pole. The mmf distribution contains only odd terms of the infinite Fourier series:

$$F(x) = \frac{4}{\pi}wI \sum_{n=1}^{\infty} \frac{1}{n} \sin\frac{\pi n}{2} \sin\frac{\pi n y}{2\tau_p} \sin\frac{\pi n x}{\tau_p} \tag{2.27}$$

Each harmonic of the mmf distribution in Eq. (2.27) has twice the amplitude of the corresponding harmonic in Eq. (2.25), since there are two times more current carrying conductors per pole pair than in the former case.

The mmf distribution created by the windings in a machine can be derived by using basically the same procedure as for a single coil. However, the results are influenced by the type of the winding.

Concentric windings are often used in the stators of single phase A.C. machines, as well as in the rotors of turbo generators. Each coil of the concentric winding can have its own number of turns, w_i, where i is the index of the coil varying from 1 to the total number of coils per pole, q. Each coil carries the same current I. An example of a concentric winding with three coils per pole with the number of turns $w_1 > w_2 > w_3$ is shown in Fig. 2.12.

Denoting by τ_s the slot pitch, and by y_1 the pitch of the outermost coil, one can express the pitch of the i-th coil, y_i, as

$$y_i = y_1 - 2(i-1)\tau_s \tag{2.28}$$

The coil pitch is always an integer multiple of the slot pitch, since the coils are placed in the slots. Introducing an integer r, such that

$$y_1 = \tau_s r \tag{2.29}$$

one can rewrite Eq. (2.28) as

$$y_i = \tau_s [r - 2(i-1)] \tag{2.30}$$

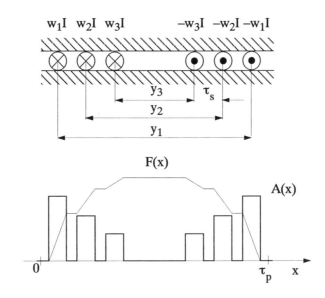

Fig. 2.12 *Three concentric coils and their air gap quantity distributions (F(x), A(x))*

The total mmf created by q coils is found by adding the mmf of each coil calculated by using Eq. (2.27). This is a consequence of the superposition law which may be applied here for the mmfs, since the driving force in this case consists of the currents in the coils. If the driving force consisted of the applied fluxes, the superposition law would have been valid for the flux densities. However, for low values of flux density when the magnetization curve of iron is linear, the superposition law is valid both for the mmfs and for the flux densities.

Applying Eq. (2.27) for a concentric winding with q coils per pole, one obtains

$$F(x) = \frac{4}{\pi} I \sum_{i=1}^{q} w_i \sum_{n=1}^{\infty} \frac{1}{n} \sin\frac{\pi n}{2} \sin\left\{\frac{\pi n}{2\tau_p} \tau_s [r-2(i-1)]\right\} \sin\frac{\pi n x}{\tau_p} \quad (2.31)$$

The amplitude of the n-th harmonic created by the i-th concentric coil, $F_{n,i}$, is equal to

$$F_{n,i} = \frac{4}{\pi} I w_i \frac{1}{n} \sin\frac{\pi n}{2} \sin\left\{\frac{\pi n}{2\tau_p} \tau_s [r-2(i-1)]\right\} \quad (2.32)$$

The expression for the mmf created by a concentric winding is different from zero only for n odd. Therefore, concentric windings produce only *odd harmonics*. Replacing n by 1 in Eq. (2.32) and considering an equal number of turns in all coils, one obtains that the amplitude of the fundamental harmonic decreases in the inner coils, since their coil pitch becomes shorter. By substituting Eq. (2.32) into Eq. (2.31), one can express the equation for the mmf distribution created by the current in q concentric coils as

$$F(x) = \sum_{i=1}^{q} \sum_{n=1}^{\infty} F_{n,i} \sin\frac{\pi n x}{\tau_p} \quad (2.33)$$

Very often electric machines are built to produce an air gap mmf waveform as close to the sinusoidal as possible. When the machine has concentric windings, the shape of the mmf distribution depends on the number of turns in each coil. Therefore, one can choose the number of turns in the coils, with the result that one or more harmonics of the mmf are suppressed.

The criterion for choosing the number of turns in coils which minimize the higher harmonics can be illustrated by means of Fig. 2.13, in which the mmf distribution of two concentric coils is shown. The difference between the real mmf distribution F(x), defined in Eq. (2.33) and its fundamental harmonic, $F_{max} \sin \pi x/\tau_p$, is denoted by the shaded areas.

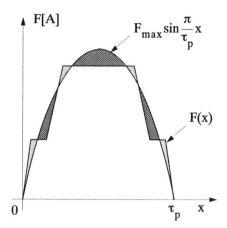

Fig. 2.13 *Illustrating the definition of the minimization curve for the higher spatial harmonics*

These areas are shaded in a different manner for the positive and for the negative difference of the two curves. The amplitudes of the higher harmonics are obviously decreased as the total shaded area is closer to zero. The shaded area in Fig. 2.13 can be defined as

$$S = \int_0^{\tau_p} \left[F(x) - F_{max} \sin \frac{\pi}{\tau_p} x \right] dx \qquad (2.34)$$

The area defined in Eq. (2.34) can be both positive and negative. This means that the total area can be equal to zero even when the difference between the rectangular and the sinusoidal distribution of the mmfs is very high. Therefore, it is much better to minimize the absolute value, or the square of the difference of the two curves in Fig. 2.13. If the square of the difference of the curves is to be minimized, one can write

$$\int_0^{\tau_p} \left[F(x) - F_{max} \sin \frac{\pi}{\tau_p} x \right]^2 dx \to \min \qquad (2.35)$$

Denoting by n_{max} the order of the highest spatial harmonic which has to be minimized, the Fourier series of the air gap mmf can be written as

$$F(x) = F_1 \sin\frac{\pi}{\tau_p}x + F_3 \sin\frac{3\pi}{\tau_p}x + \ldots + F_{n_{max}} \sin\frac{n_{max}\pi}{\tau_p}x \qquad (2.36)$$

Substituting Eq. (2.36) into Eq. (2.35), one can express the square of the expression defining the shaded area in Fig. 2.13 as

$$\frac{\tau_p}{2}\left[(F_1 - F_{max})^2 + \sum_{n=3}^{n_{max}} F_i^2\right] \qquad (2.37)$$

An obvious, but trivial, solution which has no physical meaning is to set the number of turns in each coil to zero. To obtain a physically useful solution, additional conditions for the number of turns in each coil have to be applied. One of the choices is that the number of turns in the coils do not differ too much from one another, in order to utilize the slot area as well as possible. In this case one can define the function which has to be minimized as

$$A = f_1 \sum_{i=2}^{q} (w_1 - w_i)^2 + f_2 \sum_{i=1}^{q} \sum_{n=n_{min}}^{n_{max}} F_{n,i}^2 \qquad (2.38)$$

The minimum of the function A in the previous equation is searched for unit current and for a given number of turns in the first coil, w_1. The weighing factors f_1 and f_2 are introduced here to tune the demand for an equal number of conductors per slot (f_1) and the amplitude of the higher harmonics (f_2) on the final result.

The lower index of the first sum in Eq. (2.38) is equal to 2, since the difference between the number of turns in the first and all other q coils has to be found. The order of the higher spatial harmonic n in the second double sum in the same equation varies from the order n_{min} of the lowest harmonic to be minimized to the order n_{max} of the highest harmonic to be minimized. If n_{min} is equal to 1, the amplitude F_{max} has to be defined; otherwise, the relationship $F_1 = F_{max}$ is utilized. The term $F_{n,i}$ is defined in Eq. (2.32).

Previous considerations can be illustrated by an example in which the number of turns of a winding with three coils are optimized. Assuming that the outermost coil has 100% turns, the results in Table 2.1 are obtained by executing the statements from the Appendix.

Table 2.1 The percentile number of turns in the second and third coils as a function of weighing factors

f_1	1	0.3	0.7
f_2	1	0.7	0.3
w_2	91.3%	81.8%	96%
w_3	89.1%	77.8%	95%

Distributed windings are manufactured either as lap or wave windings. All the coils in a distributed winding have an equal pitch y, which is often shorter than the pole pitch. A lap winding with three coils per pole, having an equal pitch and number of turns, is shown in Fig. 2.14.

When adjacent coils have equal pitch, the geometric shift between them is equal to the slot pitch τ_s. The geometric shift between the coils is copied into the phase angle between the electromagnetic quantities in the coils. Therefore, the fundamentals, as well as higher harmonics, of the mmfs created by different coils are out of phase.

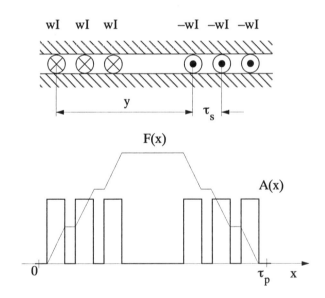

Fig. 2.14 *The current sheet and mmf distribution of a lap winding with three coils*

Windings, Currents and Air Gap Magnetomotive Force

The mmf distribution of a distributed winding can be evaluated by using Eq. (2.27) for each coil. However, one has to allow for the fact that the argument x in this equation is not equal for all the coils, due to geometric shift between them. The peripheral coordinate x_i in the i-th coil is related to the peripheral coordinate x in the first coil as

$$x_i = x - (i-1)\tau_s \quad (2.39)$$

The total mmf created by q coils is obtained by summation of the expression (2.27) for each of the q coils, which gives

$$F(x) = \frac{4}{\pi}wI \sum_{i=1}^{q} \sum_{n=1}^{\infty} \frac{1}{n} \sin\frac{\pi n}{2} \sin\frac{\pi n y}{2\tau_p} \sin\frac{\pi n x_i}{\tau_p} \quad (2.40)$$

Substituting Eq. (2.39) into (2.40) and defining the electrical angle between the slots α_{el} as

$$\alpha_{el} = \pi \frac{\tau_s}{\tau_p} \quad (2.41)$$

one can rewrite Eq. (2.40) as

$$F(x) = \frac{4}{\pi}wI \sum_{i=1}^{q} \sum_{n=1}^{\infty} \frac{1}{n} \sin\frac{\pi n}{2} \sin\frac{\pi n y}{2\tau_p} \sin\left[\frac{\pi n}{\tau_p}x - (i-1)n\alpha_{el}\right] \quad (2.42)$$

Utilizing trigonometric identity

$$\sum_{i=1}^{q} \sin[\alpha + (i-1)\beta] = \frac{\sin(\alpha + \frac{q-1}{2}\beta) \sin\frac{q\beta}{2}}{\sin\frac{\beta}{2}} \quad (2.43)$$

Eq. (2.42) can be finally written as

$$F(x) = \frac{4}{\pi}wI \sum_{n=1}^{\infty} \frac{1}{n} \sin\frac{\pi n}{2} \sin\frac{\pi n y}{2\tau_p} \frac{\sin\frac{nq\alpha_{el}}{2}}{\sin\frac{n\alpha_{el}}{2}} \sin\left[\frac{\pi n}{\tau_p}x - \frac{n\alpha_{el}}{2}(q-1)\right] \quad (2.44)$$

Again, as in the case of a concentric winding, the resultant mmf of q distributed coils exists only for n odd. As a general conclusion, one can state that, independent of its type, a winding creates an air gap mmf which has only odd harmonics. The mmf distribution is symmetrical with respect to its maximum or minimum value, and, therefore, it does not contain even harmonics. The symmetry of the mmf function is a consequence of the trivial fact that in any kind of winding the number of conductors in the air gap which conduct the current in one direction is equal to the number of conductors in the air gap which conduct the same current in the opposite direction.

The coil magnetomotive forces of a distributed winding, defined by Eq. (2.44), have equal amplitudes, but they are shifted to each other in space. This is different from the case of the coil mmfs of a concentric winding, which are all in phase, but have different amplitudes, despite an equal number of turns in the coils.

When the inductance and mutual inductance of a winding are evaluated, very often the *winding function* is utilized. The winding function $F_w(x)$ of a winding which creates the mmf distribution $F(x)$ is defined as

$$F_w(x) = \frac{F(x)}{i} \qquad (2.45)$$

The dimensionless winding function can be interpreted as an mmf distribution created by a unit current.

2.4 Winding Factor

It was shown in the previous section that when the winding has more than one coil, the total mmf for a given harmonic cannot be found simply by multiplying the value of the mmf produced by one coil with the number of coils in the winding. When the winding is concentric, the amplitude of the fundamental component of the mmf produced by a coil decreases as the coil pitch decreases. The total mmf amplitude for a given harmonic is found by adding the amplitudes of the mmfs of that harmonic in the coils *algebraically*, since they are in phase in all the coils. When the winding is distributed, the mmfs of adjacent coils have equal amplitudes when the numbers of turns are equal, but they are shifted to each other in the air gap by the angle α_{el}. The total mmf for a given harmonic can only be obtained by vector addition of the corresponding harmonics produced by each coil. In any case, the relationship between the amplitude of the mmf harmonic produced by one of the coils and the total harmonic amplitude must be calculated by using the *winding factor*. The winding factor allows for the distributed character of the machine's windings. If the winding is concentrated, all of its turns w_{tot} produce the mmfs which act in the same direction in the air gap. However, when the coils are shifted in relation to each other, the resultant mmf is smaller than the product of the total number of turns and the current in the winding. Instead, the winding acts as if it had the *effective number of turns* w_{eff}, which is always less than the total number

Windings, Currents and Air Gap Magnetomotive Force

of turns.

The winding factor f_w is defined as the ratio between the effective and the total number of turns in a winding

$$f_w = \frac{w_{eff}}{w_{tot}} \tag{2.46}$$

In a *concentric winding* the resultant amplitude of the n-th harmonic produced by all q coils is obtained by adding the expression defined in Eq. (2.32) for all q coils, which gives

$$F_n = \frac{4}{\pi} I \frac{1}{n} \sum_{i=1}^{q} w_i \sin\frac{\pi n}{2} \sin\left\{\frac{n\alpha_{el}}{2}[r-2(i-1)]\right\} \tag{2.47}$$

Since the argument of the sum in Eq. (2.47) is equal to the effective number of turns in the winding, the winding factor for the n-th harmonic, $f_{w,n}$, can be written as

$$f_{w,n} = \frac{\sum_{i=1}^{q} w_i \sin\left\{\frac{n\alpha_{el}}{2}[r-2(i-1)]\right\}}{\sum_{i=1}^{q} w_i} \tag{2.48}$$

which is not a very useful result. However, if all the coils have an equal number of turns, the winding factor in Eq. (2.48) can be written as

$$f_{w,n} = \frac{1}{q}\sum_{i=1}^{q} \sin\left\{\frac{n\alpha_{el}}{2}[r-2(i-1)]\right\} \tag{2.49}$$

or, after substituting Eqs. (2.43) and (2.29),

$$f_{w,n} = \sin\left\{n\frac{\pi}{2}\frac{1}{\tau_p}[y_1 - (q-1)\tau_s]\right\} \frac{\sin q \frac{n\alpha_{el}}{2}}{q \sin \frac{n\alpha_{el}}{2}} \tag{2.50}$$

Defining by y in a concentric winding the number of slots between one side of the outermost coil and the opposite side of the innermost coil, as shown in Fig. 2.15,

$$y = y_1 - (q-1)\tau_s \tag{2.51}$$

one can further write Eq. (2.50) as

$$f_{w,n} = \sin\left(n\frac{\pi}{2}\frac{y}{\tau_p}\right) \frac{\sin q \frac{n\alpha_{el}}{2}}{q \sin \frac{n\alpha_{el}}{2}} \tag{2.52}$$

The winding factor for the n-th harmonic, $f_{w,n}$, has two components: the *pitch factor* $f_{p,n}$, defined as

$$f_{p,n} = \sin\left(n\frac{\pi}{2}\frac{y}{\tau_p}\right) \tag{2.53}$$

and the *distribution factor*, $f_{d,n}$

$$f_{d,n} = \frac{\sin q \frac{n\alpha_{el}}{2}}{q \sin \frac{n\alpha_{el}}{2}} \tag{2.54}$$

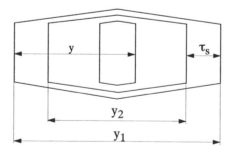

Fig. 2.15 *Illustrating Eq. (2.51)*

Substituting for n=1 in the equations above, one obtains the pitch factor for the fundamental, f_p

$$f_p = \sin\left(\frac{\pi}{2}\frac{y}{\tau_p}\right) \tag{2.55}$$

and the distribution factor for the fundamental, f_d

$$f_d = \frac{\sin q\frac{\alpha_{el}}{2}}{q\sin\frac{\alpha_{el}}{2}} \tag{2.56}$$

By utilizing the winding factor definition, one can rewrite the equation for the amplitude of the n-th harmonic of the mmf created by a concentric winding having an equal number of turns in each coil, Eq. (2.47), as

$$F_n = \frac{4}{\pi}Iqwf_{w,n}\frac{1}{n}\sin\frac{\pi n}{2} \tag{2.57}$$

The pitch factor allows for various electromagnetic conditions in the air gap at the points where the left and right sides of the coil are placed. When the pitch y is equal to the pole pitch, τ_p, the pitch factor for the fundamental harmonic is equal to one. The distribution factor allows for various electromagnetic conditions in the air gap at adjacent slots. Both the pitch and distribution factors are always *less than one*, except for a single coil with diametral pitch, where both factors are equal to one. The winding always acts as if it had fewer turns than it has in reality, as an effect of its spatial distribution.

The same considerations are valid for the distributed winding. The expression for the resultant mmf of a distributed winding, Eq. (2.44), already contains the winding factor term. The amplitude of the n-th harmonic of the mmf created by a distributed winding, defined in Eq. (2.44), can be written by utilizing the winding factor as

$$F_n = \frac{4}{\pi}Iqwf_{w,n}\frac{1}{n}\sin\frac{\pi n}{2} \tag{2.58}$$

and the mmf distribution from Eq. (2.44) can be written as

$$F(x) = \frac{4}{\pi}wqI\sum_{n=1}^{\infty}\frac{1}{n}\sin\frac{\pi n}{2}f_{w,n}\sin\left[\frac{\pi n}{\tau_p}x - \frac{n\alpha_{el}}{2}(q-1)\right] \tag{2.59}$$

The definition of the winding factor obviously does not depend on the type of winding. This is a logical consequence of the fact that the concentric, lap, and wave windings have only various forms of end windings; in the air gap, where the electromechanical energy conversion takes place, all three types of windings act in the same manner.

Since the n-th harmonic of the mmf has the meaning of a sine wave with a period n times shorter than $2\tau_p$, an intuitive visualization of the winding factor is possible by means of vector representation. The electrical phase shift between the adjacent slots for the n-th harmonic is obtained from Eq. (2.41) as

$$\alpha_{el,n} = n\alpha_{el} = n\pi \frac{\tau_s}{\tau_p} \qquad (2.60)$$

If the n-th spatial harmonic of the mmf produced by the current in the i-th coil is represented by the vector $F_{n,i}$, the vectors of the mmf produced by a distributed winding look as those shown in Fig. 2.16 (a), and in a concentric winding as in Fig. 2.16 (b). The amplitude of the vector of the total mmf in a distributed winding, denoted by $F_{n,\Sigma}$, is smaller than the algebraic sum of the amplitudes of the mmfs of the coils, since these are out of phase.

The variation of the coil distribution and pitch factor as functions of the coil parameters and harmonic order for the winding with the parameters defined in the Appendix are shown in Fig. 2.17 and Fig. 2.18.

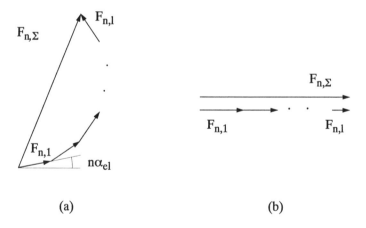

Fig. 2.16 *The vector sum of the mmfs for the n-th harmonic in a distributed (a) and concentric winding (b)*

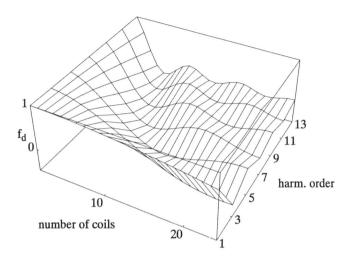

Fig. 2.17 *The dependence of the distribution factor f_d on the number of in-series connected coils and the order of the spatial harmonic*

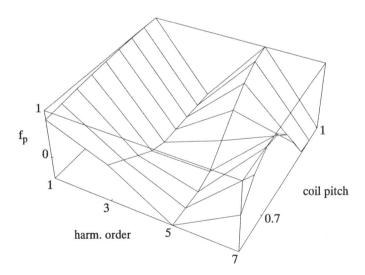

Fig. 2.18 *The dependence of the pitch factor f_p on the harmonic order and the ratio between the coil and diametral pitch*

The distribution factor for the fundamental harmonic decreases as the number of in-series connected coils increases. In an extreme case, when all the slots are filled with in-series connected coils, the distribution factor for the fundamental falls to $2/\pi = 0.6366$. The more phases the machine has, the smaller the portion of the air gap belonging to one phase, and higher distribution factors can be obtained. A single phase machine, which theoretically can have the complete air gap periphery wound by in-series coils, is never manufactured in that manner due to the poor distribution factor. Instead, usually only two thirds of the slots are filled by the coils belonging to one phase. The contribution of the coils which would occupy the rest of the slots would have been so small as compared to the amount of copper necessary to manufacture them that there is no economic reason to utilize them.

A way to minimize, or completely eliminate, particular harmonics in electric machines is to shorten the coil pitch. In the following example the effects of shortening the coil pitch to 4/5 of the pole pitch on the value of the induced voltage are demonstrated. Let the air gap flux density contain only the fundamental and the fifth harmonic, as shown in Fig. 2.19.

When the relative speed between the coils and the air gap flux density is constant, the spatial distribution of the flux density is copied into the time variation of the induced voltage. The voltage induced by the fundamental term of the air gap flux density in the left conductor of the coil with full pitch e_1 is equal to the negative induced voltage in the right conductor of the coil. The angle between the two voltages is $180°$. Summing the two induced voltages around the coil, one obtains the amount of $2e_1$. The same is valid for the voltages induced by the fifth harmonic of the flux density e_5 so that the total induced voltage in the coil with full pitch is $2(e_1 + e_5)$.

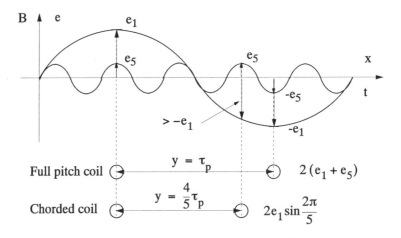

Fig. 2.19 *Illustrating the effects of shortening the coil pitch*

Windings, Currents and Air Gap Magnetomotive Force

The voltages induced by the fundamental term of the flux density in the left and right conductors of a chorded coil are shifted by an angle less than $180°$. Applying the definition of the pitch factor for the 4/5 chorded coil in Fig. 2.19, one can find the total voltage induced by the fundamental term of the flux density

$$2e_1 \sin \frac{4}{5}\frac{\pi}{2} = 2e_1 \sin \frac{2\pi}{5} \qquad (2.61)$$

which is less than $2e_1$. The voltages induced by the fifth harmonic of the flux density in the left and right conductors of the chorded coil in Fig. 2.19 are in phase and have equal amplitudes. Therefore, their sum in the coil is equal to zero. This can be proven by evaluating the pitch factor of the coil in this example for the fifth harmonic, $f_{p,5}$, defined in Eq. (2.53) as

$$f_{p,5} = \sin\left(5\frac{\pi}{2}\frac{y}{\tau_p}\right) = \sin\left(5\frac{\pi}{2}\frac{4}{5}\right) = 0 \qquad (2.62)$$

Shortening of the coil pitch decreases the fundamental, and in some cases completely eliminates certain higher harmonics.

The n-th harmonic is completely eliminated if the coil pitch y is

$$y = \frac{2k}{n}\tau_p; \qquad k = 0, 1, 2, \ldots \qquad (2.63)$$

When the coils are chorded, the index k in Eq. (2.63) is always chosen to provide that $y/\tau_p < 1$ and $y/\tau_p \Rightarrow \max$.

2.5 Matrix Representation of the Winding Magnetomotive Force

Although in previous discussions it has been assumed that the windings are placed in the machine's slots, this fact has not been exploited since the mmf distribution in the air gap does not depend on the radial position of the winding in the gap. However, it is often important to know the flux distribution in the machine's teeth. The flux distribution in the teeth can only be calculated if the mmf drop in each tooth or in the air gap under the teeth is known. A simple procedure to obtain the mmf distribution in the teeth as a function of the winding's parameters and the currents will be demonstrated in this section.

Reducing the integral of the current sheet in Eq. (2.14) to a discrete sum of ampere-conductors, and assuming that there is no change of the mmf along a tooth in a tangential direction, one can write

$$F_{i+1} = F_i + (Iw)_i; \quad i = 1, 2, ..., N \tag{2.64}$$

where F_i is the magnetomotive force which acts in the i-th tooth in a radial direction, $(Iw)_i$ are the total ampereconductors in the i-th slot, and N is the total number of teeth (slots). The relationship between these quantities is shown in Fig. 2.20.

Eq. (2.64) states that the mmf in the (i+1)-th tooth is equal to the mmf in the i-th tooth increased by the value of i-th slot ampereconductors. Due to the machine's circular geometry, after the N-th tooth the first one comes again. Therefore, the equations for all N teeth, based upon Eq. (2.64), express only the relative change of the mmf, without specifying their absolute values. In terms of mathematics, these equations are undetermined, which means that an additional physical condition must be applied, in order to obtain a unique set of numbers which represents the tooth mmfs.

This additional physical condition can be written in the form of Eq. (2.13) discretized as

$$\sum_{i=1}^{N} F_i = 0 \tag{2.65}$$

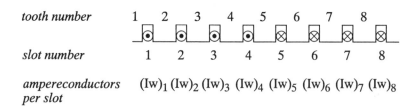

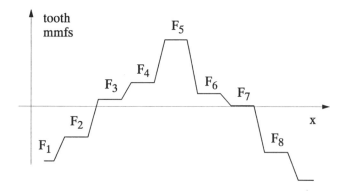

Fig. 2.20 *The relationship between the slot ampereconductors and the tooth mmfs*

The system of equations, the solution of which is the air gap mmf distribution in the teeth, consists of N–1 equations (2.64), which are written for N–1 teeth, and Eq. (2.65). Utilizing the following matrix and vector notation

$$\underline{C} = \begin{bmatrix} -1 & 1 & 0 & \ldots & 0 \\ 0 & -1 & 1 & \ldots & 0 \\ 0 & 0 & -1 & \ldots & 0 \\ \ldots & \ldots & \ldots & \ldots & \ldots \\ 0 & 0 & 0 & \ldots & 1 \\ 1 & 1 & 1 & 1 & 1 \end{bmatrix} ; \quad \underline{w} = \begin{bmatrix} w_{1a} & w_{1b} & \ldots & w_{1m} \\ w_{2a} & w_{2b} & \ldots & w_{2m} \\ \ldots & \ldots & \ldots & \ldots \\ w_{N-1,a} & w_{N-1,b} & \ldots & w_{N-1,m} \\ 0 & 0 & \ldots & 0 \end{bmatrix} ; \quad (2.66)$$

$$\underline{I} = \begin{bmatrix} I_a \\ I_b \\ \ldots \\ I_m \\ -I_a \\ -I_b \\ \ldots \\ -I_m \end{bmatrix} ; \quad \underline{F} = \begin{bmatrix} F_1 \\ F_2 \\ \ldots \\ F_N \end{bmatrix} ; \quad (2.67)$$

the N–1 equations (2.64) and Eq. (2.65) can be written in matrix form as

$$\underline{C}\,\underline{F} = \underline{w}\,\underline{I} \quad (2.68)$$

the solution of which is, obviously,

$$\underline{F} = \underline{C}^{-1}\,\underline{w}\,\underline{I} \quad (2.69)$$

The number of rows of the matrix $\underline{w}$ defined in Eq. (2.66) is equal to the number of slots N, and the number of columns is twice the number of phases, 2m. The phases are denoted by a, b, ..., m. The element w_{1a} of the matrix $\underline{w}$ denotes the number of conductors in the first slot which belong to the phase a. The information about the number of conductors in the N-th slot is redundant, because it is already included in the data for the previous N–1 slots. Therefore, this row of the matrix $\underline{w}$ is replaced by the row containing the conditions expressed in the right hand side of Eq. (2.65), i.e. by zeros. The vector of the phase currents $\underline{I}$ has 2m entries, where m denotes the number of phases in the winding. Negative signs of the phase currents in the lower half of the vector $\underline{I}$ allow for the opposite directions of currents in the conductors on the left- and right-hand side of each coil.

The vector $\underline{I}$ defined in Eq. (2.67) is obtained from the vector of phase currents $\underline{I}_{ph}$

$$\underline{I}_{ph} = \begin{bmatrix} I_a \\ I_b \\ \ldots \\ I_m \end{bmatrix} \qquad (2.70)$$

by multiplying it by the matrix of signs, $\underline{S}$

$$\underline{S} = \begin{bmatrix} 1 & 0 & \ldots & 0 \\ 0 & 1 & \ldots & 0 \\ \ldots & \ldots & \ldots & \ldots \\ 0 & 0 & \ldots & 1 \\ -1 & 0 & \ldots & 0 \\ 0 & -1 & \ldots & 0 \\ \ldots & \ldots & \ldots & \ldots \\ 0 & 0 & \ldots & -1 \end{bmatrix} \qquad (2.71)$$

The matrix of signs has m columns and 2m rows.

A simple way to generate the matrix $\underline{w}$ is illustrated in the following two examples. In the first example, a two phase concentric winding is placed into 12 slots according to the winding distribution shown in Fig. 2.21. The spatial mmf distribution is calculated for given values of phase currents. In the second example, a three phase lap double layer winding is also placed in twelve slots, and, again, the spatial distributions of the mmf are evaluated for given values of currents.

The matrix $\underline{w}$ in the first example is represented as a sum of two submatrices, $\underline{w}_a$ and $\underline{w}_b$. The entries in the i-th row of the matrix $\underline{w}_a$ are different from zero if a winding belonging to phase a has one of its sides in the i-th slot. This is illustrated in Fig. 2.22, in which the conductors of the coils belonging to phase a are drawn as entries of the matrix $\underline{w}_a$. The same is repeated for the matrix $\underline{w}_b$.

Denoting by w_{a1} the number of turns in the outer coil of phase a, by w_{a2} the number of turns in the inner coil of phase a, by w_{b1} the number of turns in the outer coil of phase b, by w_{b2} the number of turns in the inner coil of phase b, and utilizing the

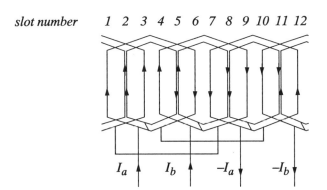

Fig. 2.21 *A schematic of a two phase winding*

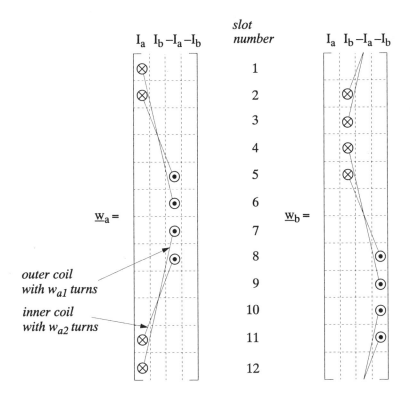

Fig. 2.22 *The submatrices of the matrix $\underline{w}$ for the winding in Fig. 2.21*

matrices $\underline{w}_a$ and $\underline{w}_b$ from Fig. 2.22, one can define the matrix $\underline{w}$ for the two phase winding in Fig. 2.21 as

$$\underline{w}^T = \begin{bmatrix} w_{a1} & w_{a2} & 0 & 0 & 0 & 0 & 0 & 0 & 0 & 0 & w_{a2} & 0 \\ 0 & w_{b2} & w_{b1} & w_{b1} & w_{b2} & 0 & 0 & 0 & 0 & 0 & 0 & 0 \\ 0 & 0 & 0 & 0 & w_{a2} & w_{a1} & w_{a1} & w_{a2} & 0 & 0 & 0 & 0 \\ 0 & 0 & 0 & 0 & 0 & 0 & 0 & w_{b2} & w_{b1} & w_{b1} & w_{b2} & 0 \end{bmatrix} \quad (2.72)$$

In the entries of the matrix $\underline{w}$ in Eq. (2.72) one can recognize how the distribution of the coils of phases a and b is copied first into the structure of the submatrices $\underline{w}_a$ and $\underline{w}_b$, and then into the structure of the matrix $\underline{w}$. The matrix $\underline{w}$ can be generated in a similar simple manner for any other winding distribution.

By executing the statements from the Appendix for the windings with numbers of turns per coil $w_{a1} = w_{b1} = 30$, $w_{a2} = w_{b2} = 25$ and for the phase currents $I_a = 1A$, $I_b = 0$, one obtains the tooth mmf distribution as shown in Fig. 2.23 (a). The tooth mmf distribution created by the same winding carrying the currents $I_a = 0.707$ A and $I_b = -0.707$ A is shown in Fig. 2.23 (b).

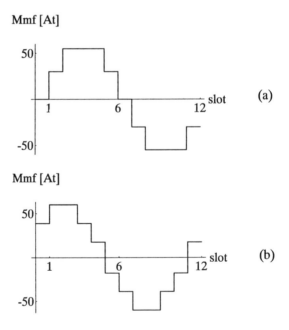

Fig. 2.23 *The tooth mmf distribution created by the winding in Fig. 2.22 for the currents Ia=1A, Ib=0 (a) and for Ia=0.707 A, Ib=-0.707A (b)*

Comparing the two mmf distributions in Fig. 2.23 one notices that the tooth mmf can be shifted along the peripheral coordinate (here represented by the slot number) solely by changing the values of the currents in the windings. Although the windings are fixed in the slots, an appropriate combination of the currents in them can create the mmf distribution which has its maximum at an arbitrary tooth along the periphery. This mechanism is utilized to create the rotating field in the air gap of electric machines.

In the second example, the mmf created by the currents in a three phase lap double layer winding is calculated. The number of turns per coil is equal in the phases w_a, w_b, and w_c. The winding is placed in twelve slots per pole pair and it has a ratio of $y/\tau_p = 5/6$ between the coil and diametral pitch. The winding distribution in this case is shown in Fig. 2.24. The matrix $\underline{w}$, defined in Eq. (2.66), in the case of a double layer winding can be represented as the difference of two submatrices: the submatrix of the number of conductors in the lower layer, $\underline{w}_l$, and the submatrix of the number of conductors in the upper layer, $\underline{w}_u$. Here the difference between the two matrices is utilized, because the mmf produced by a current in the conductors in the lower layer has opposite sign from the mmf produced by the conductors belonging to the same coil in the upper layer. The matrix $\underline{w}_u$ can be obtained from the matrix $\underline{w}_l$ by shifting the rows of the latter for the coil pitch y.

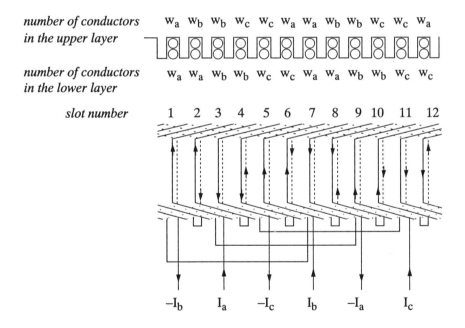

Fig. 2.24 *Schematic representation of a double layer, three phase winding with twelve slots per pole pair and $y/\tau_p = 5/6$*

The structure of the matrix w, which is created on the basis of the placement of the coils in the slots, along with the distribution of the entries in the matrix w, is defined in Eq. (2.73):

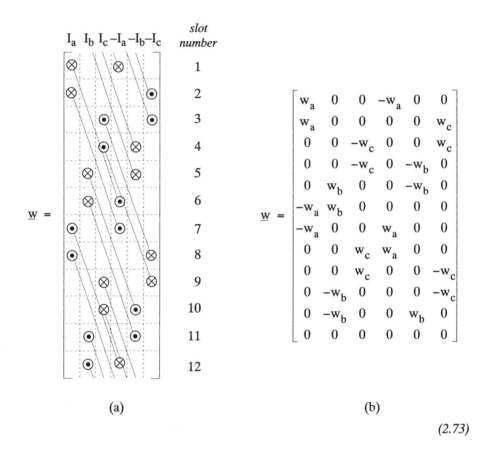

(a)   (b)

(2.73)

Since the columns of the matrix w are linearly dependent on one another, there exists a certain freedom in defining its entries. Therefore, the entries of the matrix w in Eq. (2.73) (b) do not have to follow strictly the structural distribution based on the coil placement shown in Eq. (2.73) (a). However, the given form of the matrix w is simple to define mathematically, as shown in the Appendix.

The action of the winding analyzed in this example can be illustrated by calculating the spatial mmf distribution for two sets of currents, as shown in Fig. 2.25 (a) and (b). The number of turns per coil is equal to ten for all three phases. As in the previous example of two phase winding, one can see again that the tooth mmf distribution is moving along the air gap periphery solely by varying the winding currents, without moving the windings.

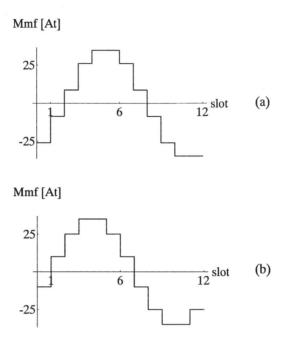

Fig. 2.25 *The tooth mmf distribution created by the winding in Fig. 2.24 for the currents $I_a=0.866$ A; $I_b=0$ and $I_c=-0.866$ A (a) and for $I_a=1$ A, $I_b=-0.5$ A and $I_c=-0.5$ A (b)*

2.6 Time Varying Excitation

The currents which produced the spatially distributed mmfs calculated in the previous sections were assumed to be constant. As will be proven in the fifth chapter, if a permanent electromechanical energy conversion has to take place in an electric machine, at least one of the currents has to be alternating. Therefore, the mmf distribution in the air gap is also time varying. The waveforms of the currents in the windings can be arbitrary; hence a general equation for the current can be written

$$I = \sum_{k=1}^{\infty} I_k \sin k\omega t \qquad (2.74)$$

Besides the fundamental term with angular frequency ω, the current waveform contains an infinite spectrum of *higher time harmonics*. The higher time harmonic of order k has a frequency k times higher than the fundamental. Each of the higher current harmonics creates its own mmf distribution when flowing through the machine's windings. The effects of the time varying excitation will be analyzed here on a chorded coil. However, the results obtained are applicable to any kind of winding. When the current is alternating, the mmf distribution of a chorded coil, expressed in Eq. (2.27), becomes a function not only of the spatial coordinate x, but also of time t

$$F(x,t) = \frac{4}{\pi} w \sum_{k=1}^{\infty} I_k \sin k\omega t \sum_{n=1}^{\infty} \frac{1}{n} \sin \frac{\pi n}{2} \sin \frac{\pi n y}{2\tau_p} \sin \frac{\pi n x}{\tau_p} \qquad (2.75)$$

With substitution

$$A_n = \frac{1}{n} \sin \frac{\pi n}{2} \sin \frac{\pi n y}{2\tau_p} \qquad (2.76)$$

Eq. (2.75) can be written as

$$F(x,t) = \frac{4}{\pi} w \sum_{k=1}^{\infty} I_k \sin k\omega t \sum_{n=1}^{\infty} A_n \sin \frac{\pi n x}{\tau_p} \qquad (2.77)$$

For a given time harmonic of order k and a spatial harmonic of order n, the time–spatial distribution of the mmf, $F_{k,n}(x,t)$, can be expressed as

$$F_{k,n}(x,t) = \frac{4}{\pi} w I_k \sin k\omega t A_n \sin \frac{\pi n x}{\tau_p} = F_{k,n} \sin k\omega t \sin \frac{\pi n x}{\tau_p} \qquad (2.78)$$

where the amplitude $F_{k,n}$ of the k-th time and n-th spatial harmonic is equal to

$$F_{k,n} = \frac{4}{\pi} w I_k \frac{1}{n} \sin \frac{\pi n}{2} \sin \frac{\pi n y}{2\tau_p} \qquad (2.79)$$

Since

$$\sin\alpha \sin\beta = \frac{1}{2} [\cos(\alpha - \beta) - \cos(\alpha + \beta)] \qquad (2.80)$$

one can rewrite Eq. (2.78) as

$$F_{k,n}(x,t) = F_{k,n}\sin k\omega t \sin\frac{\pi n x}{\tau_p} = \frac{F_{k,n}}{2}\cos\left(k\omega t - \frac{\pi n x}{\tau_p}\right) - \frac{F_{k,n}}{2}\cos\left(k\omega t + \frac{\pi n x}{\tau_p}\right)$$

(2.81)

The arguments of the cosine functions on the right hand side of Eq. (2.81) can be written as

$$k\omega t \pm \frac{\pi n x}{\tau_p} \qquad (2.82)$$

When an argument of a periodic function depends on both time and space, the latter being expressed by the peripheral coordinate x, such a periodic function describes a *wave motion*. Since there is only one spatial coordinate, x, these waves travel in only one dimension – along the rotor periphery. The cylindrical geometry of the rotor periphery causes the travelling waves to be *rotating*. It is a convention to consider the direction of rotation of the wave described by expression

$$\frac{F_{k,n}}{2}\cos\left(k\omega t - \frac{\pi n x}{\tau_p}\right) \qquad (2.83)$$

as *positive*, whereas the direction of rotation of the wave

$$\frac{F_{k,n}}{2}\cos\left(k\omega t + \frac{\pi n x}{\tau_p}\right) \qquad (2.84)$$

is taken to be *negative*. The mmf wave described by Eq. (2.83) is called the *positive sequence of the mmf*, and that of (2.84), the *negative sequence of the mmf*. These two waves form a *standing wave*, since their amplitudes are equal and since they travel in opposite directions. The nodes and peaks of a standing wave do not change position within time, as opposed to travelling waves where a complete waveform is shifted in space as time goes on.

The *speed of rotation* of the mmf wave can be found by analyzing the argument of the harmonic wave functions defined in Eq. (2.82). This speed is always expressed relative to the winding. It is obvious that the observer, who is rotating at the same speed as the mmf wave, is fixed to the wave; hence the phase shift φ between the observer and the wave is constant. This can be expressed as

$$k\omega t \pm \frac{\pi n x}{\tau_p} = \varphi = \text{const} \Rightarrow x = \pm \frac{\tau_p}{n\pi}(\varphi - k\omega t) \tag{2.85}$$

By differentiating the peripheral coordinate x from Eq. (2.85) with respect to time, one obtains the linear speed along the air gap periphery of the n-th spatial harmonic of the mmf wave, created by the k-th time harmonic of the current as

$$v_{k,n} = \mp \frac{k}{n} \frac{\tau_p}{\pi} \omega \tag{2.86}$$

whereas the angular speed of the same harmonic is

$$\omega_{k,n} = \mp \frac{k}{n} \frac{\omega}{p} \tag{2.87}$$

The mechanical speed of the n-th spatial harmonic of the wave which is created by the k-th time harmonic of the current is k times greater and n times smaller than the speed of the fundamental time and spatial harmonic ω/p.

So far, only the spatial harmonics generated by discrete winding distribution have been analyzed. These harmonics appear already in the mmf distribution, and are copied into the flux density distribution, following Eq. (2.15). These spatial harmonics are generated by the *fundamental* of the winding current; hence, their speed is, according to Eq. (2.87), equal to

$$\omega_{1,n} = \mp \frac{1}{n} \frac{\omega}{p} \tag{2.88}$$

Each of the speeds $\omega_{1,n}$ is *smaller* than the speed ω/p of the fundamental spatial component which travels in the same direction. The higher harmonics, the speed of which is given in Eq. (2.88), *travel* relatively to the fundamental, and to each other. Therefore, the complete spatial waveform, which contains higher harmonics generated by the fundamental of the winding current, changes its shape as time goes on.

Besides the higher spatial harmonics caused by the mmf distribution, the flux density in Eq. (2.15) contains the harmonics which travel at the speed of the fundamental. These harmonics are caused by a variation of reluctances through which passes the air gap flux. A reluctance which faces the air gap flux can vary due to either:

a) saturation in iron, or
b) variable air gap width

but never due to both causes simultaneously.

As will be analyzed later, the *saturation in iron* decreases the flux density in such a manner that the air gap flux density curve is flattened, independently of the rotor position. This means that the nonlinear B–H curve of the iron core generates higher harmonic components of the air gap flux density which *travel at the same speed as its fundamental spatial component*. This is only possible for the harmonics, the time order k of which is equal to their spatial order n, satisfying Eq. (2.87) so that

$$\omega_{k,n} = \mp \frac{\omega}{p} = \omega_{1,1} \qquad (2.89)$$

The various shapes of the air gap waveforms which contain higher harmonics travelling at their own speeds, in one case, and at the speed of the fundamental, in the other, are illustrated in Fig. 2.26.

In Fig. 2.26 (a) a spatial distribution containing the fundamental and the fifth harmonic is shown. The amplitude of the fifth harmonic equals 20% of the amplitude of the fundamental.

When the fifth spatial harmonic is generated by the fundamental current harmonic, it travels at one fifth of the speed of the fundamental. After one sixth ($\pi/3$) of the period of the winding current fundamental, the fifth harmonic has passed only one fifth of the path of the fundamental spatial harmonic. Such a fifth spatial harmonic is, therefore, shifted to the fundamental, causing the waveform distortion shown in Fig. 2.26 (b). If, however, the fifth spatial harmonic is generated by the fifth time harmonic of the current(s), it travels at the speed of the fundamental, and the resultant spatial distribution does not change its shape. This is illustrated in Fig. 2.26 (c), in which the air gap distribution after one sixth ($\pi/3$) of the period of the winding current fundamental is shown.

The variable air gap width means that either stator, or rotor, or both of them have salient poles and/or teeth. Therefore, the value of δ_{x0} in Eq. (2.15) at a fixed point on the stator side of the air gap varies as the rotor rotates. This can be illustrated by an example of a salient pole machine, in which a rectangularly distributed mmf generated by the windings placed on the salient poles produces a smoothed air gap flux density, due to variation of the air gap width under the main pole and inter-polar space. The distorted air gap flux density travels at the rotor speed. Therefore, all higher harmonics of the air gap flux density travel at the speed of the fundamental. The voltage which induces such a flattened air gap flux density in the stator windings contains the fundamental term, and higher harmonics caused by the nonuniform air gap.

In order to illustrate the results obtained earlier in this section, an mmf distribution of the concentric winding defined in Fig. 2.22 is evaluated for in-phase sinusoidal currents in phases a and b. The statements which generate the mmf distribution in Fig. 2.27 are given in the Appendix. The x coordinate in this figure is the spatial coordinate represented by the tooth number, and the y coordinate is time, expressed in fractions of the period of the applied currents.

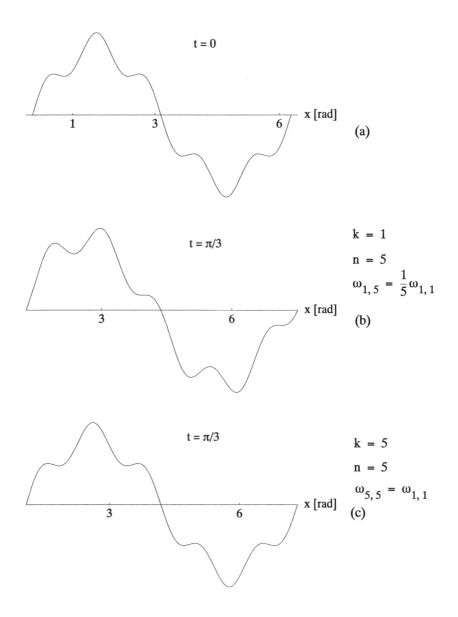

Fig. 2.26 *The variation of shape of the air gap waveform, as a function of the order of time harmonic in the winding(s) current(s): the waveform at t=0, when the fundamental and fifth harmonic are in phase (a); π/3 of the period later, if the fundamental time harmonic of current drives the fifth spatial harmonic (b), and π/3 of the period later, if the fifth time harmonic of current drives the fifth spatial harmonic (c)*

Windings, Currents and Air Gap Magnetomotive Force 117

The *spatial* mmf distribution at time t=0 in Fig. 2.27 corresponds to the mmf curve in Fig. 2.23 (a), since both phase currents at this time instant have the same values in these two figures. As time goes on, the air gap mmf changes in such a manner that the nodes, as well as the crests and troughs, are always fixed to the same position in space. This behaviour is typical for a standing wave.

The *time* distribution of the mmf at a fixed position of the peripheral coordinate x in Fig. 2.27 is always sinusoidal, since the currents in both phases are sinusoidal.

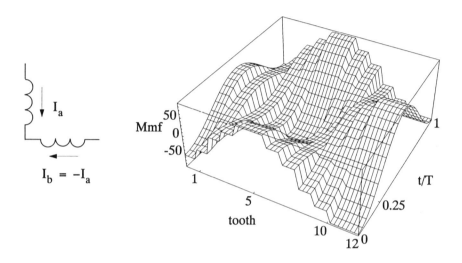

Fig. 2.27 *The time and spatial distribution of the air gap mmf created by the two phase concentric winding described in Fig. 2.22 carrying in-phase sinusoidal currents*

2.7 Generation of the Rotating Field

As demonstrated in the previous section, a coil or a group of coils carrying the same current produce the mmf which is pulsating in space. Such a pulsating mmf, which has the form of a standing wave, can be represented as the sum of two travelling waves. Each of the travelling waves has an amplitude equal to one half of the amplitude of the standing wave. These two waves travel at the same velocity, but in opposite directions.

An electric machine usually has more than one winding, each of them being fed from its own source. The windings can have a different number of turns and their spatial distributions are not necessarily identical. The frequency of currents in some windings is often equal. The amplitudes of the currents, as well as the phase shift between them, can be different. Such a large number of possibilities obviously leads to a counterproductive analysis. Therefore, a particular set of conditions which most machines have to satisfy will be analyzed here: a symmetrical, m-phase winding, in which each phase is carrying a current with equal amplitude and frequency. In a symmetrical winding all m phases have an equal number of turns and the electrical angle between the winding axes is equal in all the phases.

As shown in the previous section, the k-th time harmonic of current produces an infinite number of spatial harmonics. The n-th spatial harmonic of the mmf in the first phase is defined in Eq. (2.81) as

$$F_{1,k,n}(x,t) = \frac{F_{k,n}}{2}\cos\left(k\omega t - \frac{\pi n}{\tau_p}x\right) - \frac{F_{k,n}}{2}\cos\left(k\omega t + \frac{\pi n}{\tau_p}x\right) \quad (2.90)$$

The spatial electrical angle between the axes of adjacent phases in a symmetrical machine is equal to $2\pi/m$. The phase shift between the currents in adjacent phases is also $2\pi/m$ if a symmetrical machine is fed from symmetrical sources. Therefore, the electrical angle between the axes of the first and the i-th phases, as well as the phase shift between the current in the first and i-th phases, is equal to $2\pi(i-1)/m$. Since equation (2.90) describes the mmf created by the first phase, the n-th spatial harmonic of the mmf created by the k-th time component of current in the i-th phase is

$$F_{i,k,n}(x,t) = \frac{F_{k,n}}{2}\cos\left\{k\left[\omega t - (i-1)\frac{2\pi}{m}\right] - n\left[\frac{\pi}{\tau_p}x - (i-1)\frac{2\pi}{m}\right]\right\} -$$
$$- \frac{F_{k,n}}{2}\cos\left\{k\left[\omega t - (i-1)\frac{2\pi}{m}\right] + n\left[\frac{\pi}{\tau_p}x - (i-1)\frac{2\pi}{m}\right]\right\} \quad (2.91)$$

or, after rearrangement,

$$F_{i,k,n}(x,t) = \frac{F_{k,n}}{2}\cos\left[k\omega t - n\frac{\pi}{\tau_p}x - (i-1)\frac{k-n}{m}2\pi\right] -$$
$$- \frac{F_{k,n}}{2}\cos\left[k\omega t + n\frac{\pi}{\tau_p}x - (i-1)\frac{k+n}{m}2\pi\right] \quad (2.92)$$

An analysis of the expression (2.92) will be carried out first for the fundamental spatial and time harmonics, i.e. for k=n=1. In this case one can write for the fundamental spatial component of the mmf created by the fundamental time component of the current in the i-th coil

$$F_{i,1,1}(x,t) = \frac{F_{1,1}}{2}\cos\left(\omega t - \frac{\pi}{\tau_p}x\right) - \frac{F_{1,1}}{2}\cos\left[\omega t + \frac{\pi}{\tau_p}x - (i-1)\frac{4\pi}{m}\right] \quad (2.93)$$

Since the currents in all phases have equal amplitudes, the amplitudes of the mmfs in the phases of a symmetrical machine are equal. The total mmf, produced by all m phases, is obtained by summing the positive and negative sequence mmf components in each phase. The total positive sequence mmf, $F_{pos}(x,t)$, is equal to

$$F_{pos}(x,t) = \sum_{i=1}^{m} \frac{F_{1,1}}{2}\cos\left(\omega t - \frac{\pi}{\tau_p}x\right) = \frac{m}{2}F_{1,1}\cos\left(\omega t - \frac{\pi}{\tau_p}x\right) \quad (2.94)$$

The amplitude of the positive sequence mmf is always m/2 times greater than the amplitude of the mmf created by one phase. The total negative sequence mmf, $F_{neg}(x,t)$, is equal to

$$F_{neg}(x,t) = \sum_{i=1}^{m} \frac{F_{1,1}}{2}\cos\left[\omega t + \frac{\pi}{\tau_p}x - (i-1)\frac{4\pi}{m}\right] \quad (2.95)$$

The minimum number of phases that a machine must have in order to generate a rotating air gap field, m in Eq. (2.95), is obviously 2. However, considering that the electrical spatial angle between the axes of the windings, as well as the phase shift between the currents in this case, is equal to $2\pi/m=\pi$, one can conclude that the case in which m=2 represents a *pseudo two phase machine*. In reality, this is a one phase machine in which the two windings, shifted by the electrical angle π, and carrying the currents which are π radians out of phase, form only one phase. Therefore, the term *one phase,* instead of *single phase* machine is utilized here. A machine which is usually called single phase is, in reality, a *pseudo four phase* machine. The discrepancy between the machine's name and its structure comes primarily from the fact that the machine is connected to a single phase source but has two windings shifted by $\pi/2$ electrical radians. These windings carry currents which are shifted by an angle close to $\pi/2$ radians. Only a four phase machine, i.e. the one for which m=4 has to be substituted in the equations for the phase shift between the currents, fulfils the conditions of such a single phase machine.

Following the previous considerations, it is physically meaningful to analyze Eq. (2.95) only for m ≥ 3. For a number of phases greater than three, the vectors of negative sequence mmfs in all the phases form a *vector star*, in which the total sum of all the vectors is equal to zero. This can be illustrated with an example of a symmetrical, three phase winding for which the sum of the negative sequence mmfs is shown in Fig.

2.28 (a), and the sum of the positive sequence mmfs in Fig. 2.28 (b). $F_{pos,i}$ in this figure denotes the amplitude of the positive sequence component created by the i-th phase. $F_{neg,i}$ is the amplitude of the negative sequence component of the mmf, created by the same phase.

Now one can return to Eq. (2.92), which can be analyzed in the following manner: if the k-th time and n-th spatial harmonics of a positive or a negative sequence component of the mmf in all phases are in phase, the resultant mmf has an amplitude which is m/2 times greater than the amplitude of the k-th time and n-th spatial harmonic in each phase. If the k-th time and n-th spatial harmonics of a positive or a negative sequence component of the mmf are out of phase in all phases, the resulting mmf is equal to zero, since the winding is symmetrical. From Eq. (2.92) one can write the positive sequence component of the n-th spatial harmonic of the mmf created in the i-th phase by the k-th harmonic of the current as

$$F_{pos,i,k,n}(x,t) = \frac{F_{k,n}}{2} \cos\left[k\omega t - n\frac{\pi}{\tau_p}x - (i-1)\frac{k-n}{m}2\pi\right] \qquad (2.96)$$

If the positive sequence components in all the phases have to be collinear, the phase shift between them must be an integer multiplier of 2π. This is possible only if

$$\frac{k-n}{m} = \text{integer} \qquad (2.97)$$

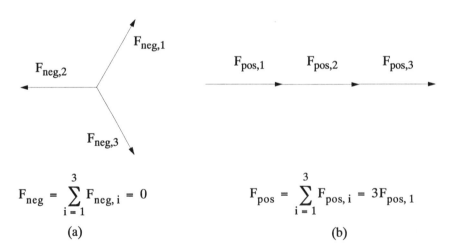

$$F_{neg} = \sum_{i=1}^{3} F_{neg,i} = 0 \qquad\qquad F_{pos} = \sum_{i=1}^{3} F_{pos,i} = 3F_{pos,1}$$

(a)   (b)

Fig. 2.28 *The vector sums of the negative (a) and positive (b) sequence mmfs in a symmetrical three phase winding carrying a symmetrical set of currents*

Windings, Currents and Air Gap Magnetomotive Force 121

Similarly, the negative sequence component of the n-th spatial harmonic of the mmf created in the i-th phase by the k-th harmonic of the current can be written as

$$F_{neg, i, k, n}(x, t) = \frac{F_{k, n}}{2} \cos\left[k\omega t + n\frac{\pi}{\tau_p}x - (i-1)\frac{k+n}{m}2\pi\right] \quad (2.98)$$

The total negative sequence component of the mmf is different from zero only if

$$\frac{k+n}{m} = \text{integer} \quad (2.99)$$

For all other combinations of k, m and n the negative sequence component of the mmf vanishes.

One can conclude from the previous analysis that for a given number of phases only certain combinations of the spatial and time harmonics generate either the positive, negative, or sometimes both components of the mmf. This is illustrated in Table 2.2, which is created for a three phase machine by utilizing conditions expressed in Eqs. (2.97) and (2.99). This table gives possible components of the total mmf for different harmonics, along with their speed of rotation. It has been assumed that only odd spatial and time harmonics exist. The existence of the positive sequence mmf is denoted by the sign +; the existence of the negative by the sign –. If both positive and negative sequence mmfs exist, the combination ± is used. The zero is used in Table 2.2 for a particular combination of the spatial and time harmonics if the windings create neither a positive, nor a negative sequence of the mmf for those harmonics.

The mechanical speed of rotation of the positive sequence component of the fundamental time and spatial harmonic, $\omega_{1,1}$, follows from Eq. (2.87) as

$$\omega_{1, 1} = \frac{\omega}{p} \quad (2.100)$$

For sinusoidal currents only the first column in Table 2.2 is significant, since there exists only the fundamental time harmonic. In a three phase symmetrical machine fed from symmetrical sinusoidal sources the fifth spatial harmonic travels at one fifth the speed of the fundamental, but in the opposite direction. The seventh harmonic travels at one seventh the speed of the fundamental in the same direction as the fundamental. A similar conclusion can be derived for all other spatial harmonics.

When the machine is fed from nonsinusoidal current sources, the spectrum of the higher harmonics of the mmf is enriched in the manner shown in Table 2.2. Some of these harmonics travel at the speed of the fundamental, $\omega_{1,1}$.

Table 2.2 The combinations of time (k) and spatial (n) harmonics which create positive (+) or negative (–) sequence components of the mmf in a symmetrical three phase winding fed from symmetrical sources, and the speeds of rotation of the mmf components

n \ k	1	3	5	7	9	11	13
1	$+\omega_{1,1}$	0	$-5\omega_{1,1}$	$+7\omega_{1,1}$	0	$-11\omega_{1,1}$	$+13\omega_{1,1}$
3	0	$\pm\omega_{1,1}$	0	0	$\pm 3\omega_{1,1}$	0	0
5	$-\dfrac{\omega_{1,1}}{5}$	0	$+\omega_{1,1}$	$-\dfrac{7}{5}\omega_{1,1}$	0	$+\dfrac{11}{5}\omega_{1,1}$	$-\dfrac{13}{5}\omega_{1,1}$
7	$+\dfrac{\omega_{1,1}}{7}$	0	$-\dfrac{5}{7}\omega_{1,1}$	$+\omega_{1,1}$	0	$-\dfrac{11}{7}\omega_{1,1}$	$+\dfrac{13}{7}\omega_{1,1}$
9	0	$\pm\dfrac{\omega_{1,1}}{3}$	0	0	$\pm\omega_{1,1}$	0	0
11	$-\dfrac{\omega_{1,1}}{11}$	0	$+\dfrac{5}{11}\omega_{1,1}$	$-\dfrac{7}{11}\omega_{1,1}$	0	$+\omega_{1,1}$	$-\dfrac{13}{11}\omega_{1,1}$
13	$+\dfrac{\omega_{1,1}}{13}$	0	$-\dfrac{5}{13}\omega_{1,1}$	$+\dfrac{7}{13}\omega_{1,1}$	0	$-\dfrac{11}{13}\omega_{1,1}$	$+\omega_{1,1}$

The condition which the combination of orders of higher time and spatial harmonics has to fulfill in order to travel at speed $\omega_{1,1}$ is that the order of the time harmonic is equal to the order of the spatial harmonic, i.e. k=n. In this case one can write

$$\omega_{k,k} = \omega_{n,n} = \frac{\omega}{p} \qquad (2.101)$$

To illustrate the mechanism for generating the air gap mmf, several three dimensional plots are created for various types of windings, their connections and the current waveforms. On the x axis of these plots the spatial argument of the mmf distribution is represented as the tooth number. The y coordinate is time, expressed in fractions of the period of the applied currents.

If a two phase machine is wound in the manner shown in Fig. 2.22 and if it carries currents which are 90 degrees out of phase and have amplitude 1 A, the generated mmf looks as shown in Fig. 2.29. There is no negative sequence fundamental component in this waveform, and the wave amplitude is almost constant. The crests in the mmf wave are caused by the higher harmonics, which rotate at different speeds from the fundamental. The difference in speeds between the fundamental and higher harmonics also means that the higher harmonics travel relative to the fundamental, so that the phase shift between the higher harmonics and the fundamental changes permanently. Therefore, the mmf waveform at every time instant varies slightly. In three dimensional plots of the mmf distribution the motion of the higher harmonics relative to the fundamental is observed as crests on the wave representing the fundamental component of the mmf.

As opposed to the waveform shown in Fig. 2.27 which reveals the air gap mmf in the same machine fed from the in-phase currents with amplitude 1 A, one can clearly notice the wave motion in Fig. 2.29 in the direction from bottom left towards top right.

In the following three dimensional plots the air gap mmf distribution in a three phase machine are shown, the distributed winding of which is revealed in Fig. 2.24.

When the winding carries symmetrical sinusoidal currents, the generated air gap mmf contains only the positive sequence component, as represented in Fig. 2.30.

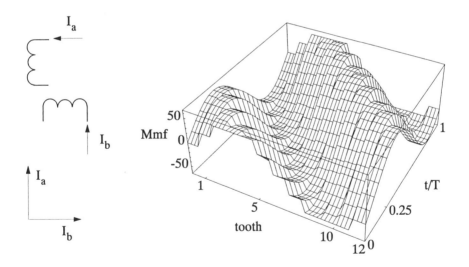

Fig. 2.29 *The time and spatial distribution of the air gap mmf created by the two phase concentric winding shown in Fig. 2.22. The winding carries sinusoidal currents which are $\pi/2$ radians out of phase.*

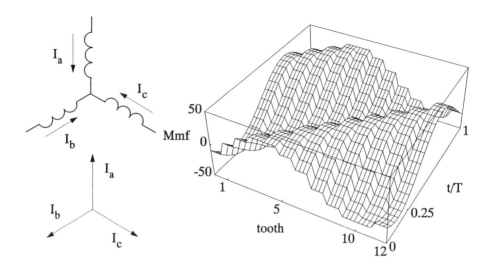

Fig. 2.30 *The time and spatial distribution of the mmf created by the winding shown in Fig. 2.24, carrying symmetrical sinusoidal currents*

Higher harmonics of the mmf are observed in this figure as crests superimposed on the fundamental component, as in the two phase machine analyzed in the previous example. One can also see by comparing Fig. 2.29 and Fig. 2.30 that the number of phases has no influence on the winding's capability to create a rotating field.

By exchanging the connections to the sources of any two phases of a three phase machine, one changes the currents and the direction of rotation of the air gap mmf. This is illustrated in Fig. 2.31, in which the currents in phases b and c of the same winding as in Fig. 2.30 are traded. Comparing the mmf waveforms in these two figures, one can see that the mmf distributions, when projected on the spatial axis, travel in opposite directions. The three dimensional mmf plots visualize the frozen mmf distributions in one period of the applied currents. Therefore, the waves in Fig. 2.30 and Fig. 2.31 carry the information about the direction of rotation of the mmf in the form of directions of propagation which are perpendicular to each other.

Very often the currents in a voltage fed electric machine are not symmetrical. There are various causes for the lack of symmetry: asymmetric windings, inappropriate connection of the windings, phase and/or amplitude unsymmetry of the applied voltages, etc. The asymmetry of the currents and/or windings is reflected in the amplitude and/or phase asymmetry of the mmfs created by each phase. Independently of its origin, every asymmetry in an electric machine creates *negative sequence mmfs*. The negative sequence mmf always deteriorates a machine's performance; hence, it is minimized or eliminated whenever possible.

Windings, Currents and Air Gap Magnetomotive Force 125

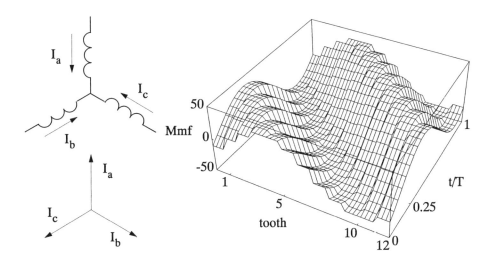

Fig. 2.31 *The air gap mmf in the same machine as in Fig. 2.30, but with exchanged currents in phases b and c*

The asymmetric mode of operation of a three phase electric machine is illustrated here by two examples. In the first example, one of the phase windings is reversely connected, so that the current through it flows in a direction which is opposite to the current direction in a symmetrical machine. The air gap mmf distribution in such a machine is shown in Fig. 2.32.

Instead of the sequence $120° - 120° - 120°$ of the spatial angles between the phases, which exists in a symmetrical machine, in this case one has the angles $60° - 60° - 240°$. When the currents in the windings are symmetric, the asymmetry of the windings is reflected on the air gap mmf distribution in the manner shown in Fig. 2.32. Although the mmf here is changing periodically both in time and space, no wave motion occurs, since the nodes and the peaks of the mmf wave are permanently fixed to a few points on the rotor periphery. One can see that the amplitude of the fundamental component of the mmf in this figure is not constant. This is always a sign of the existence of the negative sequence component. The air gap mmf is pulsating instead of rotating around the air gap periphery. The wave motion exists only if the nodes and peaks of the wave appear at every point in the air gap, or if the mmf value at every point in the three dimensional diagram in Fig. 2.32 varies periodically between the same maximum and minimum values.

In the following example, shown in Fig. 2.33, an extreme asymmetry is illustrated, caused by in-phase currents in the windings of a symmetric machine. These currents create the fundamental components of the positive and negative sequence mmfs, which form vector stars, similar to the one shown in Fig. 2.28 (a).

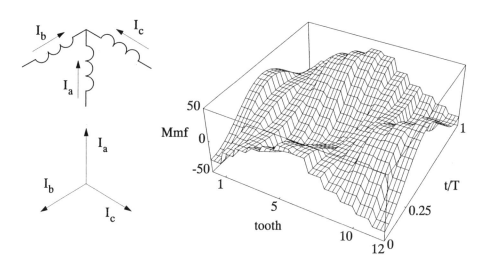

Fig. 2.32 *The air gap mmf distribution in a three phase machine in which one phase is reversed*

In a symmetric machine the positive sequence components of the fundamental of the mmf created in various phases are always collinear in space, and they support each other. When the currents are in phase as in this example, one finds, by applying Eq. (2.90), that the vectors of the positive sequence components of the fundamentals of the mmfs created by the phases are 120° out of phase, and their sum is equal to zero. The same is valid for the negative sequence components of the fundamentals of the mmf.

The *third* spatial harmonics in a three phase symmetrically wound machine are collinear in space, independent of the current waveform in its windings. By utilizing Eq. (2.90) for n=3 and for all three phases, one obtains that both positive and negative sequence components of the third spatial harmonic of the mmf are different from zero. The third spatial harmonic in this example creates a pulsating field, which is shown in Fig. 2.33. Based upon the previous considerations, one can state that a three phase, 2p pole machine carrying in-phase currents acts as a single phase, 6p pole machine.

In contemporary electric drives the machines are very often fed from power electronics sources, which supply them with nonsinusoidal currents. If the current waveform is rectangular, as shown in Fig. 2.34, and if the phase currents are shifted by 120° from each other, the three phase winding in this example generates the air gap mmf as shown in Fig. 2.35. Instead of travelling continuously around the rotor periphery, as illustrated in previous examples of the same machine fed from sinusoidal sources, the air gap mmf changes its position stepwise when fed from a rectangular current source.

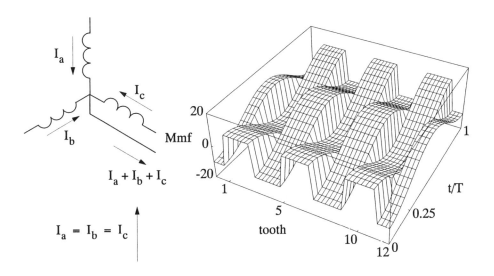

Fig. 2.33 *The air gap mmf distribution in a three phase machine in which all phases carry the same current*

Since the currents in all three phases change values every 60°, or one sixth of the period, the resultant air gap mmf jumps one sixth of the double pole pitch every one sixth of the period of the applied currents. The motion of the air gap mmf is smooth when the currents in the windings change smoothly, as in the case of sinusoidal currents.

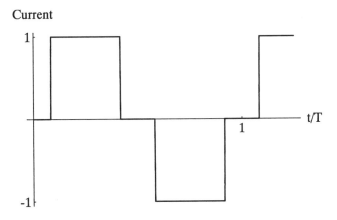

Fig. 2.34 *Rectangular current waveform*

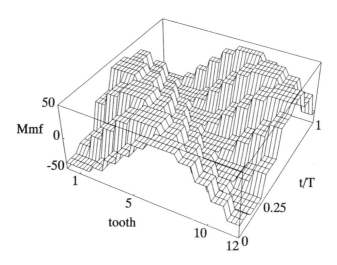

Fig. 2.35 *The air gap mmf distribution created by the three phase winding in Fig. 2.24 carrying rectangular symmetric currents*

When the currents change stepwise in time, the position of the mmf in the air gap also changes stepwise. By increasing the number of phases, the portion of the rotor periphery across which the mmf jumps when the currents change values becomes shorter, and the motion of the mmf is smoother.

The motion of the higher harmonics mmfs relative to the fundamental can again be observed in Fig. 2.35, as in the previous examples, when the machine was fed from sinusoidal current sources. However, according to the results presented in Table 2.2, the spectrum of the higher harmonics which one can observe in the air gap is enriched by additional terms when the currents are rectangular.

2.8 Representation of the Air Gap Magnetomotive Force in a Rotating Reference Frame

The waves of the rotating air gap mmf, which have been analyzed in the previous sections, are always described in the system of the observer who is fixed relative to the windings which produce the waves. Therefore, the speed of the waves, calculated in Section 2.4, is expressed relative to the winding. The observer is said to be in the *stationary reference frame*. However, sometimes a better insight into the air gap mmf behaviour can be obtained if the observer is *rotating* relative to the winding. In this case it is said that the observer is placed in a *rotating reference frame*. The speed of rotation of the reference frame can be chosen arbitrarily, depending on the particular needs of the problem being analyzed. In most cases, the speed of rotation of the reference frame is equal to the speed of either the positive or the negative sequence mmf. The freedom of motion of the observer relative to the winding is formally introduced by untying the tooth mmfs from the slot ampereturns.

By introduction of the matrix representation, the air gap mmf is discretized so that it is constant along the complete tooth width in the air gap. The minimum unit on which the changes of the air gap mmf are observable is the slot pitch. In the world of matrices only an integer notation can be used. There exists, for example, no π-th row of a vector or a matrix, but a third, or a fourth one. On the other hand, the nature of motion lies in the change of position in time, which can be performed only continuously. A discontinuous change of position, i.e. the one which would happen instantaneously, demands an infinite power to be carried out, therefore it is not possible.

In order to employ vector and matrix descriptions of the winding mmf quantities to define the reference frame(s) in a machine, one has first to convert time, a continuous quantity, into a discrete value which corresponds to the tooth number at which the maximum of the mmf is placed at a given time instant. After executing the statements given in the Appendix, one obtains a transform matrix which allows one to fictitiously shift the winding relative to the slots in which it is really placed. This way, the effect of the motion of the observer relative to the winding is obtained.

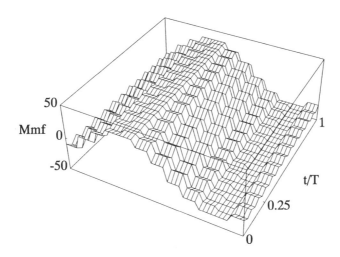

Fig. 2.36 *The air gap mmf created by a symmetric three phase winding, as seen from a reference frame which rotates at a synchronous speed in the direction of the fundamental*

If a reference frame is placed into a three phase machine with the windings shown in Fig. 2.24, and if it rotates at the speed of the fundamental of the mmf produced by symmetric currents, the air gap mmf, as seen from the reference frame, looks as shown in Fig. 2.36.

The fundamental component of the air gap mmf is stationary relative to the reference frame. Despite the stationary fundamental, the total mmf as seen from the reference frame is not smooth because the higher harmonic mmfs move relative to the fundamental, and thus relative to the reference frame, too.

When the reference frame travels at a negative synchronous speed, i.e. opposite to the direction of rotation of the fundamental, the relative speed between the two is twice the synchronous speed. In one period of rotation, a fixed point in the reference frame meets twice the maximum of the mmf. The mmf wave seen in this case from the reference frame is shown in Fig. 2.37.

If the machine is fed from rectangular current sources, as shown in Fig. 2.34, and if the reference frame rotates at the speed and in the direction of the fundamental of the mmf, the air gap mmf seen from the reference frame looks as shown in Fig. 2.38. The stepwise change of position of the mmf in the air gap can again be observed in this figure, as in Fig. 2.35, although in a different manner. The air gap mmf time–spatial distribution seen from the reference frame in this case is predominantly constant, with relatively high additional pulsating components. The pulsating components in the air gap mmf originate in the stepwise mode of motion of the air gap mmf wave and by the higher harmonics which rotate at their own speeds relative to the fundamental.

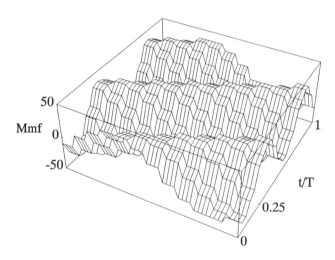

Fig. 2.37 *The air gap mmf in the same machine as in the previous figure, but seen from the reference frame rotating at a negative synchronous speed relative to the fundamental*

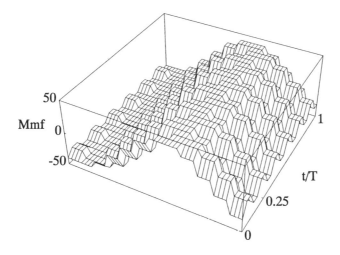

Fig. 2.38 *The air gap mmf created by a three phase winding fed from rectangular current sources, as seen from the reference frame rotating at synchronous speed in the direction of the fundamental*

2.9 Commutator Windings

The windings which have been considered so far are almost exclusively made of more than one coil. The coils are first connected in series, and then to a source. Each winding made in this way usually has only two terminals. A small number of terminals simplifies the manufacture of a machine, at the same time decreasing the ability to control the machine. One can imagine that the air gap mmf can follow a desired spatial distribution easier if the current in each coil, instead of the total current in a group of windings, is controlled.

The winding in which the connections of each coil are accessible is the *commutator winding*. A commutator winding with eight coils is shown in Fig. 2.39. Both ends of each coil are connected to segments of the *commutator*. The coils in the commutator winding are accessed via the brushes. The commutator along with the brushes forms a mechanical inverter, which converts A.C. quantities into D.C. ones, and vice versa. In order to convert the electrical energy from one form to another, the commutator with the coils has to move (rotate) relative to the brushes. Due to constructional reasons, the commutator winding is always placed on the rotor side, while the brushes are fixed to the stator side. The commutator winding is also often called the *armature winding* in a commutator machine.

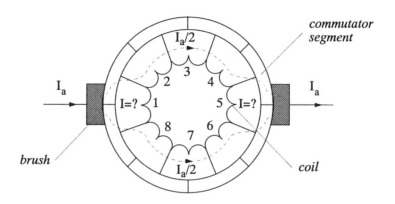

Fig. 2.39 *A commutator winding with eight coils*

The brushes scan the commutator segments during rotation, so that a rotating coil is connected directly to a brush via the commutator segment only at a particular angle between the rotor and the stator. When the commutator segments are passing beneath a brush, the coils whose ends are connected to those segments are short-circuited by the brush, and the current in the coils changes value. The current is said to be commutating in these coils. In Fig. 2.39 the current is commutating in coils 1 and 5 at a given position of the commutator winding. Coils 2, 3, and 4 in Fig. 2.39 carry one half of the positive armature current, $I_a/2$, whereas coils 6, 7, and 8 carry one half of the negative armature current, $-I_a/2$.

The commutator winding, as opposed to the windings discussed in the previous sections, is *closed*. This does not mean that only short-circuit currents can flow through the commutator winding, but that one always returns to the starting point when travelling from any point of the winding in any of the two directions.

The brushes connected to a commutator winding are used to pick up the current from the winding in the generator mode of operation, or to bring the external voltage to the winding in the motor mode of operation. The minimum number of brushes, two, is obvious only in the case of a two pole machine. Here the two brushes split the commutator winding into halves. The commutator winding can be split into various numbers of segments, called *parallel paths*, which are connected parallel to the brushes. The number of parallel paths depends on the type of commutator winding, which can be either lap or wave, on the multiplicity m of the winding, and on the number of poles, p. The winding with multiplicity one is called *simplex*, with multiplicity two *duplex*, etc. Simplex and duplex *lap* windings are shown in Fig. 2.40.

In a simplex lap winding the coil ends are connected to adjacent segments of the commutator, as shown in Fig. 2.40 (a). One can pass all the coils of a simplex lap winding by going only once around the periphery.

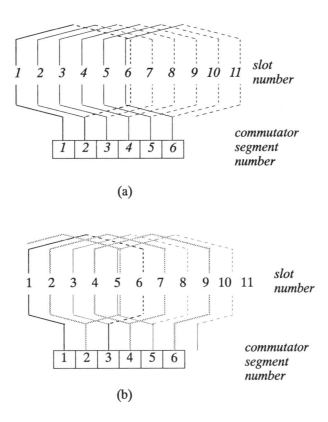

Fig. 2.40 *Simplex (a) and duplex (b) lap commutator windings*

In a duplex lap winding the ends of a coil are connected to the commutator segments in such a way that one segment remains free, as shown in Fig. 2.40 (b). The ends of the coils belonging to the second group of the coils of the duplex winding, drawn gray in this figure, are connected to these free segments. Only one half of the coils can be reached by going once around the periphery along a duplex lap winding. As the multiplicity m of the winding increases, the number of commutator segments between the ends of a coil, the so-called *commutator pitch*, also increases. At the same time, by keeping the total number of coils constant, the number of coils belonging to the group of a multiplex winding decreases. Similar conclusions can be derived for the wave winding.

The commutator pitch y_c, of a lap winding is obviously equal to the degree of multiplicity m

$$y_c = m \qquad (2.102)$$

whereas the number of parallel paths a is

$$a = 2pm \qquad (2.103)$$

The commutator pitch y_c of a wave winding can be expressed as

$$y_c = \frac{C \pm m}{2p} \qquad (2.104)$$

where C denotes the total number of commutator segments. Since the commutator pitch must be an integer, one can see that for a given number of poles p and multiplicity m a wave winding can be manufactured only with certain numbers of the commutator segments which satisfy Eq. (2.104). The number of parallel paths a in a wave winding is equal to the degree of multiplicity m

$$a = m \qquad (2.105)$$

The choice of the type of the winding is important for mitigating the effects of commutation, especially in medium and large size commutator machines. However, it has absolutely no influence on the qualitative shape of the air gap mmf curve. This conclusion is based on the fact that the same current flows in all coils between two adjacent brushes, independent of the type of winding. This is illustrated in Fig. 2.41, in which a commutator winding in the rotor slots is shown.

The air gap time–spatial mmf distributions created by a commutator winding are given in the following examples, which are created by executing the statements from the Appendix.

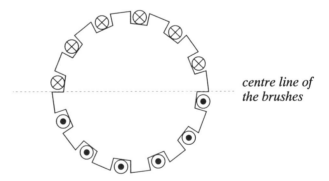

Fig. 2.41 *Conductors in the rotor slots of a two pole commutator machine and the directions of the currents*

It is assumed in these examples that the current in the commutating coil changes its value instantaneously. This assumption simplifies the analysis, but it does not affect the qualitative accuracy of the evaluated air gap mmf curves.

The air gap mmf distribution created by the commutator winding shown in Fig. 2.42 is stationary in space. This mmf distribution is seen from the stator side, or from the reference frame which is fixed to the brushes. The current through the commutator winding is D.C., so that the amplitude of the air gap mmf is constant. One should note here that, for a given armature current, the air gap mmf spatial distribution as seen from the stator side *does not depend on the rotor speed.*

The same mmf distribution seen from the reference frame which rotates at a negative rotor speed is shown in Fig. 2.43. One can see the travelling wave character of the commutator winding mmf. The speed of rotation of this travelling wave summed with the speed of the rotor gives zero. This means that the commutator winding mmf is fixed to the stator, as shown in Fig. 2.42.

Commutator motors used in appliances are often fed from sinusoidal voltage sources. Since they can work connected to either A.C. or D.C. sources, they are usually called *universal motors*. The spatial and time distribution of the air gap mmf in the stator reference frame, created by a commutator winding carrying sinusoidal currents, is shown in Fig. 2.44. One should note here the typical standing wave form of the air gap mmf. The speeds of rotation of the components of this standing wave have nothing in common with the rotor speed. The standing wave form of the rotor mmf is observed as an additional pulsating component of torque, as will be shown in the fifth chapter.

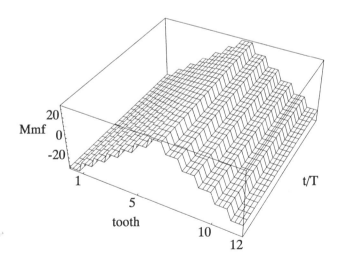

Fig. 2.42 *The time–spatial distribution of the air gap mmf created by a commutator winding, seen from the stator. The winding is placed in 12 slots, has ten turns per coil and carries 1 A.*

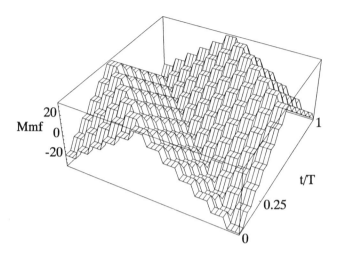

Fig. 2.43 *The time–spatial distribution of the air gap mmf created by a commutator winding, seen from the reference frame rotating at the speed of the rotor. The winding is placed in 12 slots, has ten turns per coil and carries 1 A.*

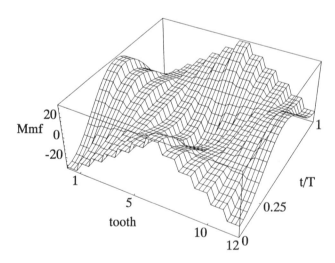

Fig. 2.44 *The time–spatial distribution of the mmf created by a commutator winding, as seen from the stator. The winding carries sinusoidal current with unit amplitude.*

One can see from Fig. 2.44 that the air gap mmf distribution of the commutator winding, when projected on the spatial axis, preserves its triangular character at any time instant, i.e. for any value of the applied current. At the same time, the projection of the air gap mmf on the time axis is a sine wave, since the current in the winding is sinusoidal.

Commutator machines are often fed from rectifiers, which supply a D.C. current with lower or higher levels of harmonics, depending on the number of phases in the source and on the type of rectifier. If the commutator winding is connected to a single phase source via a full wave bridge, as in the case of traction motors for electrical railways, the armature current waveform is similar to the absolute sine or cosine curve. The air gap mmf distribution of the commutator winding fed from a rectified sinusoidal source is shown in Fig. 2.45. Here one can again follow the spatial triangular and time rectified sinusoidal distribution of the mmf along the corresponding axes.

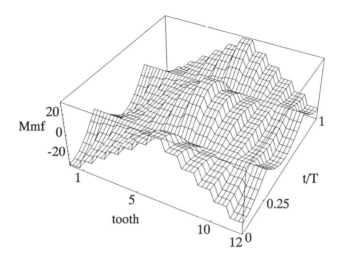

Fig. 2.45 *The time–spatial distribution of the mmf created by a commutator winding, as seen from the stator. The winding carries sinusoidal current with unit amplitude.*

2.10 Squirrel Cage Winding

In the first section of this chapter a squirrel cage winding is shown in Fig. 2.3. It is closed, as is the commutator winding, but, unlike the commutator winding, it is never connected to an external source. The currents in the squirrel cage winding can thus only be driven by the voltages induced in the bars and rings. A squirrel cage winding developed in plane is shown in Fig. 2.46.

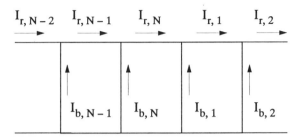

Fig. 2.46 *A plane view of a squirrel cage winding*

The squirrel cage winding in Fig. 2.46 is placed in N rotor slots. Hence it has N bars. Each ring consists of N segments. The relationship between the bar currents and the ring segment currents can be established by utilizing the assumed directions of the currents in Fig. 2.46.

$$I_{r,N} + I_{b,1} = I_{r,1}$$

$$I_{r,1} + I_{b,2} = I_{r,2}$$

$$\ldots \quad (2.106)$$

$$I_{r,N-2} + I_{b,N-1} = I_{r,N-1}$$

$$I_{r,N-1} + I_{b,N} = I_{r,N}$$

The system of Eqs. (2.106) is linearly dependent, since the last equation does not carry any new information about the ring and bar segment currents. Therefore, one of the equations in this system has to be replaced by an equation which expresses an additional physical condition which determines the performance of the squirrel cage. In this case, the additional physical condition is given by Kirchhoff's voltage law, expressed as the voltage equation for the ring segments

$$\sum_{j=1}^{N} (i_{r,j} R_{r,j} + e_{r,j}) = 0 \quad (2.107)$$

with $R_{r,j}$ denoting the resistance of the j-th ring segment, and $e_{r,j}$ the induced voltage

Windings, Currents and Air Gap Magnetomotive Force 139

in the same segment. The sum of the induced voltages and voltage drops along the ring is equal to zero, because there is no applied voltage on the ring. Since the machine is heteropolar, there is no unipolar flux linking the rings of the squirrel cage, and the sum of the induced voltages in Eq. (2.107) is equal to zero. Allowing for this condition, as well as for the fact that the resistances of the ring segments are equal, Eq. (2.107) can be further written as

$$\sum_{j=1}^{N} i_{r,j} = 0 \qquad (2.108)$$

Comparing the system of equations (2.106) and the additional condition (2.108) with system (2.64) and condition (2.65) which describe the mmf distribution in the air gap, one can conclude that the tooth mmfs created by a squirrel cage winding have the same meaning as the ring segment currents.

Eq. (2.108), along with the first N–1 equations of the system (2.106) can be written in matrix form as

$$\underline{I}_r = \underline{C}^{-1} \underline{U}' \underline{I}_b \qquad (2.109)$$

where the matrix $\underline{C}$ is defined in Eq. (2.66) and the matrix $\underline{U}'$ is the incomplete identity matrix whose last row (column) has all zero entries, given in Eq. (2.110). The vector of the bar currents, $\underline{I}_b$, and the vector of the ring currents, $\underline{I}_r$, are equal to

$$\underline{U}' = \begin{bmatrix} 1 & 0 & 0 & \dots & 0 & 0 & 0 \\ 0 & 1 & 0 & \dots & 0 & 0 & 0 \\ \dots & \dots & \dots & \dots & \dots & \dots & \dots \\ 0 & 0 & 0 & \dots & 1 & 0 & 0 \\ 0 & 0 & 0 & \dots & 0 & 1 & 0 \\ 0 & 0 & 0 & \dots & 0 & 0 & 0 \end{bmatrix} \qquad \underline{I}_b = \begin{bmatrix} I_{b,1} \\ I_{b,2} \\ \dots \\ I_{b,N} \end{bmatrix} \qquad \underline{I}_r = \begin{bmatrix} I_{r,1} \\ I_{r,2} \\ \dots \\ I_{r,N} \end{bmatrix} \qquad (2.110)$$

The relationship between the ring segment and bar currents is illustrated in the following example, which has been created by executing the statements listed in the Appendix. In this example, the fundamental spatial component of the ring segment currents of a squirrel cage winding having forty-eight bars for a given distribution of bar currents will be found.

Assuming that the maximum of the bar currents is 1 A, one obtains by applying Eq.

(2.109) a maximum ring current 7.63 A. The ring current is always greater than the bar currents. Therefore, the cross-sectional area of the rings in a squirrel cage is always greater than the cross-sectional area of the bars, if the same current density has to be obtained in both.

Another conclusion which one can derive from Fig. 2.47 is that the fundamental spatial components of the bar and ring segment currents are out of phase by 90 electrical (spatial) degrees. This is also valid for any other spatial harmonic. The bar currents have the maxima and minima at those points on the rotor periphery where the ring currents are equal to zero, and vice versa.

The ratio between the fundamental components of the ring and bar segment currents can be found analytically in the manner shown in the following procedure. The fundamental components of the bar currents are out of phase for a spatial angle α_{el}, defined by the number of bars (slots) N as

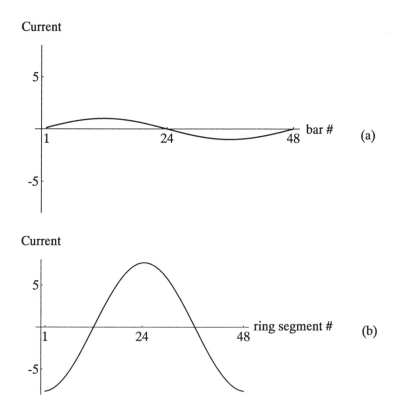

Fig. 2.47 *The relationship between the fundamental components of the bar (a) and ring segment (b) currents in a squirrel cage having forty-eight bars*

Windings, Currents and Air Gap Magnetomotive Force

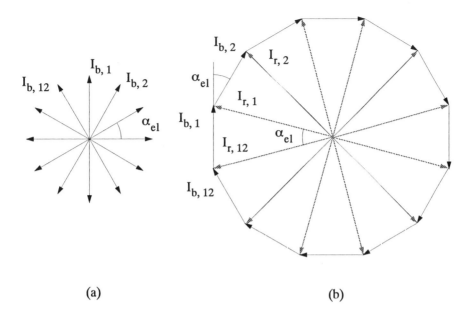

(a)   (b)

Fig. 2.48 *A representation of the bar currents in the form of a star (a) and polygon (b)*

$$\alpha_{el} = \frac{2p\pi}{N} \qquad (2.111)$$

where p is the number of pole pairs of the machine. The set of spatial vectors of the bar currents can be represented in an arbitrary manner, either in the form of a star or a polygon, as shown in Fig. 2.48. Here it is assumed that the squirrel cage winding has twelve bars per pole pair.

The relationship between the bar and ring segment currents established in Eq. (2.106), can be interpreted in vector form as shown in Fig. 2.48 (b): the vector of the bar current is equal to the difference between the vectors of the adjacent ring segment currents. This relationship enables one to define the ratio between the amplitude of the fundamental spatial components of the bar and ring segment currents.

From the vector diagram in Fig. 2.48 (b) one can write

$$\sin\frac{\alpha_{el}}{2} = \frac{I_b}{2I_r} \qquad (2.112)$$

which gives

$$I_r = \frac{I_b}{2\sin\frac{p\pi}{2N}} \qquad (2.113)$$

As long as the number of bars per pole is greater than three, the ring segment currents have a greater magnitude than the bar currents. This condition is almost always fulfilled, because squirrel cages are regularly built with the number of bars per pole greater than three.

3 Air Gap Magnetomotive Force and Flux Density

It has been shown in the previous chapter that the air gap magnetomotive force of several coils can be found simply by adding together the magnetomotive forces of each coil. The superposition law could be applied in this case, since the total effect (the air gap mmf) was obtained as the sum of partial effects (mmf of each coil). It is valid to apply the law of superposition on currents when calculating the mmf of a winding from the mmfs of each coil, since the relationship between the coil mmf and the current is independent of the current. In mathematical terms, the total mmf can be evaluated as a linear combination of the currents in the coils.

The magnetic medium in the air gap of a machine is linear, since its permeability is constant. In a linear magnetic medium the relationships between the flux and flux density, on the one hand, and magnetomotive force and field strength, on the other, are linear. Therefore, the fundamental postulate of the superposition law, expressed as $f(\alpha+\beta) = f(\alpha) + f(\beta)$, may be applied. In an electric machine, this means that the flux which corresponds to the sum of two or more mmfs can be evaluated as the sum of the fluxes produced by each mmf, or

$$\Phi(F_1 + F_2) = \Phi(F_1) + \Phi(F_2) \tag{3.1}$$

Analogously, the mmf which corresponds to the sum of two or more fluxes can be expressed as the sum of the mmfs produced by each flux, or

$$F(\Phi_1 + \Phi_2) = F(\Phi_1) + F(\Phi_2) \tag{3.2}$$

The relationships between the fluxes and mmfs in a linear magnetic medium are shown in Fig. 3.1.

When the electromagnetic response of a complete machine is calculated, the B–H curves of the stator and rotor irons have to be allowed for. Due to the nonlinear character of the B–H curve, the air gap flux density which corresponds to the resultant mmf cannot be found simply as the sum of the flux densities created by each coil's mmf.

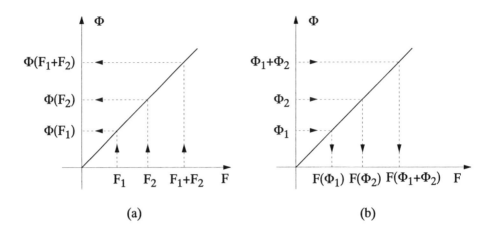

Fig. 3.1 *The resultant flux for the sum of applied mmfs (a) and the resultant mmf for the sum of fluxes (b) in a linear magnetic medium*

Analogously, the resultant mmf which corresponds to the sum of the fluxes in the air gap is not equal to the sum of the mmfs corresponding to each particular value of the fluxes. When the magnetic medium is nonlinear, one must consider that $f(\alpha+\beta) \neq f(\alpha)+f(\beta)$. The resultant flux density and mmf in a nonlinear magnetic medium are illustrated in Fig. 3.2 (a) and (b), respectively.

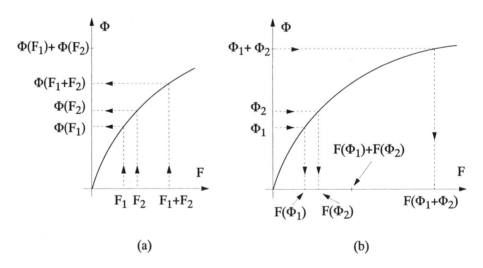

Fig. 3.2 *The resultant flux for the sum of applied mmfs (a) and the resultant mmf for the sum of fluxes (b) in a nonlinear magnetic medium*

The flux which corresponds to the sum of two or more mmfs is, due to the character of the nonlinear B–H curve, *always less* than the sum of the fluxes produced by each coil's mmf, or

$$\Phi(F_1 + F_2) < \Phi(F_1) + \Phi(F_2) \tag{3.3}$$

which is shown in Fig. 3.2 (a). On the other hand, the mmf which corresponds to the sum of two or more fluxes, is *always greater* than the sum of the mmfs produced by each flux

$$F(\Phi_1 + \Phi_2) > F(\Phi_1) + F(\Phi_2) \tag{3.4}$$

as shown in Fig. 3.2 (b).

3.1 The d–q Representation of Spatial Quantities

As a rule, an electric machine is analyzed assuming that its axial dimension (length) is infinitely long. The boundary effects are neglected, which is valid as long as the magnetic field distribution in the end windings is not the focus of interest. In such an infinitely long machine the magnetic field varies in plane, i.e. only in two dimensions – radial and tangential in radial coordinate systems, or x- and y-axis in rectangular (Cartesian) coordinate systems. Since the point in a plane is fully defined by only two coordinates, one can replace the action of the mmfs and fluxes created in an m-phase winding by only two mmf and two flux components, respectively. The two components of the mmf are perpendicular to each other. The same is valid for the two components of fluxes. When certain conditions in the machine are satisfied, the two perpendicular components of the air gap quantities are *orthogonal*, too. Orthogonal vectors are linearly independent, which means that the variation of one vector has absolutely no influence on the value of the other.

The first condition which has to be satisfied in order to represent the air gap mmf or flux in a machine by two orthogonal vectors is that the iron core B–H curve is linear. As shown in the previous section, only when the magnetization curve is linear may the superposition law be applied and the resultant flux represented as a sum of two or more fluxes.

The second condition which orthogonal vectors must satisfy is that the reluctance of the machine's magnetic circuit is independent of the rotor to stator angle. This condition is often expressed as a demand that the air gap reluctance is constant, as seen from the windings which produce the mmfs and fluxes. As illustrated in Fig. 3.3 (a), (b), and (c), the flux linkage (mutual inductance) between perpendicular coils a–a and b–b is zero, independently of the rotor angle, because the two coils face constant air gap in these three types of machines.

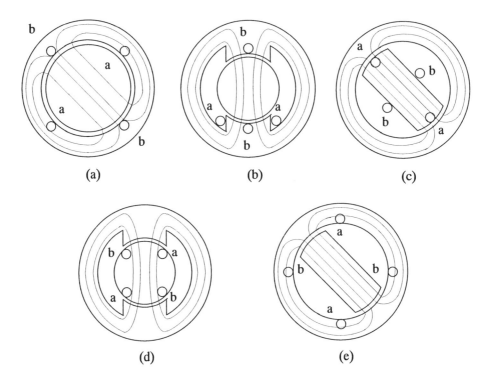

Fig. 3.3 *Approximate distribution of flux lines created by two perpendicular coils in various air gap configurations*

Since the fluxes produced by the currents in coils a–a and b–b are perpendicular and independent of each other, they are orthogonal, too. However, when the reluctance of the air gap is a function of the rotor shift, as shown in Fig. 3.3(d) and (e), the fluxes created by two perpendicular coils are independent of each other only when the axis of the minimum reluctance is collinear with the axis of one of the windings. At all other rotor positions the flux and mmf created by one coil influences the mmf and flux in the other coil. Therefore, the mmf and flux in the air gap of the machines shown in Fig. 3.3 (d) and (e) *cannot* be represented as sums of orthogonal mmfs and fluxes, respectively.

The two orthogonal components of the air gap quantities are said to lie in the d- and q-axes. Therefore, they are called the d- and q-components of the air gap mmf, flux, current, voltage, etc. When the machine has salient poles, the d-axis is always collinear with the line of minimum reluctance, whereas the q-axis is electrically perpendicular to it, i.e. shifted by 90° electrical.

3.2 The Influence of Slotting on the Air Gap Magnetomotive Force and Flux Density – Carter Factor

The number of slots on the stator side of an electric machine is almost always different from the number of slots on the rotor side. Therefore, if one wants to show the total mmf in the air gap in the way in which the mmf was represented in the previous chapter, it is necessary to define the unit length on the rotor periphery, along which the mmf is constant. Such a unit length is equal to the slot pitch if the winding which creates the mmf is placed on one side of the air gap. However, in reality the stator and the rotor pitches are different from each other, and the unit length must be shorter than both the stator and the rotor slot pitch, in order to fit an integer number of times in both of them. Such a unit length is calculated from the least common multiple (LCM) of the number of stator and rotor slots. If the rotor periphery is divided by the LCM of the number of stator and rotor slots, the obtained number fits an integer number of times in both the stator and rotor pitches. The calculation becomes more complex if not only the mmf, but also the flux density in the air gap of a machine has to be evaluated. In this case, additional subdivisions within the stator and the rotor slot pitches have to be performed, in order to take into account the variations of the air gap under the tooth and slot openings.

The calculation of the resultant air gap mmf created in a machine with a different number of stator and rotor slots is illustrated in the following example, in which the total mmf in the air gap of a machine having 12 stator and 18 rotor slots per pole is calculated. The pitch of the stator slots in this example is 15 mm, and the stator slot opening is 5.25 mm. The rotor slot pitch is 10 mm, and the rotor slot opening is 1.5 mm. The air gap width is 1 mm.

The machine carries a three phase winding on the stator side and a three phase winding on the rotor side. The spatial mmf distributions of the stator and rotor windings are obtained by executing the statements given in the Appendix. The tooth mmfs have been calculated in the previous section by assuming that the conductors are concentrated in one point within the slot, so that the air gap mmf changes stepwise in the slots. However, if one wants to calculate the air gap flux density at different points of the periphery, the slot openings cannot be neglected, since the flux density under the slot openings is smaller than under the teeth. In order to include the points on the rotor periphery which are placed within the slot openings, one has to split the complete interval of double pole pitch into the pieces which divide both the slot opening and the tooth an integer number of times. The procedure which enables one to calculate the mmf at each point of the air gap periphery, at which either the stator or the rotor change configuration from tooth to slot, and vice versa, is given in the Appendix. The stator, rotor and resultant mmf in the machine in this example are shown in Fig. 3.4. The computation of the resultant mmf was carried out at 720 points, although the mmf does not necessarily change the value between points.

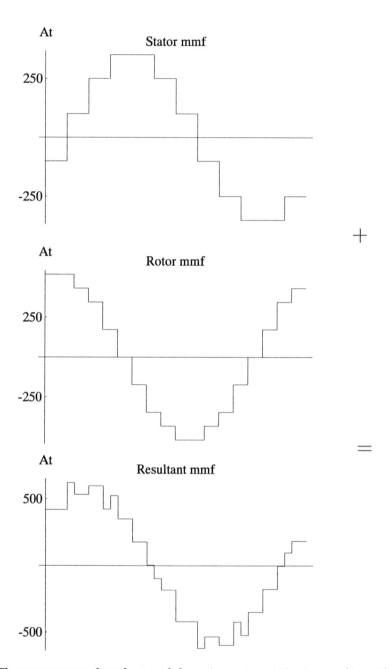

Fig. 3.4 *The stator, rotor and resultant mmfs for a given rotor position in a machine with 12 stator and 18 rotor teeth*

Air Gap Magnetomotive Force and Flux Density

The relationship between the air gap current sheet, magnetomotive force and flux density has been derived in the previous section, considering that all slot ampereturns are spent to cover the mmf drop across the air gap. The mmf drop in iron was neglected there for reasons of clarity of representation of the concept of current sheet and air gap mmf. In reality, however, the mmf drop in iron cannot be neglected. It is usually allowed for by the factor k_{sat}, which reflects the decreased mmf drop and, consequently, the decreased flux density in the air gap at point x_o

$$B_{xo} = \mu_o \frac{k_{sat} F_{xo}}{\delta_{xo}} \qquad (3.5)$$

The factor k_{sat} is always less than one. When the B–H curve of the magnetic material is nonlinear, the factor k_{sat} becomes a complex function of saturation at all points of the machine's magnetic circuit.

Eq. (3.5) is valid only if the lines of flux pass through the air gap radially. This is true when the air gap is smooth. However, if there are slots and/or poles on either the stator, or the rotor side of the air gap, the lines of flux contain additional tangential components in the vicinity of the points at which the air gap suddenly changes its width, as shown in Fig. 3.5. Therefore, the air gap width δ_{xo} in Eq. (3.5) in a slotted machine is not equal to the geometrical width of the air gap.

The air gap width δ_{xo} allows for the fact that the flux lines follow the general minimum resistance (reluctance) law, which, in magnetic circuits, can be interpreted as a tendency of the lines of flux to pass through the air along the shortest possible route, demanding the minimum magnetomotive force for that action. The lines of flux tend to change their direction and enter the slot surface laterally, as shown in Fig. 3.5 (b), instead of keeping the radial direction and entering the furthest point at the bottom of the stator slots, as shown in Fig. 3.5 (a). The distribution of the lines of flux shown in Fig. 3.5 (b) is maintained as long as the flux density in iron is low.

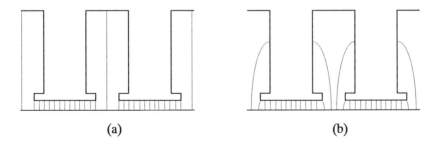

Fig. 3.5 *The lines of flux distribution in the air gap and slots: (a) non-realistic case, in which there is only a radial component of the lines of flux; (b) possible distribution, showing the fringing flux in the vicinity of the tooth–slot boundary*

Assuming a constant air gap mmf along the slot pitch, one obtains the value of the air gap flux density by dividing the constant value of the air gap mmf by the air gap width, according to Eq. (3.5). The air gap is much wider under the slot than under the teeth; hence one can expect large differences in the values of flux along the peripheral coordinate. However, this negative influence of slotting is mitigated by the fact that the flux is distributed as shown in Fig. 3.5 (b). The fringing flux in the slot openings makes the notches in the air gap flux much less pronounced than the ones expected from Eq. (3.5). The air gap flux density in a slotted, unsaturated machine can be calculated in the manner shown in [8].

By utilizing the symbols from Fig. 3.6, one can define the functions u, β, and γ as

$$u = \frac{o}{2\delta} + \sqrt{1 + \left(\frac{o}{2\delta}\right)^2} \qquad \beta = \frac{B_0}{B_{max}} = \frac{(1-u)^2}{2(1+u^2)}$$

$$\gamma = \frac{4}{\pi}\left[\frac{o}{2\delta}\text{atan}\frac{o}{2\delta} - \ln\sqrt{1 + \left(\frac{o}{2\delta}\right)^2}\right]$$

(3.6)

The effective slot opening, o′ in Fig. 3.6, is always wider than the real slot opening o. The slot fringing flux is located within the effective slot opening. The effective slot opening is defined as

$$o' = \frac{\gamma}{\beta}\delta \qquad (3.7)$$

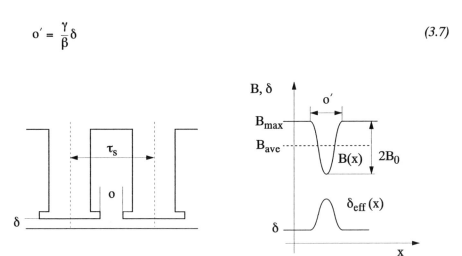

Fig. 3.6 *The variation of the air gap flux density and effective air gap width due to slotting of one side of the air gap*

Air Gap Magnetomotive Force and Flux Density

The variation of the air gap flux density within the effective slot opening can be described by a constant term and fundamental harmonic as

$$B(x) = B_{max} - B_0 \left(1 - \cos\frac{2\pi}{o'}x\right) = B_{max}\left[1 - \beta\left(1 - \cos\frac{2\pi}{o'}x\right)\right] \quad (3.8)$$

When the mmf along a slot pitch is constant, the variation of the air gap flux can be caused only by the variable air gap width. Therefore, the effective air gap width δ_{eff} within the slot opening o' can be expressed as

$$\delta_{eff} = \frac{\delta}{1 - \beta\left(1 - \cos\frac{2\pi}{o'}x\right)} \quad (3.9)$$

because the product of the flux density in Eq. (3.8) and effective air gap width in Eq. (3.9), which has the dimension of the mmf, is constant. For values of the argument x within slot pitch τ_s, the air gap width is constant and equal to δ.

The qualitative change of the air gap flux density and the effective air gap width along the peripheral coordinate are shown in Fig. 3.6. The average value of the flux density under the effective slot opening is equal to $B_{max} - B_0$, and under the rest of the slot pitch is B_{max}. Since the flux is equal to the product of the average flux density and area, one can write

$$B_{ave}\tau_s = (B_{max} - B_0)o' + B_{max}(\tau_s - o') \quad (3.10)$$

from which one can express the average flux density within the slot pitch, B_{ave}, as

$$B_{ave} = B_{max}\frac{\tau_s - \beta o'}{\tau_s} \quad (3.11)$$

The ratio between the maximum and the average flux density under one slot pitch is called the *Carter factor* k_c:

$$k_c = \frac{B_{max}}{B_{ave}} = \frac{\tau_s}{\tau_s - \beta o'} = \frac{\tau_s}{\tau_s - \gamma\delta} \qquad o' < \tau_s \quad (3.12)$$

When the effective slot opening is larger than the slot pitch, i.e. $o' > \tau_s$, the average value of the flux density within the slot pitch is

$$B_{ave} = B_{max} - B_0 = B_{max}(1 - \beta) \quad (3.13)$$

which means that the Carter factor in this case is

$$k_c = \frac{1}{1 - \beta} \qquad o' > \tau_s \quad (3.14)$$

The Carter factor, as defined in Eqs. (3.12) and (3.14), describes how much the air gap width is fictitiously increased if there are slots on one side of the gap, which is smooth on the other side. If both the stator and rotor have slots, the resultant Carter factor is equal to the product of the Carter factor for the stator slots and the Carter factor for the rotor slots. The Carter factor k_c, as a function of the ratio between the slot opening and gap width, and the ratio between the slot pitch and the gap width, is shown in Fig. 3.7. This plot is obtained by executing the statements from the Appendix. As one can see in Fig. 3.7, the Carter factor increases along with the increase of the slot opening. The air gap is apparently wider for longer than for shorter slot openings, because for a constant air gap mmf, the average value of the flux density decreases by increasing the slot opening.

The variation of the effective air gap width δ_{eff} as a function of the rotor peripheral coordinate in a machine with twelve stator slots, slot pitch equal to 15 mm, stator slot opening 5.25 mm and a smooth rotor is shown in Fig. 3.8.

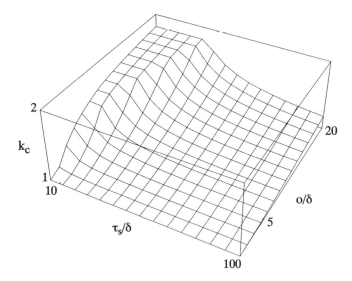

Fig. 3.7 *Carter factor k_c as a function of the ratio between the slot opening and the gap width, o/δ, and the ratio between the slot pitch and the gap width, τ_s/δ*

Air Gap Magnetomotive Force and Flux Density

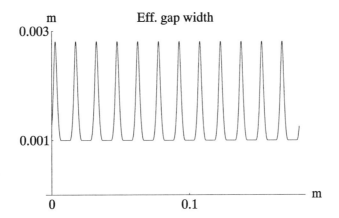

Fig. 3.8 *The effective air gap width along the double pole pitch, as a function of the peripheral coordinate in a machine with a smooth rotor and slotted stator*

This figure was obtained by executing statements given in the Appendix. The number of peaks in the effective air gap width distribution is equal to the number of stator slots.

Knowing the effective air gap width around the periphery, one can calculate the air gap flux density distribution for a given mmf curve. This is illustrated in Fig. 3.9, in which the air gap flux density for the stator mmf distribution in Fig. 3.4 and the effective air gap width given in Fig. 3.8 is calculated by executing statements from the Appendix. One can see clearly in this figure how the air gap modulates the air gap mmf, finally giving the air gap flux density.

When both stator and rotor are slotted, the effective air gap width becomes a function of the rotor to stator shift, besides the stator and rotor slot dimensions. This is illustrated in the following example, in which the effective air gap width in the machine, the parameters of which were utilized in the first example in this section, is calculated. The computational procedure is given in the Appendix. The stator in this example has much wider slot openings than the rotor. Therefore, the higher peaks in Fig. 3.10 are caused by the stator slot openings. For a given rotor position some stator slots coincide with the rotor slots. This, however, has no influence on the maximum value of the air gap width.

The resultant air gap flux density created by the resultant mmf from Fig. 3.4 and for the air gap effective width in Fig. 3.10 is given in Fig. 3.11.

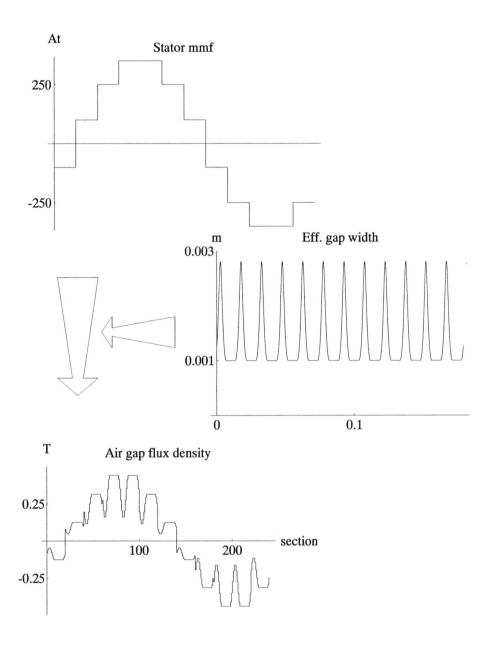

Fig. 3.9 *Illustration of the procedure for obtaining the air gap flux density from a given mmf distribution and known effective air gap width*

Air Gap Magnetomotive Force and Flux Density

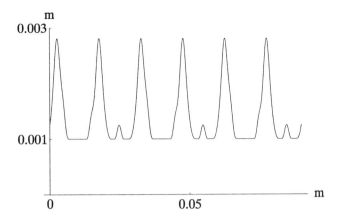

Fig. 3.10 *The effective air gap width in a machine with twelve stator and eighteen rotor slots as a function of the rotor peripheral coordinate along one pole*

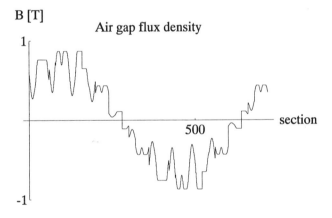

Fig. 3.11 *The resultant air gap flux density distribution for a given rotor position and resultant mmf*

3.3 The Total Air Gap Magnetomotive Force and the Air Gap Flux Density in a Commutator Machine

The generation of the magnetomotive force by the armature winding of a commutator machine has been elaborated in the previous chapter. Here we will see how the total mmf and the air gap flux density in a commutator machine are created as a result of the mmfs of all the windings.

A cross-section of a two pole commutator machine is shown in Fig. 3.12 (a). Besides the armature winding on the rotor, a commutator machine can have several windings on the stator. The *main field winding*, placed on the main poles, carries the current which builds the main magnetic flux in the air gap.

The main field winding can be performed either as a *shunt field, series field* or a *compound field winding*, depending on its connection to the armature winding. Very often the *commutating poles*, or *interpoles*, are placed in the so-called *magnetic neutral*. The magnetic neutral is the line which connects the two points on the rotor periphery, where the flux density is equal to zero. The magnetic neutral is placed in the middle of the distance between the adjacent main poles. The brushes of a commutator machine should always be placed so as to touch the commutator segments to which the coils sited in the magnetic neutral are connected. The commutating poles carry the *commutating pole winding*, which is utilized to compensate for the negative effects of the armature reaction in the zone of commutation. The main poles of medium and large size machines can also carry the *compensating winding*, whose purpose is to compensate for the negative effects of the armature reaction under the main poles and in the neutral zone. The resultant air gap mmf distribution is obtained as a sum of the mmf distributions of each winding.

The mmf distribution of the main field winding is obtained by assuming that its current sheet is concentrated at infinitesimally narrow bands in the inter-polar space on the sides of the main poles. The mmf distribution created by the field winding carrying unit ampereturns is shown in Fig. 3.13. The width of the mmf zone in this figure is proportional to the pole width, or to the parameter α in Fig. 3.12. Typical values of α vary from 0.6 to 0.7.

The air gap flux density created by the mmf distribution in Fig. 3.13 can be found after the effective air gap width is calculated for a given geometry. Assuming that the rotor has a smooth surface and that on the stator there are only main poles, without slots for the compensating winding, one obtains the effective air gap width as shown in Fig. 3.14. The effective air gap width is maximum in the centre line between the main poles. The air gap width beneath the main poles in this example is equal to 1 mm, and between the main poles it reaches a value almost thirty times higher. However, one has also to have in mind that this high value of the maximum effective air gap width is still significantly less than the real distance between the rotor surface and the stator yoke in the inter-polar space. The air gap flux density created only by the main field winding in a machine with a smooth rotor is shown in Fig. 3.15.

Air Gap Magnetomotive Force and Flux Density

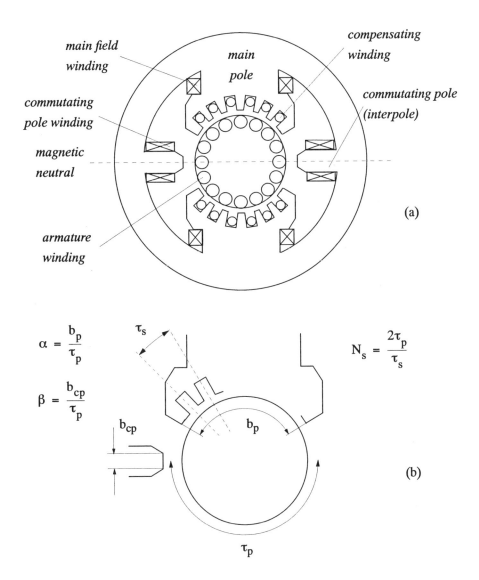

Fig. 3.12 *Cross-section of a two pole commutator machine (a), and pole dimensions (b)*

This curve is obtained by dividing the main field mmf in Fig. 3.13, created by 1000 ampereturns, by the effective gap width in Fig. 3.14. The air gap flux density in Fig. 3.15 follows the mmf distribution along the rotor periphery, except in the vicinity of the pole edges. Here the flux starts to decrease gradually, turning into the fringing flux in the inter-polar space.

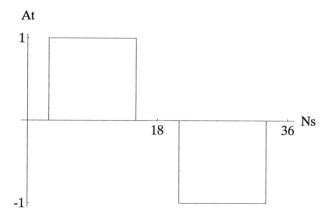

Fig. 3.13 *The air gap mmf created by unit ampereturn of the main field winding*

The commutating poles carry a winding, through which the armature current flows. The mmf of the commutator pole winding compensates for the negative effects of the armature current in the zone where the commutation of the armature current is performed. The unit ampereturns created by the commutating pole winding are shown in Fig. 3.16.

The effective value of the air gap width in a machine with commutating poles differs from the one shown in Fig. 3.14. The effective width of the air gap decreases under the commutating poles.

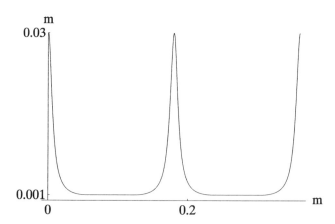

Fig. 3.14 *The effective air gap width under two poles of a commutator machine having only main poles without slots for compensating winding. The effective air gap width is maximum in the centre of inter-pole space.*

Air Gap Magnetomotive Force and Flux Density

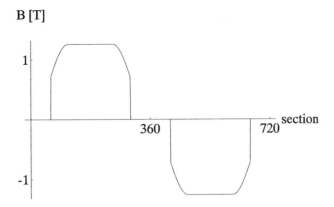

Fig. 3.15 *The air gap flux density along two poles of a commutator machine having only main poles without slots for compensating winding. Here 720 points correspond to the double pole pitch.*

However, if the commutating poles are too narrow, the effective width of the air gap under the commutator poles can be larger than the real width.

The effective air gap width in a machine with smooth rotor, commuting poles and the main poles without the compensating winding, is represented in Fig. 3.17. One can see in Fig. 3.17 that the commutating poles contribute significantly to the decrease of the maximum value of the effective air gap width. Comparing Fig. 3.17 with Fig. 3.14, one can see that the insertion of the commutating poles decreases the maximum value of the effective air gap width significantly.

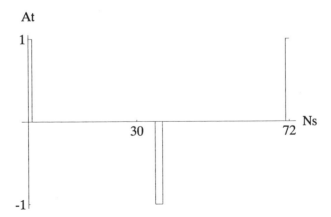

Fig. 3.16 *The unit ampereturns created by the commutating pole winding*

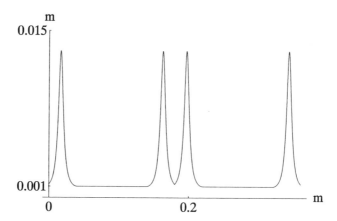

Fig. 3.17 *The effective air gap width of a commutator machine having main and commutating poles*

In this example, the decrease of the maximum value of the flux density is equal to approximately one half of the value of the maximum air gap width in a machine without commutating poles.

When a commutator machine has compensating winding, the slots into which this winding is inserted modify the value of the effective air gap width under the main poles. Within the zone of the main poles, the air gap width is influenced by the compensating winding poles, and described in the manner shown in the Appendix. The effective air gap width in a commutator machine with smooth rotor, commutating poles and main poles with the compensating winding slots, is shown in Fig. 3.18.

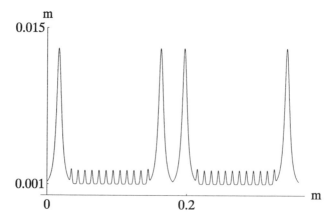

Fig. 3.18 *The effective air gap width in a commutator machine with smooth rotor, commutating poles and main poles with the compensating winding*

One can clearly see in Fig. 3.18 the influence of the compensating winding slots, which increase the effective air gap width under the main poles.

The air gap flux density, created by the main field winding, is influenced by the compensating winding slotting. The air gap flux density decreases under the compensating winding slots, as shown in Fig. 3.19. The average value of the flux density in Fig. 3.19 is less than the one in a machine without the compensating winding slots, shown in Fig. 3.15. The decrease of the average value of the flux density can be calculated by utilizing the Carter factor for the compensating winding slots.

The compensating winding is used to neutralize the effects of the armature reaction in the air gap beneath the main pole and in the neutral zone. The compensating winding carries the armature current, as the commutating pole does. The air gap mmf distribution created by a compensating winding having unit ampereconductors in the slots is shown in Fig. 3.20.

The spatial distribution of the air gap flux density created by the compensating winding of a machine with commutating poles and with a smooth rotor is shown in Fig. 3.21.

The air gap mmf and flux density created by all stator windings will be represented next. The calculation procedure to obtain these two spatial distributions, along with the parameters of the windings, are given in the Appendix. The stator created mmf distribution is shown in Fig. 3.22, whereas the corresponding air gap flux density is given in Fig. 3.23.

One should again notice that all the air gap flux density distributions shown in this section have been derived by assuming that the mmf drop on the machine's iron is equal to zero. Therefore, the complete ampereturns of the windings are spent on the air gap, and the relationships between the mmfs and flux densities are magnetically linear, however, modulated by various width of the air gap.

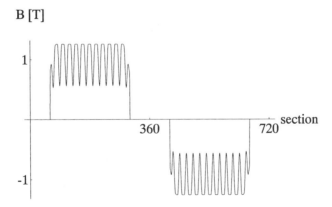

Fig. 3.19 *The air gap flux density along two poles of a commutator machine, which has twelve compensating winding slots on each main pole. Only the main field winding is energized, creating an air gap mmf distribution, such as the one shown in Fig. 3.13.*

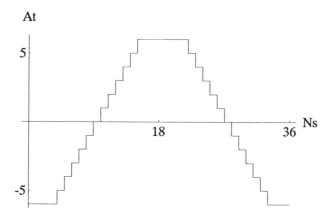

Fig. 3.20 *The air gap mmf created by unit ampereconductors of the compensating winding. The maximum ampereturns are located in the inter-pole section of the machine.*

The armature winding in a commutator machine creates the mmf which has the maximum and minimum colinear with the axis of the commutating poles, i.e. in the q-axis of the machine. The air gap mmf distribution created by the armature winding is shown in Fig. 3.24.

One can notice that the armature winding mmf, shown in Fig. 3.24, and the compensating winding mmf, Fig. 3.20, have the same linearly increasing or decreasing character in the zone of the main poles. The ampereturns of the compensating winding act against the armature winding ampereturns, cancelling their effects in the air gap.

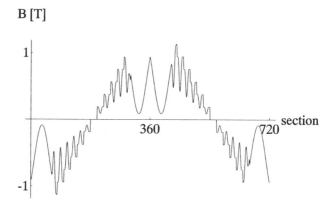

Fig. 3.21 *The air gap flux density created by the compensating winding in a machine with commutating poles and smooth rotor. The corresponding air gap mmf distribution is generated by utilizing the unit mmf distribution in Fig. 3.20.*

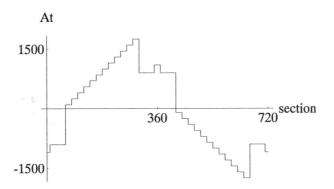

Fig. 3.22 *The air gap mmf created by the common action of all stator windings*

The positive effects of the compensating winding on the resultant air gap mmf can be seen in Fig. 3.25. The air gap mmf under the main poles does not substantially differ from the mmf created solely by the main field winding.

This effect has been obtained by a careful choice of the number of turns of the compensating winding. The higher harmonics of the mmf under the main poles, shown in Fig. 3.25, are a consequence of the difference between the number of slots in which the compensating winding is placed and the number of rotor slots. The amplitudes and phase shifts of these harmonics vary as a function of the rotor shift.

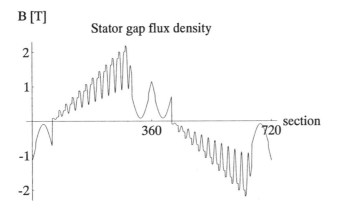

Fig. 3.23 *The air gap flux density created by the common action of all stator windings. The corresponding value of the total stator mmf is shown in Fig. 3.22.*

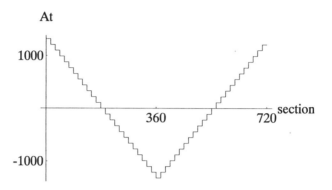

Fig. 3.24 *The air gap mmf distribution created by the armature winding placed in 48 rotor slots*

The effective air gap width in a commutator machine with slotted main poles and rotor, and having inter-poles is a function of the rotor to stator shift. This curve is shown in Fig. 3.26. One can see that the effective air gap width under the main poles has several local maxima, at the points where the stator and rotor slots align. This curve is also influenced by the variable rotor surface geometry in the inter-pole region.

The total air gap mmf, shown in Fig. 3.25, creates in the air gap with the effective width as in Fig. 3.26 the flux density represented in Fig. 3.27. The rotor slots introduce the holes in the flux density distribution, the final effect of which is a lower average flux density, or a higher Carter factor.

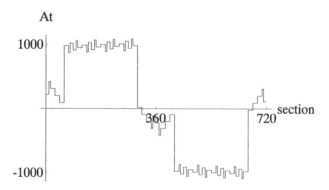

Fig. 3.25 *The total air gap mmf in a commutator machine*

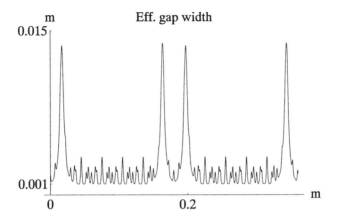

Fig. 3.26 *The effective air gap width in a commutator machine for a given stator to rotor shift*

For the purpose of savings, large size commutator machines are sometimes built without the compensating winding. Such a construction decreases the performance of the machine, as illustrated in the following example in which the mmf distribution created in a loaded commutator machine without the commutator winding is calculated. This curve, shown in Fig. 3.28, has been obtained by executing the statements given in the Appendix. One can clearly see in this figure how the armature current distorts the air gap mmf from its value in an unloaded machine. This effect is called the *armature reaction*.

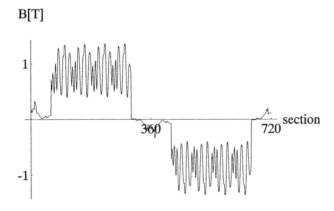

Fig. 3.27 *The air gap flux density created by the stator and rotor windings in a commutator machine*

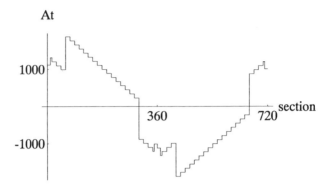

Fig. 3.28 *The resultant mmf distribution in the air gap of a loaded commutator machine which has no compensating winding*

The armature reaction is pronounced in commutator and synchronous machines, where the fundamental components of the mmf are strongly influenced by it. In induction machines, however, the armature reaction exists only for the higher air gap harmonics.

Some special purpose D.C. machines can have an additional pair of brushes besides the ones in the magnetic neutral of the machine. The additional brushes are usually placed perpendicular to the regular ones, i.e. in the d-axis, and d-axis currents flow through them.

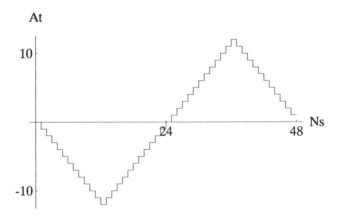

Fig. 3.29 *The d-axis mmf produced by unit ampereconductors in the rotor slots*

Air Gap Magnetomotive Force and Flux Density

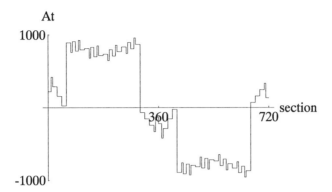

Fig. 3.30 *The resultant air gap mmf distribution in a machine, the armature winding of which generates both d- and q-axis mmfs*

The mmf produced by this current has its maximum value in the d-axis of the machine; hence it is called the d-axis mmf. The distribution of the d-axis mmf for unit ampereconductors in the slots is shown in Fig. 3.29.

When the d-axis mmf produced by the armature winding are superimposed on the ones created by other machine windings, one obtains the mmf distribution similar to the one represented in Fig. 3.30. In this figure one can see the decrease in the mmf curve under the main poles, caused by the d-axis current in the armature winding.

3.4 The Total Air Gap Magnetomotive Force and the Air Gap Flux Density in a Synchronous Machine

As a rule, the stator of a synchronous machine carries a three phase winding connected to the power system, whereas on the rotor side there is a D.C. fed coil on each pole of the machine. The rotor of a synchronous machine can be either cylindrical, when the machine is called a *turbogenerator*, or it can have salient poles, when the machine is called a *hydrogenerator*. The two possible rotor shapes are shown in Fig. 3.31.

The field winding of a turbogenerator is concentric and placed in the rotor slots, whereas the field winding of a salient pole rotor consists of one coil per each pole. The rotor of a synchronous machine rotates at the speed of the fundamental of the stator created rotating field, or at the *synchronous speed*. Therefore, in the reference frame fixed to the rotor, the fundamental stator mmf is stationary.

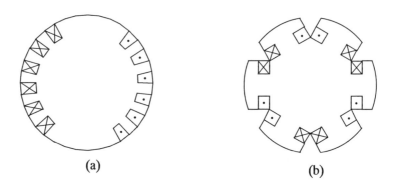

Fig. 3.31 *The cross-section of a high speed synchronous machine – turbogenerator (a), and a low speed synchronous machine – hydrogenerator (b)*

On the other hand, a field winding, which rotates at a synchronous speed and carries a constant current, is seen on the stator side as a coil carrying currents which have a frequency equal to the fundamental frequency of the stator currents. The mmf distribution created by the rotor winding in a turbogenerator is shown in Fig. 3.32.

The stator (armature) winding of a turbogenerator is double layer. The instantaneous values of currents in it produce the mmf distribution which has minimum amplitudes of the higher harmonics, due to the shortening of the coils. If the phase shift of the currents is chosen so that the stator mmf is collinear in space with the rotor mmf, the armature mmf looks as shown in Fig. 3.33.

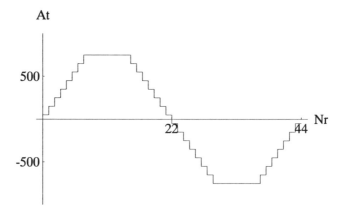

Fig. 3.32 *The air gap mmf created by the field winding in a turbogenerator*

Air Gap Magnetomotive Force and Flux Density

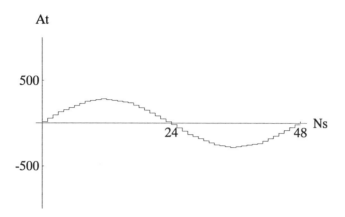

Fig. 3.33 *The stator winding mmf distribution for given values of phase currents*

The mmf waveform in Fig. 3.32 is typical for a concentric winding with equal number of turns per coil. However, very often the number of turns per coil in the field winding of a turbogenerator is not equal, in order to minimize higher mmf harmonics.

The total mmf is obtained by adding the armature and the field winding mmfs. Since the number of the rotor slots and the stator slots is always different, the stator and rotor mmfs have to be added in the manner shown in Section 3.2. After applying this procedure, as shown in the Appendix, one obtains the resultant air gap mmf for the stator and rotor mmfs, the fundamentals of which are in phase, as shown in Fig. 3.34.

The phase shift between the fundamentals of the armature F_{arm} and field F_{field} mmfs can be controlled in a synchronous machine by varying its active and/or reactive power. The phase shift obviously influences the amplitude of the fundamental of the resultant air gap mmf. The total mmf in a turbogenerator for some typical values of the shift between the stator and rotor mmf, besides the one revealed in Fig. 3.34, are shown in Fig. 3.35 (a), (b), and (c).

The air gap flux density in a turbogenerator for a given mmf distribution and effective air gap width is obtained in the same manner as was done for a D.C. machine. The air gap mmf in this example is obtained as a superposition of the stator and rotor mmf which are spatially shifted electrical $\pi/3$ one to the other, as shown in Fig. 3.36.

The effective air gap width in this example is calculated in the Appendix assuming that both sides of the air gap are slotted. With this assumption one obtains the air gap flux density distribution for the given mmf curve in Fig. 3.36, as shown in Fig. 3.37.

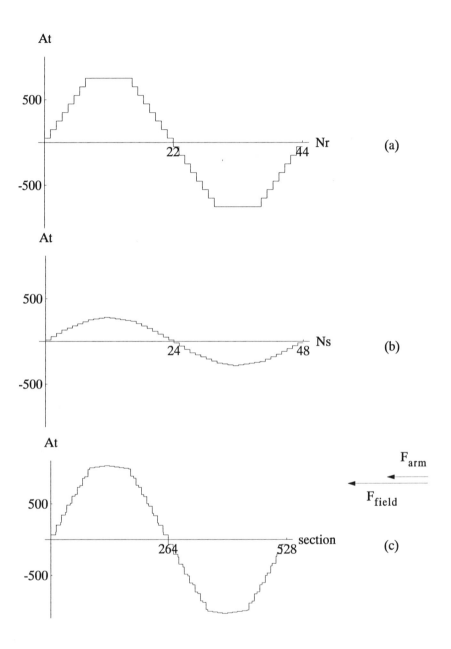

Fig. 3.34 *The rotor (a), stator (b) and resultant (c) air gap mmf distribution in a turbogenerator in which the stator and rotor mmfs are in spatial phase*

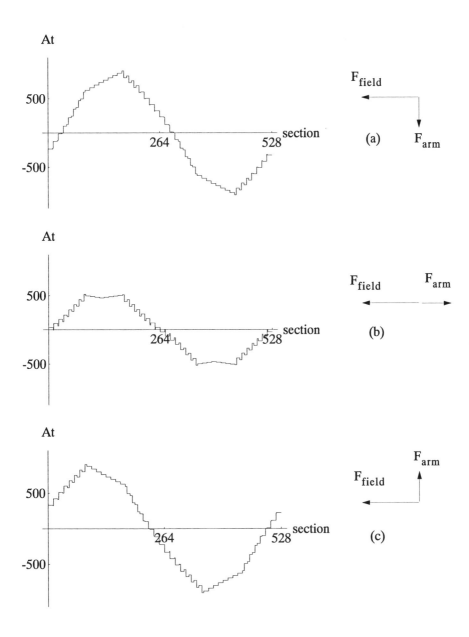

Fig. 3.35 *The total mmf as a function of the angle between the field and armature mmfs in the air gap of a turbogenerator*

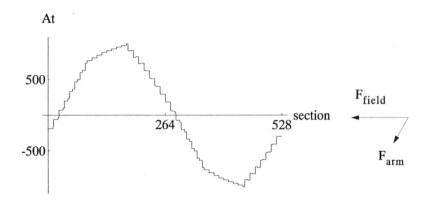

Fig. 3.36 *The resultant air gap mmf distribution in a turbogenerator for the phase shift between the stator and rotor mmfs equal to π/3*

The air gap mmf distribution created by the field winding in a *hydrogenerator* is obtained in a similar manner as the field winding mmf distribution in a D.C. machine, which is shown in the Appendix. Since the rotor salient poles are assumed smooth in this analysis, the unit length along which the field winding mmf is calculated is the stator slot pitch. The field winding mmf distribution in the air gap of a hydrogenerator which has 48 stator slots is shown in Fig. 3.39.

The resultant air gap mmfs for four different spatial phase shifts between the field and armature mmfs are shown in Fig. 3.38.

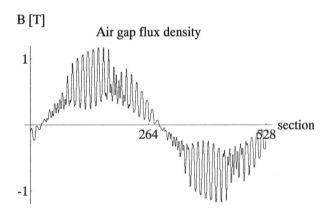

Fig. 3.37 *The air gap flux density in a turbogenerator, created by the mmf in Fig. 3.36*

Air Gap Magnetomotive Force and Flux Density 173

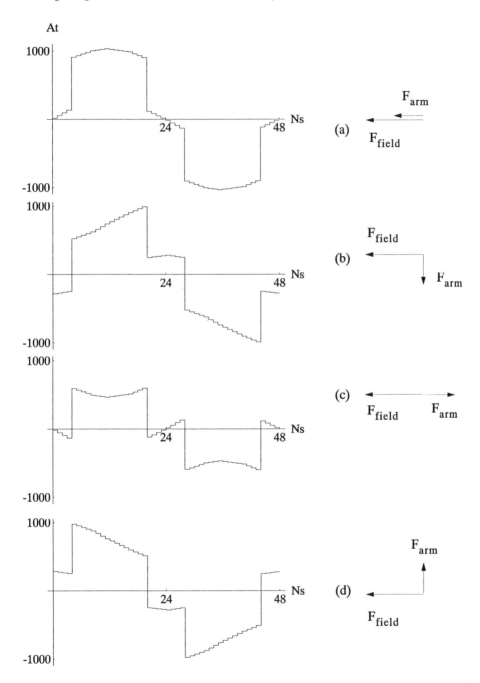

Fig. 3.38 *The total mmf as a function of the angle between the field and armature mmfs in the air gap of a hydrogenerator*

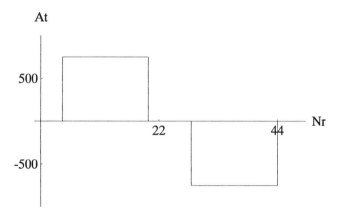

Fig. 3.39 *The air gap mmf distribution created by the field winding in a turbogenerator*

The effective air gap width of a salient pole synchronous machine is evaluated in a similar manner as for a D.C. commutator machine without slots for the compensating winding and commutating poles, as shown in the Appendix. The salient rotor poles of a turbogenerator are, however, very often manufactured in such a manner to create a variable air gap width with a cylindrical stator. The distance between the rotor pole surface and inner stator cylinder is more or less close to the inverse cosine function. Therefore, the air gap permeance under the poles follows the cosine function, with the air gap peripheral coordinate as an argument. The higher harmonics of the air gap permeance are minimized in this way.

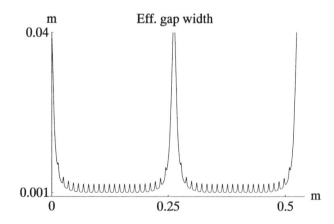

Fig. 3.40 *The effective air gap width in a hydrogenerator*

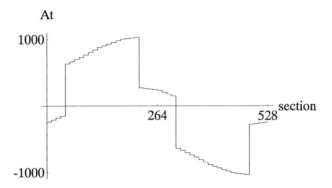

Fig. 3.41 *The air gap mmf in a hydrogenerator for the shift between the fundamentals of the stator and rotor mmf equal to π/3*

The effective air gap width in the machine, the air gap of which is constant under the poles, is shown in Fig. 3.40.

The air gap flux density created by the air gap mmf in Fig. 3.41 in the air gap, the effective width of which is given in Fig. 3.40, is represented in Fig. 3.42.

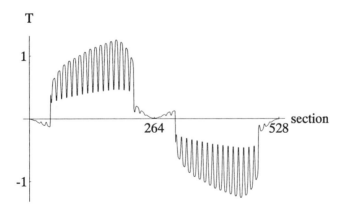

Fig. 3.42 *The air gap flux density in a hydrogenerator for the mmf distribution as shown in Fig. 3.41*

3.5 The Total Air Gap Magnetomotive Force in an Induction Machine

An induction machine has a stator identical to the stator of a synchronous machine. The rotor of an induction machine is always cylindrical, and has slots which carry the rotor winding. The rotor winding can be either a squirrel cage, or an m-phase concentric or a distributed winding. If the induction machine has a wound rotor, the number of rotor phases is usually, but not necessarily, equal to the number of stator phases. The number of phases of a squirrel cage winding is equal to the number of bars, since the bar currents are out of phase. The number of rotor slots in an induction machine is always different from the number of stator slots, in order to avoid dips in the torque–speed curve.

The armature winding in a synchronous and a D.C. machine carries an A.C. current, whereas the field winding is connected to a D.C. source. In an induction machine one cannot separate the armature from the field winding, since all the stator and rotor currents have components which create the main flux and components which generate the torque.

The stator winding of an induction machine is usually connected to an A.C. source(s). The rotor side of an induction machine does not contain any external source, except in some special connections. The only source which drives the currents in the rotor windings of a regular induction machine is the voltage induced by the air gap flux. In this section the mechanisms for inducing the voltage and driving the currents in the rotor winding will not be elaborated. Only the relationships between the stator, rotor and resultant air gap mmfs will be analyzed.

An induction machine can operate as a motor, generator, or a brake. The mode of operation of an induction machine is defined by the rotor relative speed to the air gap mmf. An indication of the induction machine's mode of operation is the rotor *slip*, or the *per unit rotor relative speed*. The slip s is defined as

$$s = \frac{\omega_s - \omega_r}{\omega_s} \qquad (3.15)$$

where ω_s is the angular speed of the stator rotating field, or the synchronous speed, and ω_r is the rotor electrical angular speed. The fundamental frequency f_2 of the rotor currents and voltages is equal to

$$f_2 = sf_1 \qquad (3.16)$$

with f_1 denoting the fundamental frequency of the stator currents and voltages.

The speed of the rotor mmf relative to the stator winding is equal to the synchronous speed, because both the fundamental spatial component of the rotor field and the fundamental spatial component of the stator field travel at the same speed relative to the stator windings. This speed is equal to the synchronous speed.

Various modes of operation of an induction machine will be illustrated here by utilizing the reference frames introduced in the previous chapter. The speed of the reference frame in the following examples is equal to the rotor speed, in order to show what the rotor winding sees at a certain slip. The statements which are used to obtain the plots in this section are listed in the Appendix.

Assume that the stator has a three phase winding, placed into twelve slots per pole. If the induction machine operates as a brake, the slip must be greater than one. The stator mmf and the rotor move in opposite directions. The air gap mmf as seen from the rotor which rotates at a negative synchronous speed, i.e. at s = 2, is shown in Fig. 3.43.

The coordinate t/T in Fig. 3.43 is so defined that it is equal to 1 when the rotor makes one turn. Since the slip in this example is equal to two, the rotor faces the air gap mmf twice within one period of rotation, i.e. for t/T = 1.

When the rotor of an induction machine is at *standstill*, the slip is equal to one. The speed of the air gap mmf relative to the rotor is equal to the speed of the air gap mmf relative to the stator. The air gap mmf seen from the rotor at s =1 is shown in Fig. 3.44.

When the induction machine operates in the motor mode, the slip is in the range 1>s>0. The rotor turns in the same direction as the air gap mmf, but at a lower speed.

The motor mode of operation is illustrated in the following example, in which the rotor speed equals 75% of the synchronous speed. At this speed the slip is equal to s = 0.25. The air gap mmf seen from the rotor in this case is generated by using statements from the Appendix, and is shown in Fig. 3.45.

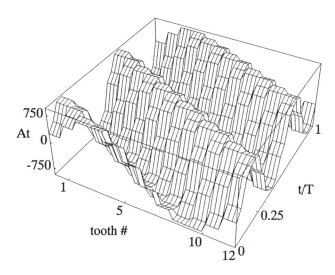

Fig. 3.43 *The air gap mmf seen from the rotor which rotates at a negative synchronous speed, i.e. at slip s = 2*

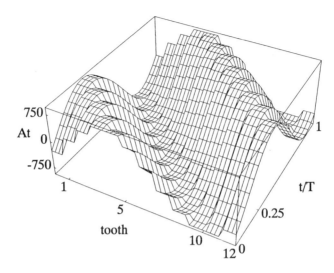

Fig. 3.44 *The air gap mmf seen from the rotor at a standstill, when s = 1*

When an induction machine is at *synchronism*, the rotor turns at the speed of the fundamental of the air gap mmf. The slip, as a measure of the rotor speed relative to the air gap mmf speed, is equal to zero in this case. The air gap mmf is fixed to the rotor at s = 0, as shown in Fig. 3.46.

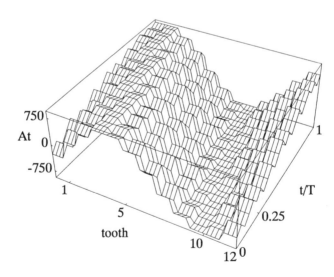

Fig. 3.45 *The air gap mmf seen from the rotor which rotates at slip s = 0.25*

Air Gap Magnetomotive Force and Flux Density   179

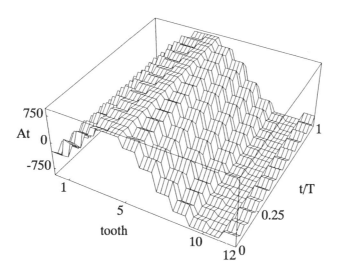

Fig. 3.46 *The air gap mmf seen from the rotor at synchronism*

When an induction machine operates as a generator, the speed of the rotor is higher than the synchronous speed and the slip is negative. The slip range in the generator mode is s<0. The air gap mmf distribution in a generator is illustrated in Fig. 3.47.

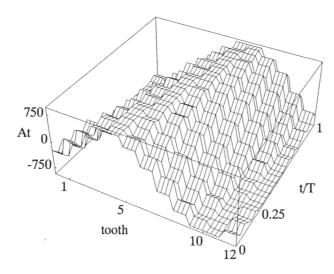

Fig. 3.47 *The air gap mmf seen from the rotor running at s = −0.25*

In Fig. 3.47 the air gap mmf seen from the rotor running at 125% of the synchronous speed, or at slip s = –0.25, is shown. The direction of rotation of the air gap mmf in this figure is opposite to the one in the previous figures, since the relative speed between the rotor and the air gap mmf, expressed by the slip, has changed sign.

When the rotor is running at twice the synchronous speed, the slip is equal to s=–1, and the air gap mmf seen from the rotor looks as shown in Fig. 3.48.

The voltage in the armature winding of a D.C. or synchronous machine is induced when its conductors cut the flux lines created by the filed current. Since in an induction machine there is no field coil, the only mechanism for inducing the voltages in the rotor (secondary) windings is through the Faraday's electromagnetic induction law. An induction machine fed from sinusoidal voltage source(s) is a flux driven device – the air gap flux in it remains constant in a wide interval of speeds around the rated speed. The fundamental component of the air gap flux is defined by the induced voltages in the phases connected to the external voltage sources. The difference between the applied voltage per phase and the induced voltage is equal to the voltage drop across the stator winding resistance and leakage reactance. The absolute value of this difference is only a few percent around the rated slip, since the stator impedance is relatively small. The induced voltage, and the fundamental component of the air gap flux, can therefore be considered constant around the rated slip. At synchronism the rotor windings of an induction machine see the air gap mmf as shown in Fig. 3.46. When the induction motor is loaded, the rotor slows to below the synchronous speed.

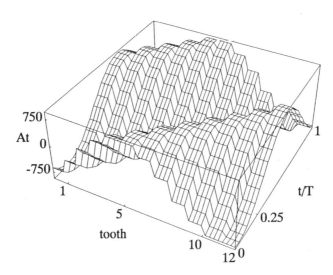

Fig. 3.48 *The air gap mmf seen from the rotor running at twice the synchronous speed, when the slip is s = –1*

Air Gap Magnetomotive Force and Flux Density

The air gap flux density becomes alternating for the rotor windings, inducing the voltages in them. When the rotor windings are closed, these induced voltages drive currents, which along with the rotor turns create the rotor mmf. The rotor mmf generates its own magnetic field in the air gap, which, following Faraday's law, acts *against* the stator air gap field, trying to decrease it. Since the fundamental component of the stator air gap field must remain unchanged, the only way to compensate for the rotor ampereturns is to draw more current from the source(s) on the stator. The additional stator currents create a component of the air gap mmf, which cancels the load created rotor ampereturns. Nevertheless, only the fundamental and some higher spatial harmonic components of the rotor ampereturns can be fully compensated by additional stator ampereturns. Different numbers of slots and different winding distributions on the rotor and stator side make it impossible to compensate for those harmonics in the rotor spatial mmf distribution which do not exist in the spectrum of the stator mmf curve. This phenomenon, illustrated in Fig. 3.49, is called the armature reaction for higher spatial harmonics in an induction machine.

One can see in Fig. 3.49 that, although the fundamental components of both mmf distributions are almost equal, the stator ampereturns do not compensate completely for the rotor ones, because these are created by a winding placed in a different number of slots, having a different distribution of coils. The difference between the two mmfs, which is shaded in Fig. 3.49, drives the so-called *differential flux* in the air gap. Assuming, for reasons of simplicity, smooth stator and rotor surfaces, the flux density distribution follows the mmf distribution in the air gap. The flux produced by the rotor currents is equal to

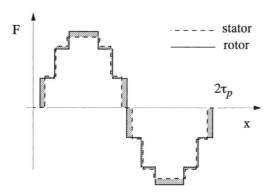

Fig. 3.49 *The stator and rotor mmf distributions in the air gap of an induction machine at a given time instant*

$$\Phi_r = \int_0^{\tau_p} 1 B_r(x)\,dx = 1\frac{\mu_0}{\delta}\int_0^{\tau_p} F_r\,dx \qquad (3.17)$$

Analogously, the flux produced by the additional stator currents is

$$\Phi_s = 1\frac{\mu_0}{\delta}\int_0^{\tau_p} F_s\,dx \qquad (3.18)$$

The difference between the two fluxes is the differential flux. In the following example, the stator mmf will be found which compensates for the rotor mmf caused by currents which draw the rotor of an induction machine when loaded. Assume that the rotor has fourteen slots per pole pair, and that at a certain load the spatial distribution of its mmf within one period of the rotor currents looks as shown in Fig. 3.50. The additional stator ampereturns, obtained by executing statements given in the Appendix, are shown in Fig. 3.51. One can see in this figure that, despite different numbers of slots and winding distributions on the stator and rotor sides, the additional stator ampereturns compensate as much as possible for the rotor ampereturns.

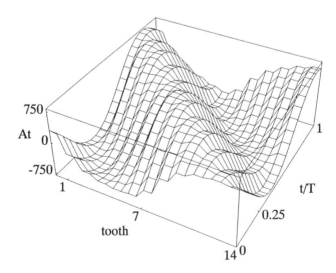

Fig. 3.50 *The rotor mmf in a loaded induction machine*

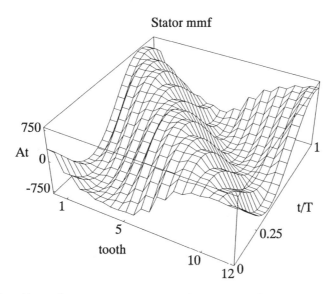

Fig. 3.51 *The additional stator ampereturns, which compensate for the rotor mmf in Fig. 3.50*

The difference between the stator and rotor mmfs is shown in Fig. 3.52. This differential mmf drives the differential flux, which acts as an armature reaction in the air gap of an induction machine.

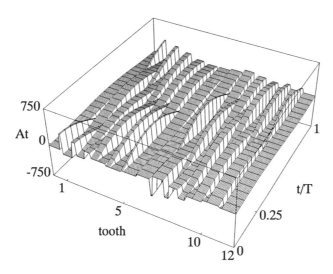

Fig. 3.52 *The difference between the stator and rotor mmfs, caused by different numbers of the stator and rotor slots, and different stator and rotor winding distributions*

One of the stator currents which compensate for the rotor load ampereturns is shown in Fig. 3.53.

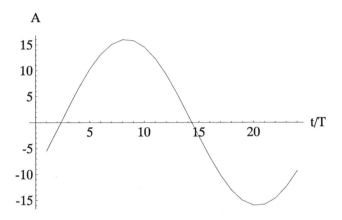

Fig. 3.53 *Additional component of stator current in phase b*

4 An Electric Machine as a Circuit Element

When a power system or a control circuit is connected to an electric machine's terminals, the machine performs as a combination of nonlinear inductances, resistances, and voltage sources. For clarity and simplicity, the number of elements which represent a machine in an electric circuit should be very small. Following this criterion, only a few elements have to reflect the whole universe of the machine's magnetic and electric circuit peculiarities: the distribution of its windings, the variable geometry of the magnetic circuit, saturation at various parts of the machine, etc. The task of completely representing such a large variety of phenomena with a small number of electrical elements is therefore impossible to fulfil. On the other hand, it would be counterproductive to deal with an electric circuit consisting of a large number of elements, in order to represent correctly every single electromagnetic phenomenon in a machine. An equivalent circuit is, therefore, a reasonable compromise between an efficient scheme with a minimum number of elements, and a tendency to include in the analysis as many physical phenomena as possible.

4.1 Main and Leakage Inductance Concept

The currents flowing through the machine's windings create magnetic flux which has several components, as illustrated in Fig. 4.1. The *total flux* Φ_{tot}, which links all the conductors of the winding passes through various parts of the machine. A component of the total flux which does not pass through the stator and rotor irons links the winding only in the end winding zone. It is called the *end winding leakage flux*, and is denoted by $\Phi_{\sigma,e}$. Another component of the total flux which links the conductors on only one side of the air gap and passes predominantly tangentially through the slots is the *slot leakage flux* $\Phi_{\sigma,s}$. The rest of the total flux leaves the irons on one side of the air gap.

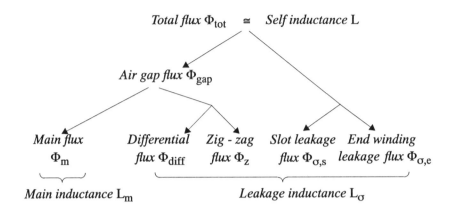

Fig. 4.1 *The components of flux created by a winding in an electric machine*

Depending on the ratio between the slot opening and air gap width, a part of this flux, called the *zig-zag flux* Φ_z goes back to the iron without linking the conductors on the other side of the air gap. Therefore, the zig-zag flux is a leakage flux.

The flux which passes through the air gap and links the conductors on the other side of the gap does not completely participate in the electromechanical energy conversion. Its fundamental spatial component, *the main flux* Φ_m, carries energy which is converted from an electrical to a mechanical form and vice versa. The higher spatial harmonics of the flux which link the conductors on the other side of the air gap travel at different speeds than does the fundamental. By definition of Fourier series, higher harmonics are orthogonal to each other and to the fundamental. This means that the higher harmonics are independent of each other, and that there is no magnetic coupling between them and the fundamental. When there is no magnetic coupling between two fluxes, each of them is, for the other, a *leakage flux*. Therefore, the higher harmonic components of flux are, by their nature, leakage for the fundamental. The higher spatial components of flux are called the *differential leakage flux*, and are denoted by Φ_{diff}.

The slot leakage flux was analyzed in Section 1.3. The analysis in this section was carried out by assuming that the machine was infinitely long. This means that the boundary effects on the iron–air border on both sides of the lamination have no influence on the slot leakage flux. Therefore, the complete analysis in Section 1.3 could be performed in a plane defined by the slot radial and tangential dimensions.

In the analysis of the end winding leakage flux, however, all three spatial components of the magnetic field must be allowed for. Due to the geometry of the end windings, it is not possible to consider any of the radial, tangential, or axial components of the end winding leakage flux constant. Since the magnetic field analysis tools have for a long time been incapable of handling three-dimensional problems, the values of the end winding leakage inductance have been determined on the basis of experience collected by measurements of manufactured machines. It was only a few years ago that progress in the calculation of the end winding leakage inductance was made.

The distribution of fluxes in the axial cross-section of a machine is shown in Fig. 4.2. The line of flux denoted by Φ_{gap} in this figure represents the common flux for the stator and rotor windings, or the air gap flux. The line of flux denoted by $\Phi_{\sigma,s}$ stands for the slot leakage flux, and Φ_z stands for the zig-zag flux. The representative of the main flux in the machine's equivalent circuit is the *main inductance*, whereas each leakage flux is represented by a corresponding *leakage inductance*.

The main inductance is a machine's parameter characteristic for no load. At no load there exists no current component covering the mechanical power. Neglecting the electrical losses, the complete winding current i creates the magnetic flux in the machine. Therefore, one may utilize the procedure illustrated in Fig. 4.3 to calculate the main inductance of a winding. Since the inductance is defined as the ratio between the flux linkage and the current which creates the flux linkage, one has to find the flux linkage for a given current, machine's geometry, and winding distribution. Therefore, the mmf distribution of the winding has to be found first, in the manner shown in the second chapter of this book. The corresponding flux density distribution results from applying the laws introduced in the third chapter. However, one has to be careful here. The flux density distribution in the air gap was derived in the third chapter by assuming that the mmf drop in the iron part of the magnetic circuit was negligible. This is true only at low values of flux density. When iron becomes saturated, the mmf drops at various parts of the machine increase significantly, and the flux density becomes a function of both the reluctance of the air gap and the iron.

When the flux density distribution is determined, the flux linkage Ψ of the winding is calculated. The main inductance of the winding is then evaluated according to its definition as the ratio between the flux linkage and the current which produced it.

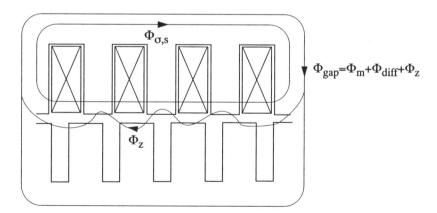

Fig. 4.2 *The distribution of fluxes in the axial cross-section of a machine*

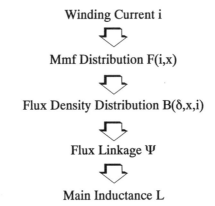

Fig. 4.3 *Steps in the calculation of the main inductance*

Another way to evaluate the main inductance of a winding is to utilize the definition of magnetic energy. As shown in the first chapter, the magnetic energy stored in a volume element which is defined by its cross-sectional area S and length l is a function of the flux Φ through the element and the mmf drop F on it

$$W_{mg} = \int F d\Phi = \int F B dS = \mu_0 l \int \frac{F^2}{\delta} dx \qquad (4.1)$$

Since magnetic energy is defined as a function of the main inductance as

$$W_{mg} = \frac{1}{2} L i^2 \qquad (4.2)$$

one can express the main inductance as

$$L = 2l\mu_0 \int \left(\frac{F}{i}\right)^2 \frac{dx}{\delta} = 2l\mu_0 \int F_w^2 \frac{dx}{\delta} \qquad (4.3)$$

where F_w is the winding function, introduced in chapter 2, and defined in Eq. (2.45).

In this chapter both analytical and numerical procedures for evaluating the machine's main and mutual inductances will be elaborated. An analytical solution will be derived by assuming that the complete mmf drop in the machine's magnetic circuit is concentrated on its air gap. In a nonlinear solution of the magnetic circuit at no load, the mmf drops on the machine's teeth and yokes on both sides of the air gap will be allowed for. In this case, the mmf drop in iron will not be considered constant, but as a function of the winding current.

4.2 The Main Inductance of a Coil and a Winding in a Slotless, Cylindrical, Unsaturated Machine

The mmf distribution of a single coil with diametral pitch which carries the current i can be derived from Eq. (2.37). By substituting q=1 for one coil per pole, $y=\tau_p$ since the coil has a diametral pitch, and n=1 because only the fundamental spatial harmonic is considered, one obtains the mmf distribution

$$F_1(x, i) = \frac{4}{\pi} w_p i(t) \sin\frac{\pi}{\tau_p} x \qquad (4.4)$$

with w_p denoting the number of turns per pole. The assumption that the machine is unsaturated means that the mmf drop in iron is proportional to the flux. Since the air gap reluctance in this case is much greater than the reluctance of iron, the mmf drop in the air gap can be considered equal to the ampereturns of the coil, given in Eq. (4.4). The fundamental term of the winding function $F_{w,1}$ may be further expressed as

$$F_{w,1}(x) = \frac{F_1(x,i)}{i} = \frac{2}{p\pi} w \sin\frac{\pi}{\tau_p} x \qquad (4.5)$$

where the number of turns per pole w_p was replaced by the total number of turns w and the number of poles 2p, i.e.

$$w_p = \frac{w}{2p} \qquad (4.6)$$

The machine is slotless and cylindrical, which means that the air gap width δ is constant. By applying Eq. (4.3) one obtains

$$L = \frac{2l\mu_0}{\delta}\int_0^{\tau_p} F_{w,1}^2(x)\,dx = \frac{2l\mu_0}{\delta}\left(\frac{2}{p\pi}\right)^2 w^2 \int_0^{\tau_p}\left(\sin\frac{\pi}{\tau_p}x\right)^2 dx \qquad (4.7)$$

with l denoting the machine's axial length. After integration, Eq. (4.7) gives the main inductance per pole pair $L_{m,p}$

$$L_{m,p} = \frac{\tau_p l \mu_0}{\delta}\left(\frac{2}{p\pi}\right)^2 w^2 \qquad (4.8)$$

The main inductance L_m is obtained by multiplying the main inductance per pole pair $L_{m,p}$ by the number of pole pairs p, which gives

$$L_m = \frac{\tau_p l \mu_0}{\delta p}\left(\frac{2}{\pi}\right)^2 w^2 \qquad (4.9)$$

Introducing the *air gap specific permeance* Λ

$$\Lambda = l\tau_p \frac{\mu_0}{\delta}\frac{4}{p\pi^2} \qquad (4.10)$$

one can express the main inductance of a coil as

$$L_m = \Lambda w^2 \qquad (4.11)$$

As expected, the main inductance is proportional to the square of the number of turns in the coil.

When several coils are distributed in the slots, thus forming a winding, the number of turns in the previously derived equations has to be replaced by the effective number of turns, i.e. the winding factor f_w has to be allowed for. The main inductance of a winding with w turns and with the winding factor f_w is then

$$L_m = \Lambda (wf_w)^2 \qquad (4.12)$$

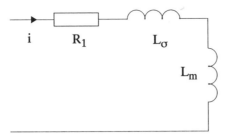

Fig. 4.4 *The equivalent circuit of a winding in an electric machine*

Denoting by L_σ the total leakage inductance of the winding, and by R_1 its resistance, one can draw the equivalent electric circuit of a winding in an unsaturated machine with a cylindrical stator and rotor in the manner shown in Fig. 4.4.

4.3 The Main Inductance of a Winding in a Slotted, Cylindrical, Saturated Machine

When the main inductance of a winding in an actual, saturated electric machine has to be calculated, the relationships become more complex than the ones derived in the previous section. A portion of the winding ampereturns in a saturated machine covers the mmf drops in the stator and rotor yokes and teeth. These ampereturns are nonlinearly dependent on the main flux. The magnetic circuit in this case cannot be solved analytically, but numerically. Since the reluctance of each part of the magnetic circuit is a nonlinear function of the flux through it, the number of unknowns of the system of equations that have to be solved numerically can be high.

The most accurate way to calculate the main inductance of a coil is by using the finite element method. The variable geometry of the teeth and core, spatial distribution of the windings, and nonlinear B–H curve of the magnetic material can be taken into account in such an analysis. However, the complexity of computation and the necessity of a high performance mainframe computer or workstation are still limiting wide application of the finite element method in solving routine problems in machine design and analysis.

A main inductance computation procedure much simpler than the finite element method, but still highly accurate, is the magnetic equivalent circuit method, described in [6]. In this approach, each stator and rotor tooth, as well as each portion of the stator and rotor yoke between adjacent teeth, is represented by a single reluctance. The winding ampereturns introduced in Section 2.5 are represented by the mmf sources placed in the machine's teeth. Since the computation of the main inductance of the winding has to be carried out at no load, only the stator or rotor windings are excited, depending on the type of machine analyzed.

The stator and rotor teeth make the air gap reluctance dependent on the stator to rotor shift. Therefore, the main inductance of a coil is a function of the angle between the stator and the rotor, and can be represented as the sum of the constant and pulsating terms. Very often, however, only the average value of the main inductance is important. The average value of the main inductance is independent of the rotor angle, which allows one to simplify the machine's magnetic equivalent circuit for the purpose of the main inductance computation. According to this simplification, both stator and rotor have the same number of teeth, equal to the number of teeth on that side of the air gap at which the excitation winding is placed. Since the number of teeth on the side of the air gap without excitation is fictitiously changed by this simplification, the width of each tooth so obtained must be altered according to the following equation

$$b_{t, \text{fict}} = b_t \frac{N_t}{N_{t, \text{wound}}} \qquad (4.13)$$

where b_t is the real tooth width, $b_{t,\text{fict}}$ is the fictitious tooth width, N_t is the number of teeth on the side of the air gap without excitation winding and $N_{t,\text{wound}}$ is the number of teeth on the side of the air gap with excitation winding.

The previous principle of reduction of the number of teeth is illustrated in Fig. 4.5, in which the reduced magnetic equivalent circuit of an induction machine is shown. The number of rotor teeth is equal to the number of stator teeth, so that the width of each rotor tooth has to be recalculated following Eq. (4.13).

The reluctance of each element which represents the iron part of the machine's magnetic equivalent circuit is a nonlinear function of the flux through it. For this reason the solution of a magnetic circuit cannot be represented in closed form, where the inductance of the winding would be expressed as a function of all the reluctances and the winding current. The solution has to be found numerically, by solving a system of nonlinear algebraic equations which relate the fluxes at various branches of the magnetic circuit to the reluctances and mmf drops. Since the number of teeth per pole pair of a machine, the main inductance of which is calculated, can sometimes be high, as is the case in two pole synchronous generators, the number of nonlinear algebraic equations which have to be solved can be very large. The solution of a large system of nonlinear algebraic equations is incomparably harder to find than the solution of a small system. Therefore, the system of nonlinear algebraic equations that has to be solved in the main inductance computation procedure is reduced, based upon some physical conditions of symmetry which are fulfilled in this case.

The first condition of symmetry is a consequence of the absence of a constant term in the spatial distribution of the air gap flux density. Therefore, the flux under an S pole is equal to the negative flux under an N pole, and it is sufficient to calculate the magnetic circuit for only one pole of the machine.

An Electric Machine as a Circuit Element

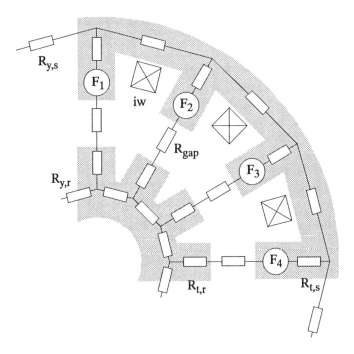

Fig. 4.5 *A portion of the magnetic equivalent circuit of a machine having excitation only on the stator side*

The second condition of symmetry which is valid in this case follows from the absence of even harmonics in the mmf distribution of the winding. If the magnetic circuit of the machine is symmetrical, the tooth flux density distribution does not contain even harmonics either. The tooth flux density distribution under one pole is therefore symmetric with respect to the maximum, so it is necessary to solve the magnetic circuit only for one half of the pole.

When both conditions of symmetry are satisfied, the number of nonlinear algebraic equations that have to be solved in order to find the main inductance of a winding in a machine with nonlinear B–H curve can be reduced to one quarter of the number of teeth per pole pair.

The principles of evaluation of the winding's main inductance in a saturated machine are illustrated in the following example, in which the main inductance of an induction machine with twelve stator slots per pole pair is calculated. The winding distribution on the stator side is given in Eq. (2.67), and is shown in Fig. 2.21. The magnetic equivalent circuit per one half of the pole pitch is shown in Fig. 4.6. The following equations can be written for the circuit in this figure:

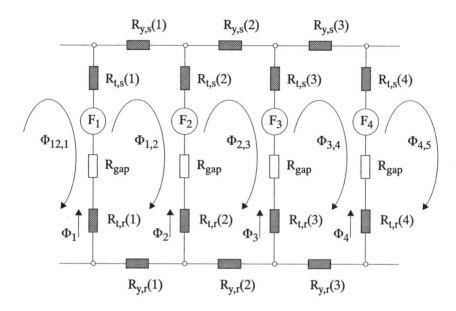

Fig. 4.6 *The magnetic equivalent circuit of one half of the pole pitch in a machine with twelve teeth per pole pair*

$$\Phi_{1,2}[2R_{gap} + R_{t,s}(1) + R_{y,s}(1) + R_{t,s}(2) + R_{t,r}(2) + R_{y,r}(1) + R_{t,r}(1)] -$$
$$- \Phi_{12,1}[R_{gap} + R_{t,s}(1) + R_{t,r}(1)] - \Phi_{2,3}[R_{gap} + R_{t,s}(2) + R_{t,r}(2)] = F_1 - F_2$$

(4.14)

$$\Phi_{2,3}[2R_{gap} + R_{t,s}(2) + R_{y,s}(2) + R_{t,s}(3) + R_{t,r}(3) + R_{y,r}(2) + R_{t,r}(2)] -$$
$$- \Phi_{1,2}[R_{gap} + R_{t,s}(2) + R_{t,r}(2)] - \Phi_{3,4}[R_{gap} + R_{t,s}(3) + R_{t,r}(3)] = F_2 - F_3$$

(4.15)

$$\Phi_{3,4}[2R_{gap} + R_{t,s}(3) + R_{y,s}(3) + R_{t,s}(4) + R_{t,r}(4) + R_{y,r}(3) + R_{t,r}(3)] -$$
$$- \Phi_{2,3}[R_{gap} + R_{t,s}(3) + R_{t,r}(3)] - \Phi_{4,5}[R_{gap} + R_{t,s}(4) + R_{t,r}(4)] = F_3 - F_4$$

(4.16)

The winding is placed in the stator slots so that the centre line of the coils lies between the twelfth and first slot. Following Ampère's circuital law, the mmf source F_1 and the flux density Φ_1 in the first tooth are therefore always equal to zero, indepen-

An Electric Machine as a Circuit Element 195

dent of the value of the current. This can be expressed as

$$\Phi_1 = \Phi_{1,2} - \Phi_{12,1} = 0 \Rightarrow \Phi_{12,1} = \Phi_{1,2} \qquad (4.17)$$

Since the tooth flux density distribution is symmetrical with respect to the maximum, one can write for this machine

$$\Phi_{4,5} = -\Phi_{3,4} \qquad (4.18)$$

Substituting Eq. (4.17) into Eq. (4.14), and Eq. (4.18) into Eq. (4.16), one obtains

$$\Phi_{1,2}[R_{gap} + R_{y,s}(1) + R_{t,s}(2) + R_{t,r}(2) + R_{y,r}(1)] -$$
$$-\Phi_{2,3}[R_{gap} + R_{t,s}(2) + R_{t,r}(2)] = F_1 - F_2 \qquad (4.19)$$

and

$$\Phi_{3,4}[3R_{gap} + R_{t,s}(3) + R_{y,s}(3) + 2R_{t,s}(4) + 2R_{t,r}(4) + R_{y,r}(3) + R_{t,r}(3)] -$$
$$-\Phi_{2,3}[R_{gap} + R_{t,s}(3) + R_{t,r}(3)] = F_3 - F_4 \qquad (4.20)$$

The branch fluxes $\Phi_{1,2}$, $\Phi_{2,3}$, and $\Phi_{3,4}$ are obtained by solving Eqs. (4.15), (4.19), and (4.20).

Utilizing the solution procedure elaborated in the Appendix for a given machine geometry and B–H curve, one obtains the characteristics of the machine's magnetic circuit shown in the following figures.

If the iron core B–H curve is linearized in the manner illustrated in Fig.1.1, the spatial distribution of the stator tooth flux densities as a function of the coil current looks as shown in Fig. 4.7. The tooth flux densities are proportional to the coil current, because the iron core reluctances are constant. However, the flux density distribution shown in this figure is unrealistic, since flux densities significantly higher than two tesla are seldom reached in a regular machine. The tooth flux density distribution shown in Fig. 4.7 follows the tooth mmf distribution, derived for this type of winding in the second chapter.

The spatial tooth flux density distribution of a machine in which the iron core has a nonlinear B–H curve is shown in Fig. 4.8. As the coil current increases, the reluctance of the iron core becomes higher and limits the flux in the magnetic circuit. Therefore, the tooth flux densities are in reality limited to values not much higher than 2 T.

The flux which links the turns of the winding is obtained by multiplying the flux per turn by the number of turns and the winding factor. In the magnetic equivalent circuit presentation of electric machines the flux in every tooth is computed.

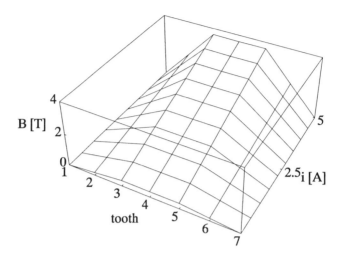

Fig. 4.7 *The spatial distribution of the tooth flux density under one pole of a machine with a linearized B–H curve of the iron core*

Therefore, the flux linked by each coil can be expressed as a sum of the fluxes in the teeth within the coil. This way one obtains the dependence of coil flux linkage on current shown in Fig. 4.9.

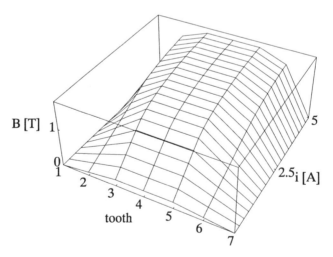

Fig. 4.8 *The spatial distribution of the tooth flux density under one pole of a machine with a nonlinear B–H curve of the iron core*

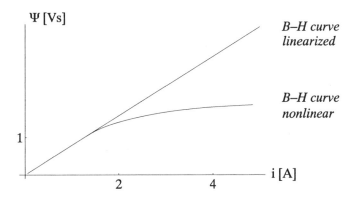

Fig. 4.9 *The flux linkage in a machine with a linearized B–H curve (line) and with a nonlinear B–H curve*

The line in this figure represents the Ψ–i characteristic for a machine having an iron core with a linearized B–H curve, whereas the other curve stands for a machine with an iron core which has a nonlinear B–H curve.

The main inductance, being equal to the ratio between the flux linkage and current in the winding, is constant as long as the flux linkage increases proportionally to the winding current. When the Ψ–i characteristic of a machine with a nonlinear B–H curve in Fig. 4.9 starts to decline from the line, the main inductance begins to decrease as shown in Fig. 4.10.

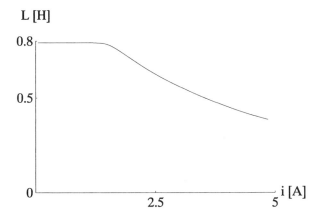

Fig. 4.10 *The main inductance of a winding in a machine with a nonlinear B–H curve of the magnetic material*

4.4 The Main Inductance of a Winding in an Unsaturated Machine with Variable Air Gap Geometry

When there are salient poles on one side of the air gap, the main inductance of a winding placed on the other side of the gap is dependent on the relative stator to rotor position. The main inductance of the winding is maximum when the winding axis is collinear with the pole axis; it is minimum when the two axes are electrically perpendicular to each other. An electric machine with salient poles on the rotor and cylindrical stator is shown in Fig. 4.11 (a). The main inductance of the stator winding shown in this figure is a function of *double* the stator to rotor angle γ : it is maximum when any of the rotor poles is aligned with the winding axis, and minimum at the rotor angles which are $\pi/2$ electrical radians shifted to the position of the maximum inductance.

Instead of using the air gap width $\delta(x)$, it is more convenient to perform further analysis with the *inverse air gap width* $\Delta(x)$, defined as

$$\Delta(x) = \frac{1}{\delta(x)} \qquad (4.21)$$

The inverse air gap width Δ_{min} in Fig. 4.11 (b) corresponds to the maximum air gap width in the inter-polar space, whereas Δ_{max} corresponds to the minimum air gap width under the poles.

Denoting by Δ_0 the average value of the inverse air gap width as shown in Fig. 4.11 (b), $\Delta(x)$ can be represented in the form of a Fourier series as

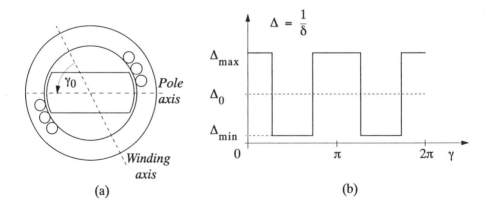

Fig. 4.11 *An electric machine with salient poles on the rotor and winding on the stator side (a) and the variation of the inverse air gap width Δ (b)*

An Electric Machine as a Circuit Element

$$\Delta(x) = \Delta_0 + \sum_{i=1,3,\ldots}^{\infty} \Delta_i \cos \frac{2\pi i}{\tau_p}(x - x_0) =$$

$$= \frac{\Delta_{max} + \Delta_{min}}{2} + \sum_{i=1,3,\ldots}^{\infty} \Delta_i \cos \frac{2\pi i}{\tau_p}(x - x_0) \qquad (4.22)$$

where rotor shift x_0 corresponds to the angle γ_0 in Fig. 4.11 (a) according to

$$x_0 = \frac{D}{2}\gamma_0 \qquad (4.23)$$

D in the equation above denotes the average air gap diameter. The spatial period of the fundamental term of the inverse air gap width $\Delta(x)$ is equal to the pole pitch τ_p. The function $\Delta(x)$ has only odd higher harmonics in period τ_p; in period $2\tau_p$ it has, therefore, the 2, 6, 10, 14,... harmonics.

The coefficients Δ_i in Eq. (4.22) can be accurately determined only by calculating the magnetic field distribution in the machine using the finite element method. These coefficients are defined by the air gap geometry and saturation of the iron, both of which determine the air gap flux density distribution. However, a detailed treatment of the machine which would result in accurate values of Δ_i is outside the scope of this book. For further analysis of machines with salient poles or teeth in which the inverse air gap width is used, the existence of the particular harmonics of inductances is of much greater importance than an accurate determination of their magnitudes.

The main inductance of a winding is calculated by applying Eq. (4.3), in which the product of the square of the winding function $F_w^2(x)$ and the inverse air gap width $\Delta(x)$ is integrated in interval $[0, \tau_p]$. Since both the functions $F_w^2(x)$ and $\Delta(x)$ are represented as an infinite Fourier series, their product is also an infinite Fourier series. Each term of this product has two components: the ν-th harmonic of $F_w^2(x)$, and the μ-th harmonic of $\Delta(x)$. Both the ν-th harmonic of $F_w^2(x)$ and the μ-th harmonic of $\Delta(x)$ fit integer numbers of times into the pole pitch τ_p. The finite integral of the product of the two harmonics in interval $[0, \tau_p]$ is equal to zero when the orders of the harmonics are not equal, i.e. for $\mu \neq \nu$. Only those terms in the series $F_w^2(x)$ and $\Delta(x)$ which have equal periods can give a result different from zero when their product is integrated along the pole pitch.

This principle is illustrated in the following example, in which the functions $F_w(x)$ and $\Delta(x)$ are represented by their first five harmonics:

$$\Delta(x) = \Delta_0 + \Delta_1 \cos\frac{2\pi}{\tau_p}(x-x_0) + \Delta_3 \cos\frac{6\pi}{\tau_p}(x-x_0) + \Delta_5 \cos\frac{10\pi}{\tau_p}(x-x_0)$$

(4.24)

and

$$F_w(x) = F_{w,1} \cos\frac{\pi}{\tau_p}x + F_{w,3} \cos\frac{3\pi}{\tau_p}x + F_{w,5} \cos\frac{5\pi}{\tau_p}x \qquad (4.25)$$

The square of the winding function $F_w(x)$ has only even terms in the interval $[0, 2\tau_p]$

$$F_w^2(x) = \frac{1}{2}(F_{w,1}^2 + F_{w,3}^2 + F_{w,5}^2) + \left(\frac{F_{w,1}^2}{2} + F_{w,1}F_{w,3} + F_{w,3}F_{w,5}\right)\cos\frac{2\pi}{\tau_p}x +$$

$$+ F_{w,1}(F_{w,3} + F_{w,5}) \cos\frac{4\pi}{\tau_p}x + F_{w,1}F_{w,5}\cos\frac{6\pi}{\tau_p}x + F_{w,3}F_{w,5}\cos\frac{8\pi}{\tau_p}x +$$

$$+ \frac{F_{w,5}^2}{2}\cos\frac{10\pi}{\tau_p}x \qquad (4.26)$$

or

$$F_w^2(x) = G_{w,0} + \sum_{n=1,2,\ldots}^{5} G_{w,n} \cos\frac{2\pi n}{\tau_p}x \qquad (4.27)$$

The term $G_{w,n}$ in Eq. (4.27) comprises the amplitudes of those components of the winding function distribution, the orders of which give n as a result of addition or subtraction. Note that each harmonic of the winding function participates in creating the main inductance, the measure of which is $G_{w,0}$ in Eq. (4.27). This principle can be verified by comparing Eqs. (4.24), (4.25), and (4.26).

Besides the constants $G_{w,0}$ and Δ_0, only the terms which have equal periods in Eqs. (4.24) and (4.27) can give a self inductance which is not identically equal to zero. In this example, the combinations of the following terms satisfy the previous condition:

An Electric Machine as a Circuit Element

$$\Delta_1 \cos\frac{2\pi}{\tau_p}(x-x_0) \quad \text{with} \quad G_{w,1}\cos\frac{2\pi}{\tau_p}x \tag{4.28}$$

$$\Delta_3 \cos\frac{6\pi}{\tau_p}(x-x_0) \quad \text{with} \quad G_{w,3}\cos\frac{6\pi}{\tau_p}x \tag{4.29}$$

and

$$\Delta_5 \cos\frac{10\pi}{\tau_p}(x-x_0) \quad \text{with} \quad G_{w,5}\cos\frac{10\pi}{\tau_p}x \tag{4.30}$$

The component of the main inductance created by the constant terms $G_{w,0}$ and Δ_0 is independent of the rotor shift. The component of the main inductance $L_2(x_0)$ created by the fundamental component of the inverse air gap width $\Delta_1\cos[2\pi(x-x_0)/\tau_p]$ and the fundamental component of the square of the winding function $G_{w,1}\cos(2\pi x/\tau_p)$ is obtained by using Eq. (4.3):

$$L_2(x_0) = 2l\mu_0 \int_0^{\tau_p} G_{w,1}\cos\frac{2\pi x}{\tau_p}\Delta_1\cos\frac{2\pi}{\tau_p}(x-x_0)\,dx = L_{2,max}\cos\left(\frac{2\pi}{\tau_p}x_0\right) \tag{4.31}$$

The main inductance represented by the constant and fundamental term of the Fourier series can now be written as

$$L_m(x_0) = \frac{L_{max}+L_{min}}{2} + \frac{L_{max}-L_{min}}{2}\cos\frac{2\pi}{\tau_p}x_0 \tag{4.32}$$

where L_{max} corresponds to the maximum inverse air gap width Δ_{max}, or the minimum air gap width δ_{min}. The minimum inductance L_{min} corresponds to the minimum inverse air gap width Δ_{min}, or the maximum air gap width δ_{max}. The variation of the main inductance described in Eq. (4.32) is shown in Fig. 4.12 (c).

Eq. (4.31) can be extended to all terms of the Fourier series which represent the winding function and inverse air gap width. The result is the main inductance of a stator winding in a machine with salient rotor poles

$$L_m(x_0) = L_0 + \sum_{n=1,3,5,\ldots}^{\infty} L_{2n,max}\cos\frac{2\pi n}{\tau_p}x_0 \tag{4.33}$$

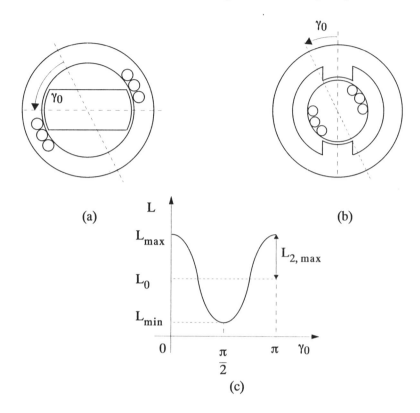

Fig. 4.12 *The representation of the self inductance of a winding in salient pole machines (a), with a constant term (b) and fundamental harmonic (c)*

The self inductance of a coil, represented by the constant term and fundamental harmonic, as a function of the rotor shift in a machine with salient poles is shown in Fig. 4.12 (c).

If the rotor has N teeth per pole pair instead of two poles, Eq. (4.33) can still be applied. The main inductance in this case is created by those harmonics of the winding function, the orders of which are multiples of the number of teeth per pole N/2. Therefore, the main inductance of a coil on the stator can be written as

$$L_m(x_0) = L_0 + \sum_{n = 1, 3, 5, \ldots}^{\infty} L_{2n, max} \cos \frac{N \pi n}{\tau_p} x_0 \qquad (4.34)$$

The variation of the main inductance of a machine which has six teeth per pole pair is shown in Fig. 4.13 (b). The fundamental term has as many periods in the double pole pitch $2\tau_p$ as there are rotor teeth per pole pair.

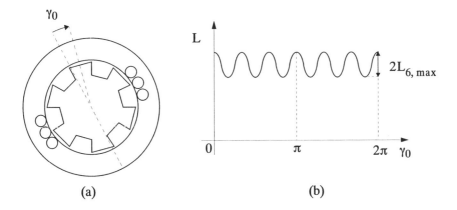

Fig. 4.13 *An electric machine with six teeth on the rotor (a) and the main inductance variation of the stator winding as a function of the rotor shift (b). The main inductance is represented by the constant term and fundamental harmonic.*

When the winding is placed into N slots per pole pair in the air gap the other side of which is cylindrical, the inverse air gap width can be written as

$$\Delta(x) = \Delta_0 + \sum_{i=1,3,\ldots}^{\infty} \Delta_i \cos \frac{N\pi i}{\tau_p}(x - x_0) \qquad (4.35)$$

The amplitude of the air gap flux density is defined as

$$B(x) = \mu_0 i F_w(x) \Delta(x) \qquad (4.36)$$

The fundamental harmonic of the winding function generates with the fundamental harmonic of the inverse air gap width in Eq. (4.35) *two* harmonics of the air gap flux density. Since the period of the fundamental harmonic of the inverse air gap width is N times shorter than the period of the fundamental of the air gap mmf, the two harmonics of the air gap flux density have orders N+1 and N−1. These harmonics of the air gap flux density are therefore called *the slot harmonics*.

The distribution factor for the slot harmonics follows from Eq. (2.50) by substituting N±1 for n:

$$f_{d, N \pm 1} = \frac{\sin q \frac{(N \pm 1)\alpha_{el}}{2}}{q \sin \frac{(N \pm 1)\alpha_{el}}{2}} = \frac{\sin q \frac{\alpha_{el}}{2}}{q \sin \frac{\alpha_{el}}{2}} = f_{d, 1} \qquad (4.37)$$

because $\alpha_{el} = 2p\pi/N$ for slot harmonics. The distribution factor for the slot harmonics is equal to the distribution factor for the fundamental. The same is valid for the pitch factor, defined in Eq. (2.49) for higher harmonics of order $n = N \pm 1$ as

$$f_{p, N \pm 1} = \sin\left[(N \pm 1) \frac{\pi}{2} \frac{y}{\tau_p} \right] = \sin \frac{y}{\tau_p} \frac{\pi}{2} = f_{p, 1} \qquad (4.38)$$

because N is an integer. The pitch factor for the slot harmonics is equal to the pitch factor for the fundamental. Therefore, *the winding factor for the slot harmonics is equal to the winding factor for the fundamental.*

If the winding in Fig. 4.11 (a) is replaced by a single coil with diametral pitch, its main inductance is constant, despite salient poles on the other side of the air gap. Such a conclusion is on first sight absurd; however, it has a simple physical explanation. The total area in the air gap through which the coil drives the flux is constant and equal to the area of one salient pole, independent of the rotor shift. Therefore, the main inductance of the coil is also independent of the rotor shift. Mathematical proof that the main inductance is constant can be derived from Eq. (4.3). The winding function of a full pitch coil is constant within one pole. The square of the winding function is therefore constant along the whole pole pair. The expression in Eq. (4.3) is thus reduced to an integral of the inverse air gap width, i.e. to an integral of a constant plus an infinite sum of harmonic functions. In this expression only the constant, when integrated, gives a term different from zero. This term has the meaning of the main inductance of the coil, and it is independent of the rotor shift. A definite integral of all harmonics of the inverse air gap width is equal to zero.

4.5 Mutual Inductance between the Windings in a Slotless, Unsaturated Machine

The mutual inductance $L_{1,2}$ between two windings is evaluated by applying the same principles as the ones used in the main inductance computation. Since in this case the accumulated magnetic energy is a function of the parameters of both windings, Eq. (4.3) has to be modified slightly: instead of the square of the winding function, the product of the winding functions of each winding has to be used. In this way, one obtains the expression for mutual inductance between two coils characterized by their winding functions $F_{1,w}$ and $F_{2,w}$

$$L_{1,2} = l\mu_0 \int F_{1,w}(x) F_{2,w}(x) \Delta(x) dx \qquad (4.39)$$

In terms of inductive coupling, the mutual inductance between the windings in a cylindrical machine exists only for those harmonics of winding functions which are of equal order. Following the principles elaborated in the previous section, a definite integral from 0 to τ_p of the product of two winding functions which have different periods is equal to zero. This means that, for example, the mutual inductance between the fundamental harmonic of a two pole stator winding function and the fundamental harmonic of a six pole rotor winding function is equal to zero. Nevertheless, the mutual inductance between the third harmonic of the two pole stator winding function and the fundamental of the six pole rotor winding function is *not* identically equal to zero.

The angle between the stator and rotor components of the air gap mmf is equal to the angle between the stator and rotor winding functions. Since the winding function of each winding is fixed to that side of the machine at which the winding is placed, the angle between the stator and rotor components of the air gap mmf is equal to the electrical angle between the stator and rotor.

The parameters utilized in deriving the dependence of the mutual inductances on the rotor shift γ_0 for various placements of the windings and air gap geometry, evaluated in accordance with Eq. (4.39), are given in Fig. 4.14. The winding functions and inverse air gap width are represented here only by their fundamental harmonics. The winding functions have an equal number of poles.

The air gap configuration analyzed in case (a) is characteristic for induction machines and turbogenerators. Applying Eq. (4.39) one finds that the mutual inductance between the stator and rotor winding is proportional to the cosine of the rotor shift, or

$$L_{1,2} = L_{max} \cos \frac{\pi}{\tau_p} x_0 \qquad (4.40)$$

The period of the mutual inductance is $2\tau_p$, and its maximum value is denoted by L_{max}. The mutual inductance can be either positive or negative, as opposed to the self inductance, which is always positive. A negative mutual inductance means that the flux produced by the current in the first coil and linked by the turns of the second coil has the opposite direction from the flux produced by the same current in the second coil.

Cases (b), (c), and (d) shown in Fig. 4.14 are characteristic of a salient pole synchronous machine. The inverse air gap width $\Delta(x)$ is represented in these cases by a constant term and the fundamental harmonic, the period of which is τ_p

$$\Delta(x_0) = \frac{\Delta_{max} + \Delta_{min}}{2} + \frac{\Delta_{max} - \Delta_{min}}{2} \cos \frac{2\pi}{\tau_p}(x - x_0) \qquad (4.41)$$

In case (b), the mutual inductance between one stator and one rotor winding is evaluated. The period of the mutual inductance is again $2\tau_p$, as in case (a). However, the maximum value of the mutual inductance is smaller than L_{max} because of the influence of the wide air gap in the inter-polar space of a salient pole machine. The mutual inductance in this case is

$$L_{1,2} = \frac{L_{max} + L_{min}}{2} \cos \frac{\pi}{\tau_p} x_0 \qquad (4.42)$$

In cases (c) and (d) the mutual inductances between two *stator* windings in a salient pole machine are analyzed. The angle between the axes of the stator windings in case (c) is $\pi/2$ electrical radians, and in case (d) it is $2\pi/3$ electrical radians. Although the windings in case (c) are perpendicular to each other, the mutual inductance between them is not identically equal to zero. As explained in Section 3.1, the reason for such behaviour of the mutual inductance is the variable air gap width in a salient pole machine. The flux linkages in two electrically perpendicular windings of a salient pole synchronous machine are not orthogonal, i.e. independent of each other. The mutual inductance between the two stator windings can be expressed as

$$L_{1,2} = \frac{L_{max} - L_{min}}{2} \sin \frac{2\pi}{\tau_p} x_0 \qquad (4.43)$$

and its period is τ_p. The fundamental component of the mutual inductance in case d) has period τ_p equal to that in case (c). The mutual inductance between the stator windings denoted by 1 and 2 in Fig. 4.14 (d) can be described as

$$L_{1,2} = -\frac{L_{max} + L_{min}}{4} - \frac{L_{max} - L_{min}}{4} \cos \frac{2\pi}{\tau_p}(x_0 + \frac{\tau_p}{6}) = L_{2,1} \qquad (4.44)$$

The mutual inductance $L_{1,2}$ in Eq. (4.44) is always negative because of the angle of $120°$ between the two stator windings. The mutual inductance reaches maximum at $x_0 = -\tau_p/6 \pm k\tau_p$, $k = 1,2,...$, which corresponds to the position of the salient pole axis in the centre between windings 1 and 2. The minimum of the mutual inductance is obtained when the salient pole axis is at positions $x_0 = \tau_p/3 \pm k\tau_p$, $k = 1,2,....$

An Electric Machine as a Circuit Element

		$F_{1,w}$	$F_{2,w}$	Δ
(a)	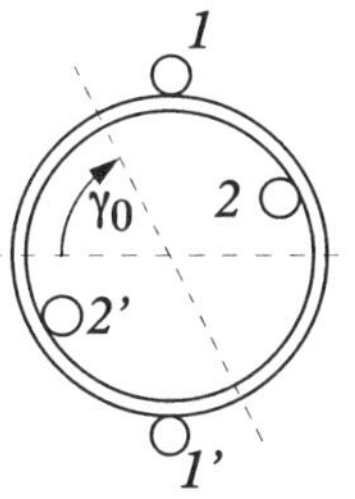	$F_{1,M}\cos\dfrac{\pi}{\tau_p}x$	$F_{2,M}\cos\dfrac{\pi}{\tau_p}(x-x_0)$	Δ_{max}
(b)	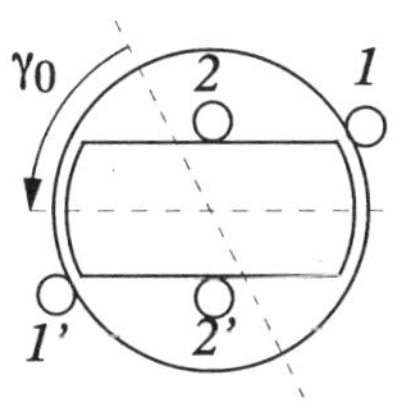	$F_{1,M}\cos\dfrac{\pi}{\tau_p}x$	$F_{2,M}\cos\dfrac{\pi}{\tau_p}(x-x_0)$	Eq. (4.41)
(c)	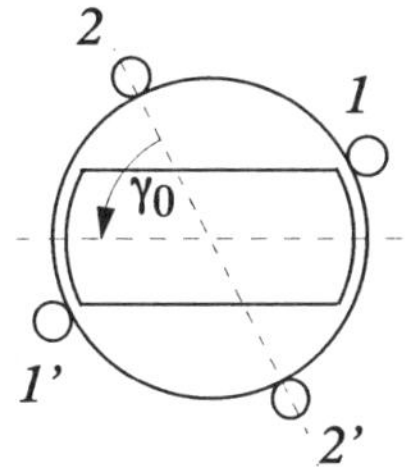	$F_{1,M}\cos\dfrac{\pi}{\tau_p}x$	$F_{2,M}\cos\dfrac{\pi}{\tau_p}\left(x-\dfrac{\tau_p}{2}\right)$	Eq. (4.41)
(d)	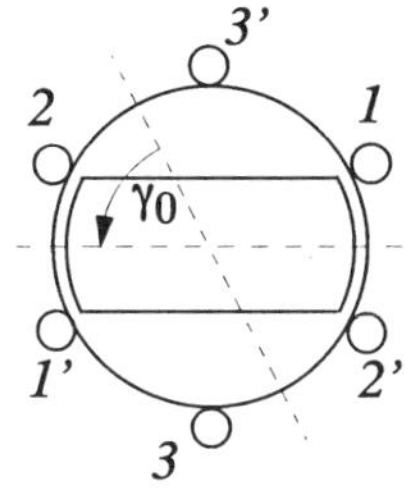	$F_{1,M}\cos\dfrac{\pi}{\tau_p}x$	$F_{2,M}\cos\dfrac{\pi}{\tau_p}\left(x-\dfrac{2\tau_p}{3}\right)$	Eq. (4.41)

Fig. 4.14 *The winding functions and inverse air gap widths for various air gap geometries and placements of the windings*

Since the relationship between the flux density and field strength in the air gap is linear, one may write $L_{2,1} = L_{1,2}$.

The mutual inductance between phases denoted by 2 and 3 in Fig. 4.14 (d) is

$$L_{2,3} = -\frac{L_{max}+L_{min}}{4} + \frac{L_{max}-L_{min}}{4}\cos\frac{2\pi}{\tau_p}x_0 = L_{3,2} \qquad (4.45)$$

whereas the mutual inductance between phases 3 and 1 is

$$L_{3,1} = -\frac{L_{max}+L_{min}}{4} - \frac{L_{max}-L_{min}}{4}\cos\frac{2\pi}{\tau_p}(x_0 - \frac{\tau_p}{6}) = L_{1,3} \qquad (4.46)$$

The slots in a machine are often axially skewed, as shown in Fig. 4.15. Skewing has no effect on the main inductance of the winding which is placed into skewed slots. It does, however, decrease the mutual inductances between the windings placed in skewed slots on one side of the air gap, and the windings in unskewed slots on the other side of the gap. Therefore, it is irrelevant on which side of the air gap the slots are skewed. If the slots in an induction machine have to be skewed, this is done on the rotor side because of the simpler manufacturing procedure. The slot leakage inductance of the winding placed into skewed slots increases with skewing, because skewed conductors are longer than unskewed, keeping the machine's length constant.

The mmf produced by the skewed winding is a function of both the peripheral coordinate x and the machine's axial coordinate. If the rotor slots are skewed, the fundamental harmonic of the winding function of the rotor winding can be written as

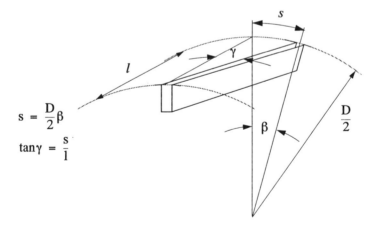

Fig. 4.15 *Illustrating the axial skewing of the rotor slots*

$$F_{2,w}(x, y) = F_{2,M} \cos \frac{\pi}{\tau_p}(x - x_0 - \frac{s}{l}y) \qquad (4.47)$$

with notation defined in Fig. 4.15. The mutual inductance is obtained by integrating the product of the stator and rotor winding functions in two dimensions: along the peripheral coordinate x, and along the axial coordinate y

$$L_{1,2} = \mu_0 \iint F_{1,w}(x) F_{2,w}(x, y) \Delta(x) \, dx \, dy \qquad (4.48)$$

The stator slots are not skewed. Therefore, the stator winding function depends only on the rotor peripheral coordinate x

$$F_{1,w}(x) = F_{1,M} \cos \frac{\pi}{\tau_p} x \qquad (4.49)$$

After integrating the product of the two winding functions within the pole pitch τ_p and along the machine's length l, one finds that the magnitude of the mutual inductance is modulated by the term called *the skewing factor*, being defined as

$$f_{sk,1} = \frac{\sin \beta}{\beta} \qquad (4.50)$$

where the angle β is defined in Fig. 4.15. Except for $\beta = 0$ (no skewing), the skewing factor is always less than one.

The effects of skewing on the air gap quantities are illustrated in Fig. 4.16. When the slots are not skewed, as shown in Fig. 4.16 (a), the positive half wave of the mmf produced by one winding is linked uniformly by the other winding along the machine's length. When the rotor slots are skewed, as shown in Fig. 4.16 (b), the rotor winding produces the air gap mmf which cannot be uniformly linked by the unskewed winding on the stator.

The stator winding links the part of the mmf with an opposite sign, which makes the total linked mmf smaller. The dependence of the skewing factor for the fundamental harmonic on the skewing angle β is shown in Fig. 4.17

The skewing factor for a higher harmonic of order n is

$$f_{sk,n} = \frac{\sin n\beta}{n\beta} \qquad (4.51)$$

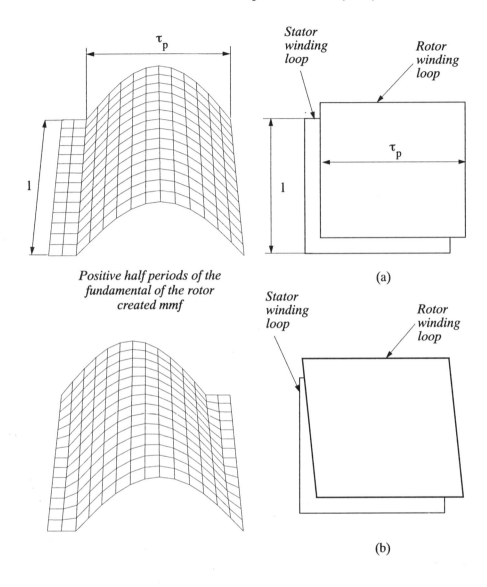

Fig. 4.16 *Illustrating the mutual inductance between the windings placed in unskewed slots on both sides of the gap (a) and between unskewed slots on one side and skewed slots on the other side of the gap (b)*

because the electrical angle β for the n-th higher harmonic is n times greater than for the fundamental. Of special importance is the skewing factor for the slot harmonics, the order of which is $N\pm 1$:

An Electric Machine as a Circuit Element

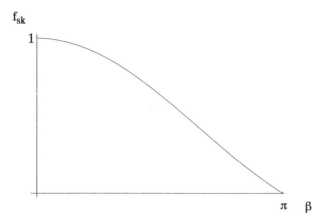

Fig. 4.17 *The skewing factor for the fundamental as a function of the skewing angle β*

$$f_{sk, N \pm 1} = \frac{\sin(N \pm 1)\beta}{(N \pm 1)\beta} \qquad (4.52)$$

The slot harmonics which are insensitive to the winding pitch shortening, as demonstrated in the previous section, can be efficiently diminished by skewing the slots.

4.6 Influence of Slotting on Both Sides of the Air Gap on the Main and Mutual Inductances

The air gap in a machine was considered in the previous sections to be either smooth or unilaterally slotted. However, many electric machines have slots on both sides of the air gap, in order to accommodate for the winding or to produce a variable reluctance torque. The flux in a doubly slotted machine has to pass through the air gap which can be divided into three regions, as shown in Fig. 4.18. The width of the part of the air gap denoted by δ_s is related to the stator slotting and it changes periodically following the distribution

$$\delta_s(x) = \sum_{k = 1, 3, 5\ldots}^{\infty} \delta_{s, k} \cos k N_s \frac{\pi}{\tau_p} x \qquad (4.53)$$

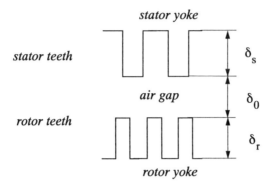

Fig. 4.18 *Illustrating the definition of various portions of the air gap*

where N_s is the number of stator slots (teeth) per pole pair. The width of the part of the air gap denoted by δ_0 is constant, whereas the width of the air gap denoted by δ_r is related to the rotor slotting. This term is described by Fourier series

$$\delta_r(x, x_0) = \sum_{k=1,3,5,\ldots}^{\infty} \delta_{r,k} \cos k N_r \frac{\pi}{\tau_p}(x - x_0) \qquad (4.54)$$

where N_r is the number of the rotor slots (teeth) per pole pair, and x_0 is the rotor shift.

The total air gap width at a point fixed to the stator is a function of the coordinate x and the rotor shift x_0:

$$\delta(x, x_0) = \delta_0 \left[1 + \frac{\delta_s(x)}{\delta_0} + \frac{\delta_r(x, x_0)}{\delta_0} \right] \qquad (4.55)$$

The inverse air gap width, introduced in Eq (4.21), is obtained by using identity

$$\frac{1}{1+x} = 1 - x + x^2 \mp \ldots \qquad (4.56)$$

Physically, Eq. (4.56) means that the real *series* combination of reluctances in the air gap, defined by the air gap widths δ_s, δ_0, and δ_r, is replaced by an equivalent *par-*

An Electric Machine as a Circuit Element

allel combination of elements, the reluctances of which are functions of δ_s, δ_0, and δ_r. Substituting Eq. (4.55) into (4.56) gives

$$\Delta(x, x_0) = \frac{1}{\delta_0}\left[1 + \frac{\delta_s(x)}{\delta_0} + \frac{\delta_r(x, x_0)}{\delta_0} - \frac{\delta_s(x)\delta_r(x, x_0)}{\delta_0^2}\right] \quad (4.57)$$

Utilizing the following substitutions:

$$\Delta_0 = \frac{1}{\delta_0} \qquad \Delta_s(x) = \frac{\delta_s(x)}{\delta_0^2}$$

$$\Delta_r(x, x_0) = \frac{\delta_r(x, x_0)}{\delta_0^2} \qquad \Delta_{r,s}(x, x_0) = \frac{\delta_s(x)\delta_r(x, x_0)}{\delta_0^3} \quad (4.58)$$

one can write for the inverse air gap width in Eq. (4.57)

$$\Delta(x, x_0) = \Delta_0 + \Delta_s(x) + \Delta_r(x, x_0) + \Delta_{r,s}(x, x_0) \quad (4.59)$$

The power series in Eq. (4.56) is limited in Eq (4.57) only to those combinations of the air gap widths, the physical significance of which can be recognized in various types of machines. The inverse air gap width in Eq. (4.57) is represented by the sum of a constant and three terms of Fourier series. The inverse air gap width Δ_0 determines the component of the air gap flux which is independent of the stator and rotor slotting; Δ_s determines the component of the air gap flux dependent only on the stator slotting, whereas Δ_r determines the component of the air gap flux dependent only on the rotor slotting. The orders o within one pole pair of harmonics of the three Fourier series in Eq. (4.58) are as follows:

$$o[\delta_s(x)] = iN_s; \quad i = 1, 2, 3, \ldots \quad (4.60)$$

$$o[\delta_r(x, x_0)] = jN_r; \quad j = 1, 2, 3, \ldots \quad (4.61)$$

$$o[\delta_s(x)\delta_r(x, x_0)] = \pm iN_s \pm jN_r; \quad i = 1, 2, 3, \ldots, \quad j = 1, 2, 3, \ldots \quad (4.62)$$

From Eqs. (4.60) – (4.62) one can see that the inverse air gap width Δ in a doubly slotted machine contains a constant term, the harmonics which are integer multiples of the number of stator teeth N_s, the harmonics which are integer multiples of the rotor number of teeth N_r, and the harmonics which are combinations of integer multiples of the numbers of stator and rotor teeth. Each of the harmonics of the inverse air gap width function contributes to the main- and mutual inductances of the windings in a machine.

The **main inductance** L_m of a winding described by the winding function F_w is evaluated following Eq. (4.3) as

$$L_m(x_0) = 2l\mu_0 \int_0^{\tau_p} F_w^2(x, x_0) \Delta(x, x_0) \, dx \qquad (4.63)$$

When the winding is fixed to the stator, the spatial shift x_0 in the argument of the winding function in Eq. (4.63) is constant; when it is fixed to the rotor, x_0 is equal to the rotor electrical angle.

The winding function is defined as an infinite sum of *odd* harmonics of different orders. The square of the winding function contains, accordingly, only *even* harmonics. As opposed to the original winding function in which each harmonic is represented by only one term, each even harmonic in the square of the winding function comprises an *infinite* number of terms. The n-th harmonic of the square of the winding function is created by the i-th and l-th harmonic of the winding function which satisfy condition

$$i \pm l = n \qquad (4.64)$$

This also means that the n-th harmonic of the main inductance is created by any two harmonics of the winding function(s) which have orders i and l that satisfy Eq. (4.64). In the special case when n/2 is odd, a component of the n-th harmonic of the main inductance is created only by the n/2-th harmonic of the winding function.

The principle of generating the harmonics of the square of the winding function is also illustrated in Eq. (4.26).

The winding function $F_w(x)$ of a stator winding creates the main inductance with:

(a) the inverse air gap width Δ_0. Since both F_w and Δ_0 are independent of x_0, this component of the main inductance is also independent of x_0. The inverse air gap width Δ_0 is constant; hence, this component of the main inductance can be created only by the harmonics of the winding function which have equal orders

An Electric Machine as a Circuit Element

$$L_{s, 0, k} = 2l\mu_0 \int_0^{\tau_p} F_{w, k}^2(x) \Delta_0 dx = \text{const} \tag{4.65}$$

(b) the inverse air gap width $\Delta_s(x)$. Substituting Eq. (4.53) into (4.58), one can write

$$\Delta_s(x) = \frac{1}{\delta_0^2} \sum_{k=1,3,5,\ldots}^{\infty} \delta_{s,k} \cos kN_s \frac{\pi}{\tau_p} x = \sum_{k=1,3,5,\ldots}^{\infty} \Delta_{s,k}(x) \tag{4.66}$$

As in the previous case, both the winding function and inverse air gap width are independent of the rotor shift x_0. Therefore, the component of the main inductance which they produce cannot be dependent on x_0. The component of the main inductance created by the k-th harmonic of $\Delta_s(x)$ is evaluated as

$$L_{s, s, k} = 2l\mu_0 \int_0^{\tau_p} F_{w, 1}(x) F_{w, i}(x) \Delta_{s, k}(x) dx = \text{const}; \quad 1 \pm i = kN_s \tag{4.67}$$

where $\Delta_{s,k}(x)$ denotes the k-th harmonic of $\Delta_s(x)$. In the special case when $kN_s/2 =$ odd, the k-th harmonic of the main inductance $L_{s,s,k}$ is created by the k-th harmonic of $\Delta_s(x)$ and the $kN_s/2$-th harmonic of the winding function, along with all the harmonics the sum or difference of which is equal to kN_s.

(c) the inverse air gap width $\Delta_r(x,x_0)$. From Eqs. (4.54) and (4.58) one can write

$$\Delta_r(x - x_0) = \frac{1}{\delta_0^2} \sum_{k=1,3,5,\ldots}^{\infty} \delta_{r,k} \cos kN_r \frac{\pi}{\tau_p}(x - x_0) = \sum_{k=1,3,5,\ldots}^{\infty} \Delta_{r,k}(x - x_0) \tag{4.68}$$

The inverse air gap width is a function of the rotor shift x_0, as is the component of the main inductance which it creates along with the harmonics of the winding function. The component of the main inductance created in this case can be expressed as

$$L_{s, r, k} = 2l\mu_0 \int_0^{\tau_p} F_{w, 1}(x) F_{w, i}(x) \Delta_{r, k}(x - x_0) dx; \quad 1 \pm i = kN_r \tag{4.69}$$

where $\Delta_{r,k}(x-x_0)$ denotes the k-th harmonic of $\Delta_r(x-x_0)$. After integration, as shown in the Appendix, one obtains

$$L_{s,r,k} = L_{s,r,k,M} \cos k N_r \frac{\pi}{\tau_p} x_0; \quad k = 1, 3, 5, \ldots \quad (4.70)$$

The kN_r-th harmonic of the main inductance $L_{s,r,k}$ in Eq. (4.70) is created by the k-th harmonic of $\Delta_r(x-x_0)$ along with all harmonics of the winding distribution which satisfy the condition that the sum or difference of their orders is equal to kN_r.

The main inductance of a rotor winding can be found by applying the same rules as for a stator winding, with one formal difference: the winding function of the rotor winding is a function of the rotor shift. However, the physical effects in both cases are equal. Therefore, only the final result for the main inductance of the rotor winding due to stator slotting $L_{r,s,k}$ is given here

$$L_{r,s,k} = L_{r,s,k,M} \cos k N_s \frac{\pi}{\tau_p} x_0; \quad k = 1, 3, 5, \ldots \quad (4.71)$$

The **mutual inductance between two stator windings** can depend on the rotor shift only if created in interaction with the inverse air gap width $\Delta_r(x,x_0)$. The kN_r-th harmonic of the mutual inductance between two stator windings is created by the i-th harmonic of the winding function of the first winding and the l-th harmonic of the winding function of the second winding if the orders of harmonics i and l satisfy

$$i \pm l = kN_r \quad (4.72)$$

The k-th harmonic of the mutual inductance $L_{s,s,r,k}$ between two stator windings can be expressed as

$$L_{s,s,r,k} = l\mu_0 \int_0^{\tau_p} F_{1,w,i} \cos i \frac{\pi}{\tau_p} x \; F_{2,w,l} \cos l \frac{\pi}{\tau_p} x \; \Delta_{r,k} \cos k N_r \frac{\pi}{\tau_p} (x - x_0) \; dx$$

$$(4.73)$$

Provided that condition (4.72) is satisfied, the mutual inductance $L_{s,s,r,k}$ can be written as

$$L_{s,s,r,k} = L_{s,s,r,k,M} \cos k N_r \frac{\pi}{\tau_p} x_0; \quad k = 1, 3, 5, \ldots \tag{4.74}$$

The components of the **mutual inductance between a stator and a rotor winding** in a doubly slotted air gap are created by the interaction of their winding functions with various components of the inverse air gap width in Eq. (4.58). The following groups of mutual inductance components between the windings denoted by indices 1 and 2 can be distinguished:

- the mutual inductance $L_{1,2,0}$, independent of the stator and rotor slotting

$$L_{1,2,0} = l\mu_0 \int_0^{\tau_p} F_{1,w}(x) F_{2,w}(x - x_0) \Delta_0 dx \tag{4.75}$$

- the mutual inductance $L_{1,2,s}$, dependent only on the stator slotting

$$L_{1,2,s} = l\mu_0 \int_0^{\tau_p} F_{1,w}(x) F_{2,w}(x - x_0) \Delta_s(x) dx \tag{4.76}$$

- the mutual inductance $L_{1,2,r}$, dependent only on the rotor slotting

$$L_{1,2,r} = l\mu_0 \int_0^{\tau_p} F_{1,w}(x) F_{2,w}(x - x_0) \Delta_r(x - x_0) dx \tag{4.77}$$

- the mutual inductance $L_{1,2,s,r}$, dependent both on the stator and rotor slotting

$$L_{1,2,s,r} = l\mu_0 \int_0^{\tau_p} F_{1,w}(x) F_{2,w}(x - x_0) \Delta_{r,s}(x - x_0) dx \tag{4.78}$$

The orders o of harmonics of functions which appear in Eqs. (4.75) – (4.78) are:

$$o(F_{1,w}) = 1, 3, 5, \ldots \qquad (4.79)$$

$$o(F_{2,w}) = 1, 3, 5, \ldots \qquad (4.80)$$

$$o(\Delta_s) = o(\delta_s) = N_s, 3N_s, 5N_s, \ldots \qquad (4.81)$$

$$o(\Delta_r) = o(\delta_r) = N_r, 3N_r, 5N_r, \ldots \qquad (4.82)$$

$$o(\Delta_{r,s}) = \pm o(\delta_s) \pm o(\delta_r) \qquad (4.83)$$

The mutual inductance $L_{1,2,0}$ defined in Eq (4.75) is independent of the stator and rotor slotting. It is created exclusively by the harmonics of the stator and rotor winding functions which have equal orders. Therefore

$$L_{1,2,0} = \sum_{k=1,3,5,\ldots}^{\infty} L_{1,2,0,M,k} \cos k \frac{\pi}{\tau_p} x_0 \qquad (4.84)$$

where index M denotes the maximum value of the mutual inductance. For $k = 1$ the equation above is identical to that in column (a) in Fig. 4.14. When the orders of the winding functions are different, the definite integral of their product within one pole pitch given in Eq. (4.75) is identically equal to zero.

In Eqs. (4.76) and (4.77) the definite integrals of the products of three harmonic functions have to be evaluated, whereas in Eq. (4.78) the product of four trigonometric functions defines a component of the inductance. These definite integrals can be expressed as functions of the rotor shift x_0 only when a sum or a difference of a combination of orders of the harmonics is equal to zero. In particular, this means that the component $L_{1,2,s}$ of the mutual inductance is not identically equal to zero if the orders of subintegral functions satisfy the following equation:

$$\pm o(F_{1,w}) \pm o(F_{2,w}) \pm o(\delta_s) = 0 \qquad (4.85)$$

As shown in the Appendix, when the condition (4.85) is satisfied, the mutual inductance $L_{1,2,s}$ defined in Eq. (4.76) can be expressed as

$$L_{1,2,s} = \sum_{k}^{\infty} L_{1,2,s,M,k} \cos k \frac{\pi}{\tau_p} x_0 \qquad (4.86)$$

where the order k of a higher harmonic of the main inductance $L_{1,2,s}$ satisfies the following equation:

$$k = o(\delta_s) \pm o(F_{1,w}) \qquad (4.87)$$

or, after substituting Eqs. (4.79) and (4.81)

$$k = N_s \pm 1, N_s \pm 3, ..., 3N_s \pm 1, 3N_s \pm 3, ... \qquad (4.88)$$

When the number of stator slots per pole pair N_s is even, the component of the mutual inductance $L_{1,2,s}$ has only odd harmonics, and vice versa.

The component of the mutual inductance $L_{1,2,r}$ in Eq. (4.77) caused by the rotor slotting can be analogously expressed as

$$L_{1,2,r} = \sum_{k}^{\infty} L_{1,2,r,M,k} \cos k \frac{\pi}{\tau_p} x_0 \qquad (4.89)$$

where the order k of the higher harmonic of the main inductance $L_{1,2,r}$ satisfies the following equation:

$$k = o(\delta_r) \pm o(F_{2,w}) \qquad (4.90)$$

or, by substituting Eqs. (4.80) and (4.82)

$$k = N_r \pm 1, N_r \pm 3, ..., 3N_r \pm 1, 3N_r \pm 3, ... \qquad (4.91)$$

When the number of rotor slots per pole pair N_r is even, the component of the mutual inductance $L_{1,2,r}$ has only odd harmonics, and vice versa.

The component of the mutual inductance $L_{1,2,s,r}$ caused by both stator and rotor slotting exists when the orders of harmonics of the inverse air gap widths and winding functions satisfy the following equation:

$$\pm o(F_{1,w}) \pm o(F_{2,w}) \pm o(\delta_s) \pm o(\delta_r) = 0 \qquad (4.92)$$

The mutual inductance $L_{1,2,s,r}$ can be expressed as

$$L_{1,2,s,r} = \sum_{k}^{\infty} L_{1,2,s,r,M,k} \cos k \frac{\pi}{\tau_p} x_0 \qquad (4.93)$$

where the order of harmonic k must simultaneously satisfy the following equations:

$$\begin{aligned} k &= iN_s \pm l = mN_r \pm n; i = 1, 3, 5, \ldots \\ l &= 1, 3, 5, \ldots \\ m &= 1, 3, 5, \ldots \\ n &= 1, 3, 5, \ldots \end{aligned} \qquad (4.94)$$

It is assumed in the analysis of the effects of higher harmonics of winding functions in this section that the number of pole pairs of all windings is equal. However, the n-th harmonic of the winding function of a two pole winding produces identical effects as a fundamental of a 2n pole winding, provided that all other parameters of the two windings are equal. Therefore, the expressions for the mutual inductances created by the higher harmonics of winding functions of windings which have an equal number of poles may also be used to evaluate the mutual inductances created by the fundamentals of winding functions of windings having different numbers of poles.

4.7 Equivalent Circuit Representation of Windings in Electric Machines

When a single winding in a machine is connected to a time varying source, it creates a pulsating air gap flux. As shown in the first chapter, each harmonic of a pulsating air gap flux density can be represented as a sum of the positive and negative sequence flux densities. The amplitude of the positive and negative sequence components of the fundamental term of the air gap flux density is one half the amplitude of the fundamental. Accordingly, the flux linkage Ψ_1 associated with the fundamental term of the pulsating flux density can be represented as a sum of the flux linkage $\Psi_{1,pos}$, related to the positive sequence component of the flux density, and $\Psi_{1,neg}$, related to the negative sequence component of the flux density.

Since both components of the flux linkage are created by the same winding current, the main inductance of the winding L_m, evaluated in Eq. (4.12), can be represented as a sum of two inductances. The first inductance $L_{m,pos}$ builds, along with the winding current, the flux linkage $\Psi_{1,pos}$. The second inductance $L_{m,neg}$ builds, along with the winding current, the flux linkage $\Psi_{1,neg}$. Therefore, the equivalent circuit of the winding in Fig. 4.4 can be redrawn as shown in Fig. 4.19.

The phases of an m-phase winding placed on one side of the air gap are magnetically coupled, since they share the same magnetic circuit. Magnetic coupling between the phases i and j is expressed by the mutual inductance coefficient $L_{i,j}$. In an unsaturated machine one can write $L_{i,j} = L_{j,i}$, because the magnetic medium is linear.

When the phases of an m-phase winding carry alternating currents, the resultant fundamental component of the positive sequence mmf in the air gap is obtained by summation of the fundamental components of the positive sequence components created by currents in each phase. The same is valid for the fundamental component of the negative sequence mmf. Depending on the relationship between the amplitudes of the resultant fundamental components of positive and negative sequence mmfs, the total air gap mmf can be either *pulsating, elliptical* or *rotating*. The three different types of the total air gap mmf are represented in Fig. 4.20, in which the vector trajectories of the fundamental component of the total air gap mmf are shown.

Depending on the level of asymmetry of the windings and/or currents which they carry, the amplitudes of the positive and negative sequence components of the fundamental term of the air gap mmf vary between zero and m/2 F_1, where F_1 is the amplitude of the fundamental component of the air gap mmf created by one phase. Introducing coefficients k_{pos} and k_{neg}, which can both vary between zero and m/2, one can draw the equivalent circuit per phase of a general m-phase winding as shown in Fig. 4.21.

In the special case when symmetrical windings carry symmetrical currents, the equivalent circuit per phase shown in Fig. 4.21 can be simplified. The fundamental component of the positive sequence mmf in this case has an amplitude which is m/2 times greater than the amplitude of the fundamental component of the mmf created by each phase. The fundamental component of the negative sequence mmf is equal to zero. The same is valid for the flux linkages.

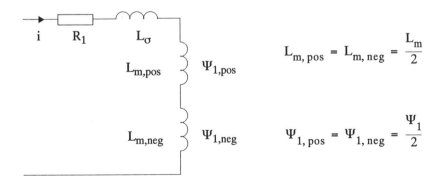

Fig. 4.19 *The equivalent circuit of a winding allowing for the positive and negative sequence*

components of the air gap flux

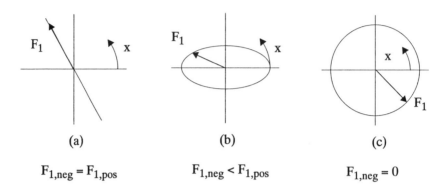

Fig. 4.20 *The trajectories of the vector of the fundamental component of the total air gap mmf for various relationships between the positive and negative sequence components: (a) pulsating; (b) elliptical and (c) rotating air gap mmfs*

Therefore, one can conclude that the positive sequence main inductance in a symmetrical m-phase machine is m times greater than the main inductance of each phase, whereas the negative sequence main inductance is equal to zero. The equivalent circuit per phase in this case is shown in Fig. 4.22.

Electric machines usually have windings on both sides of the air gap. In such a machine an m_s-phase stator winding produces in the air gap the mmf F_s, and an m_r-phase rotor winding produces in the air gap the mmf F_r. The total air gap mmf is equal to the sum of the two. If both stator and rotor windings are symmetrical, and if they carry symmetrical currents, the stator and rotor air gap mmfs contain only positive sequence components. The resultant air gap mmf is pure rotational, since there are no negative sequence components in it.

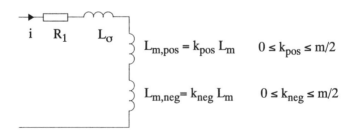

An Electric Machine as a Circuit Element 223

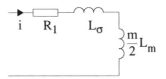

Fig. 4.22 *Equivalent circuit per phase of a symmetrical m-phase machine carrying symmetrical m-phase currents*

Applying the results from the second chapter, one can express the total fundamental component of the positive sequence mmf $F_{1,pos}(x,t)$ as

$$F_{1,pos}(x,t) = \frac{m_s}{2} F_{1,s}(x,t) + \frac{m_r}{2} F_{1,r}(x,t) \qquad (4.95)$$

where $F_{1,s}(x,t)$ is the fundamental term of the pulsating mmf created by one stator phase, and $F_{1,r}(x,t)$ is the fundamental term of the pulsating mmf created by one rotor phase. The stator and rotor positive sequence mmfs rotate at the same speed and usually in the same direction. The angle between the stator and rotor positive sequence components of the air gap mmf is independent of the stator to rotor angle. This is opposite to the case analyzed in the previous section, when the angle between the mmf created by one rotor winding and one stator winding was shown to be equal to the rotor to stator electrical angle.

At no load the fundamental component of the stator air gap mmf is aligned with the fundamental component of the rotor air gap mmf in every electric machine. Therefore, the average value of the positive sequence component of the air gap flux Φ_{pos} is created by the mmf which is equal to the algebraic sum of the stator and rotor mmfs:

$$\Phi_{pos} = 1 \int_0^{\tau_p} B_{1,pos} \sin\frac{\pi}{\tau_p} x \, dx = \frac{2}{\pi} l \tau_p \frac{\mu_0}{p \pi \delta} (\frac{m_s}{2} F_{1,s} + \frac{m_r}{2} F_{1,r}) \qquad (4.96)$$

where $F_{1,s}$ denotes the amplitude of the fundamental term of the pulsating mmf created by one stator phase, and $F_{1,r}$ is the analogous rotor quantity. The turns of the i-th stator phase link the flux $\Psi_{s,i,pos}$ which is equal to

$$\Psi_{s,i,pos} = w_s f_{w,s} \Phi_{pos} = w_s f_{w,s} \frac{2}{\pi} l \tau_p \frac{\mu_0}{p \pi \delta} (\frac{m_s}{2} F_{1,s} + \frac{m_r}{2} F_{1,r}) \qquad (4.97)$$

The flux linkage $\Psi_{s,i,pos}$ has two components: the first is created by the stator ampereturns $F_{1,s}$, and the second is generated by the rotor ampereturns $F_{1,r}$. As already shown, the main inductance $L_{m,s}$ per one stator phase in a symmetrical m_s-phase stator winding fed from symmetrical sources is equal to

$$L_{m,s} = \frac{m_s}{2} \Lambda (w_s f_{w,s})^2 = \frac{m_s}{2} L_s \qquad (4.98)$$

where the air gap specific permeance Λ is defined in Eq. (4.10), and L_s is the main inductance of one stator phase. Analogously, one can define the main inductance $L_{m,r}$ per one rotor phase in a symmetrical m_r-phase rotor winding fed from symmetrical sources as

$$L_{m,r} = \frac{m_r}{2} \Lambda (w_r f_{w,r})^2 = \frac{m_r}{2} L_r \qquad (4.99)$$

Note that the air gap specific permeance Λ is common for stator and rotor winding. The component of flux linkage $\Psi_{s,i,pos}$ created by the rotor currents is equal to

$$\Psi_{s,i,pos,r} = w_s f_{w,s} \frac{2}{\pi} l \tau \frac{\mu_0}{p \pi \delta} \frac{m_r}{2} F_{1,r} = \frac{m_r}{2} L_{s,r} I_r \qquad (4.100)$$

The coefficient of the mutual inductance between one stator and one rotor phase $L_{s,r}$ is defined as

$$L_{s,r} = \Lambda (w_s f_{w,s})(w_r f_{w,r}) = \sqrt{L_s L_r} \qquad (4.101)$$

with L_r denoting the main inductance of one rotor phase. With the previous substitutions, one can express the flux linkage $\Psi_{s,i,pos}$ as

$$\Psi_{s,i,pos} = \frac{m_s}{2} L_s I_s + \frac{m_r}{2} L_{s,r} I_r \qquad (4.102)$$

An Electric Machine as a Circuit Element

where I_s is the amplitude of the stator phase current, and I_r is the amplitude of the rotor phase current.

The air gap flux Φ_{pos} defined in Eq. (4.96) links the turns of the j-th rotor phase, giving the flux linkage $\Psi_{r,j,pos}$ which is calculated in the same manner as the flux linkage for one stator phase

$$\Psi_{r,j,pos} = \frac{m_r}{2} L_r I_r + \frac{m_s}{2} L_{s,r} I_s \qquad (4.103)$$

When comparing Eqs. (4.102) and (4.103) one can see that the mutual inductance term in them is not equal: in the stator flux linkage the mutual inductance is expressed as $m_r L_{s,r}/2$, and in the rotor flux linkage it is equal to $m_s L_{s,r}/2$. This property of the mutual inductance results from the different number of stator and rotor phases.

With previously defined values of the self and mutual inductances, one can draw a *per phase equivalent circuit* of a symmetrical m_s-phase stator winding and a symmetrical m_r-phase rotor winding, both being fed from a symmetrical source, in the manner shown in Fig. 4.23.

An air gap mmf which rotates at mechanical speed ω/p relative to the stator, where p is the number of pole pairs, can be produced:
(a) by *stator* currents, the frequency of which is ω, or
(b) by *rotor* currents, the frequency of which is also ω, and the rotor is at standstill, or
(c) by *rotor* currents, the frequency of which is $\omega - p\omega_m$, and the rotor is rotating at angular speed ω_m, or
(d) by *rotor* D.C. current, if the rotor is rotating at angular speed ω/p

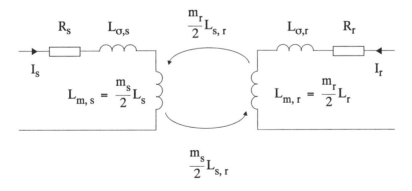

Fig. 4.23 *The equivalent circuit of a machine with symmetrical stator and rotor windings fed from symmetrical sources*

Since the same air gap mmf can be generated in all four cases, the mmf in each case can be replaced by any of the mmfs produced in a different way. This principle is utilized when the magnetic and/or electric quantities produced by one winding are substituted by fictitious equivalent quantities produced by another winding. The frequency of currents in a winding, the parameters of which are referred to another winding, has no significance. Therefore, it is valid to reflect an A.C. rotor current in an induction machine to the stator side, although the stator and rotor frequencies are not equal. The mmf produced by a D.C. current in the rotor of a synchronous machine which runs at synchronous speed may be reflected in the same manner to the stator winding which carries A.C. currents.

The effects of the rotor mmf $F_{1,r}$ which drives the air gap flux Φ_{pos} defined in Eq. (4.96), can be replaced by an *equivalent stator mmf* $F'_{1,r}$ which satisfies the condition

$$m_r F_{1,r} = m_s F'_{1,r} \tag{4.104}$$

The equivalent stator mmf $F'_{1,r}$ is created by a *fictitious* current I'_r. In order to satisfy Eq. (4.104), the fictitious stator current I'_r has to be equal to

$$I'_r = I_r \frac{m_r w_r f_{w,r}}{m_s w_s f_{w,s}} f_{sk} \tag{4.105}$$

The current I'_r is called the *reduced (reflected) rotor current to the stator winding*. The reduced rotor current produces with the stator winding function the same ampere-turns as the real rotor current with the rotor winding function.

Besides the currents, the flux linkages can be reduced from one winding to another, in accordance with the following procedure: the positive sequence component of the air gap flux Φ_{pos}, when linked with the turns of one rotor phase, gives the rotor flux linkage $\Psi_{r,pos}$ defined as

$$\Psi_{r,pos} = w_r f_{w,r} \Phi_{pos} \tag{4.106}$$

The same flux Φ_{pos} when linked with the turns of one *stator* phase gives the flux linkage

$$\Psi'_{r,pos} = w_s f_{w,s} \Phi_{pos} \tag{4.107}$$

After eliminating the flux Φ_{pos} from Eqs. (4.106) and (4.107), and substituting the flux linkages in the equation obtained by the products of the main inductances and winding currents, one can write

$$\frac{I'_r L'_{m,r}}{w_s f_{w,s}} = \frac{I_r L_{m,r}}{w_r f_{w,r}} \frac{1}{f_{sk}} \qquad (4.108)$$

from which the rotor main inductance reflected to the stator side $L'_{m,r}$ equals

$$L'_{m,r} = L_{m,r} \frac{w_s f_{w,s}}{w_r f_{w,r}} \frac{I'_r}{I_r} \frac{1}{f_{sk}} \qquad (4.109)$$

Substituting the reduced rotor current defined in Eq. (4.105) into the equation above, one obtains

$$L'_{m,r} = L_{m,r} \left(\frac{w_s f_{w,s}}{w_r f_{w,r} f_{sk}}\right)^2 \frac{m_s}{m_r} \qquad (4.110)$$

Since the rotor equivalent circuit main inductance $L_{m,r}$ is $m_r/2$ times greater than its per phase value L_r, and the stator equivalent circuit main inductance $L_{m,s}$ is $m_s/2$ times greater than its per phase value L_s, by utilizing Eq. (4.12) one can write

$$L'_r = L_r \left(\frac{w_s f_{w,s}}{w_r f_{w,r} f_{sk}}\right)^2 = \Lambda (w_r f_{w,r})^2 \left(\frac{w_s f_{w,s}}{w_r f_{w,r}}\right)^2 = L_s \qquad (4.111)$$

or, the *main inductance of the rotor winding reflected to the stator winding L'_r is identical to the main inductance of the stator winding L_s.*

When reflected to the stator winding, the rotor winding leakage reactance and the rotor resistance have to be multiplied by the factor which was used for reduction of the main inductance. Therefore,

$$L'_{\sigma,r} = L_{\sigma,r} \left(\frac{w_s f_{w,s}}{w_r f_{w,r} f_{sk}}\right)^2 \frac{m_s}{m_r} \qquad (4.112)$$

and

$$R'_r = R_r \left(\frac{w_s f_{w,s}}{w_r f_{w,r} f_{sk}}\right)^2 \frac{m_s}{m_r} \qquad (4.113)$$

The reflection of the rotor quantities to the stator side and vice versa is physically valid, because the copper loss in the resistance remains unchanged by this action. To prove this, let us express the copper loss in a total of m_s reflected resistances R'_r as

$$P'_{Cu,r} = m_s (I'_r)^2 R'_r = m_s \left(\frac{m_r w_r f_{w,r}}{m_s w_s f_{w,s}}\right)^2 I_r^2 \frac{m_s}{m_r} \left(\frac{w_s f_{w,s}}{w_r f_{w,r}}\right)^2 R_r =$$

$$= m_r I_r^2 R_r = P_{Cu,r} \qquad (4.114)$$

By utilizing the procedure for reducing the parameters of one winding to another, one replaces a set of *magnetically coupled and electrically disconnected real circuits* by a unique fictitious circuit in which all the elements are *electrically connected, but magnetically decoupled*.

The previously derived principles are illustrated in Fig. 4.24 (a), which represents the equivalent circuit of a symmetrical polyphase machine reflected to one stator phase, and in Fig. 4.24 (b), which represents the equivalent circuit of the same machine, but this time reflected to one rotor phase.

Since the equivalent circuit of a machine is defined for one winding, which in case of A.C. machines has the meaning of one phase, the voltages and currents in it are called the *per phase quantities*.

When one of the windings is skewed, the winding to which the equivalent circuit is reflected always acts as if it were not skewed, and the winding of which the parameters are reflected always acts as if it were skewed, independently of which of the windings is in reality placed in skewed slots. Therefore, the skewing factor always appears in numerators of the reflected quantities in Fig. 4.24.

An Electric Machine as a Circuit Element

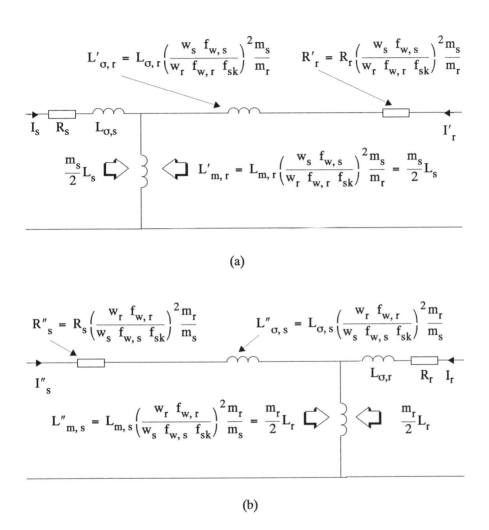

Fig. 4.24 *The equivalent circuit of a machine with symmetrical windings fed from symmetrical sources reflected (a) to the stator and (b) to the rotor*

4.8 Induced Voltages in the Windings of Electric Machines

A force **F** on an electron exists in both electric and magnetic fields. In an electric field it is expressed as

$$\mathbf{F} = q\mathbf{E} \qquad (4.115)$$

where q is the charge of an electron, and **E** is the electric field strength. The force acts in the direction of the electric field.

The force on an electron exists in a magnetic field only when the electron moves relative to the magnetic field. This force is equal to

$$\mathbf{F} = q\mathbf{v} \times \mathbf{B} \qquad (4.116)$$

where **v** is the electron's speed relative to the magnetic field, the flux density of which is **B**. The force is perpendicular to the plane defined by the vectors **v** and **B**. The force exists as long as the relative speed between the electron and the magnetic field is different from zero.

Electrons in metals form an electronic gas. When a conductor travels relative to an external magnetic field **B**, a force on the electrons in it is created. For an external observer who is fixed to the magnetic field, this force is caused by the relative motion of electrons in the magnetic field, and can be expressed by Eq. (4.116). For an observer fixed to the conductor, the electrons are at standstill. A force on electrons for the fixed observer can be created only by an electric field which is induced by the motion of the conductor in an external magnetic field. The force on electrons is, however, independent of whether an observer is fixed to the conductor or to the external field. It is equal in both cases, which means that Eqs. (4.115) and (4.116) describe the same quantity. Therefore, the electric field **E** induced by the motion of the conductor at speed **v** in an external magnetic field **B** is

$$\mathbf{E} = \mathbf{v} \times \mathbf{B} \qquad (4.117)$$

From the equation above one can express the amplitude e of the induced voltage as

$$e = Blv \qquad (4.118)$$

The conductors in the slots of electric machines are not directly exposed to the air gap flux, as illustrated in Fig. 4.2. The air gap flux is equal to the total yoke flux, independent of the order of harmonic of flux density which produced it. Denoting by $B_{y,Max}$ the amplitude of the yoke flux, and by $B_{g,Max}$ the amplitude of the air gap flux, one can write, based upon Maxwell's divergence equation (2.6), the following relationship between the air gap flux and total yoke flux

An Electric Machine as a Circuit Element

$$B_{y,Max} 2h_y = \int_0^{\tau_p} B_g \, dx \tag{4.119}$$

where h_y is the yoke height. The air gap flux density distribution B_g can be represented as a Fourier series which contains only odd terms. For each term of this series Eq. (4.119) must be satisfied. The average value of the n-th spatial harmonic of the air gap flux density in interval $[0, \tau_p]$ is equal to

$$\int_0^{\tau_p} B_{g,Max,n} \sin n \frac{\pi}{\tau_p} x \, dx = \frac{2\tau_p}{\pi n} B_{g,Max,n} \tag{4.120}$$

The amplitude of the n-th spatial harmonic of the yoke flux density $B_{y,Max,n}$ can be expressed from Eqs. (4.119) and (4.120) as

$$B_{y,Max,n} = B_{g,Max,n} \frac{\tau_p}{\pi n h_y} \tag{4.121}$$

As shown in Fig. 4.25, the yoke and air gap flux densities are shifted by $\pi/2$ radians in space. If the n-th spatial harmonic of the air gap flux is expressed by the equation

$$B_{g,n} = B_{g,Max,n} \sin n\varphi \tag{4.122}$$

then the same harmonic of the yoke flux is, following Eq. (4.121) and the flux lines distribution in Fig. 4.25,

$$B_{y,n} = B_{g,Max,n} \frac{\tau_p}{n\pi h_y} \sin n\left(\varphi - \frac{\pi}{2}\right) \tag{4.123}$$

When the flux is alternating, the k-th time harmonic of the flux density has a frequency which is k times greater than the frequency of the fundamental. Therefore,

$$B_{g,n} = B_{g,Max,n} \sin(n\varphi - k\omega t) \tag{4.124}$$

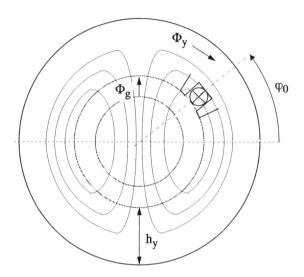

Fig. 4.25 *Illustrating the spatial relationship between the air gap flux Φ_g and yoke flux Φ_y*

and

$$B_{y,n} = B_{g,\text{Max},n} \frac{\tau_p}{n\pi h_y} \sin(n\varphi - k\omega t - n\frac{\pi}{2}) \tag{4.125}$$

The conductor which is φ_0 degrees shifted from the origin links the yoke flux Φ_y

$$\Phi_y = lh_y B_{y,n} = B_{g,\text{Max},n} \frac{l\tau_p}{n\pi} \sin(n\varphi_0 - k\omega t - n\frac{\pi}{2}) \tag{4.126}$$

where l is the machine's length. Since the conductor is linking the *yoke* flux, the induced voltage per conductor $e_{k,n}$, induced by the n-th spatial harmonic of the flux density which alters with frequency $k\omega$, follows from Faraday's law

$$e_{k,n} = -\frac{d\Phi_y}{dt} = k\omega B_{g,\text{Max},n} \frac{l\tau_p}{n\pi} \cos(n\varphi_0 - k\omega t - n\frac{\pi}{2}) \tag{4.127}$$

or

$$e_{k,n} = k\omega B_{g,\text{Max},n} \frac{l\tau_p}{\pi} \sin(n\varphi_0 - k\omega t) \quad (4.128)$$

The frequency of the voltage induced in a conductor at standstill by the n-th spatial harmonic of flux which alters with frequency $k\omega$ is *independent of the order* n *of the spatial harmonic*. The frequency of the induced voltage in Eq. (4.128) is equal to the frequency $k\omega$ of flux which induces it.

Substituting the value of the air gap flux density for $\varphi = \varphi_0$ from Eq. (4.124) into Eq. (4.128), one obtains

$$e_{k,n} = B_{g,n} l k v \quad (4.129)$$

where $B_{g,n}$ is the amplitude of the n-th spatial harmonic of the air gap flux density *at that point in the air gap where the conductor is placed*, and kv is the peripheral speed of the k-th time harmonic of the air gap flux density. Although the conductor in the slot is not directly exposed to the air gap flux density, the induced voltage in it is proportional to the value of the air gap flux density at that point of the air gap where the conductor is placed! Such a dependence of the induced voltage is determined by the fact that the yoke flux which links the conductor is in a firm relationship with the air gap flux, as shown in Eq. (4.121).

According to the general equation for the induced voltage $e = Blv$, the curve representing the spatial distribution of the flux density is identical to the curve representing the time distribution of induced voltage.

Since the n-th harmonic of the air gap flux density $B_{g,n}$ in Eq. (4.129) is varying sinusoidally, the R.M.S. value of the induced voltage per conductor can be further written as

$$e_{k,n} = \frac{1}{\sqrt{2}} k\omega B_{g,\text{Max},n} \frac{l\tau_p}{\pi} = \frac{1}{2\sqrt{2}} \Phi_{g,\text{Max},n} k\omega \quad (4.130)$$

The R.M.S. value of the n-th harmonic of the induced voltage in the winding which has 2w conductors, or w turns, is

$$E_{n,k} = 2w f_{w,n} e_{n,k} = \frac{2\pi}{\sqrt{2}} \Phi_{g,\text{Max},n} k f w f_{w,n} \quad (4.131)$$

where f is the frequency of the fundamental harmonic of the alternating flux. The

R.M.S. value of the fundamental of the induced voltage is therefore

$$E_{1,1} = 4.44 \Phi_{g,Max,1} f w f_{w,1} \qquad (4.132)$$

When the conductor travels relative to the rotating air gap flux which induces voltages in it, the frequency of the induced voltage is dependent on the relative speed between the flux and the conductor. Defining by v_{cond} the speed of the conductor, and by $v_{n,k}$ the speed of the n-th spatial and k-th time harmonic of current, one can introduce the slip $s_{n,k}$ between them

$$s_{n,k} = \frac{v_{n,k} - v_{cond}}{v_{n,k}} \qquad (4.133)$$

According to Eq. (4.131), the frequency of the voltage induced by the k-th time harmonic of flux in a conductor at standstill is kf, where f is the frequency of the fundamental of the current which creates the flux. If the conductor travels relative to the field at the speed v_{cond}, the frequency $f_{n,k,s}$ of the voltage induced by the k-th time and n-th spatial harmonic of flux is proportional to the product of slip $s_{n,k}$ and the frequency kf:

$$f_{n,k,s} = s_{n,k} k f \qquad (4.134)$$

After substitutions, one obtains the angular frequency of the induced voltage $\omega_{n,k,s}$ as

$$\omega_{n,k,s} = k\omega_1 - np\omega_m \qquad (4.135)$$

where ω_1 is the fundamental angular frequency of the flux, ω_m is the mechanical speed of rotation of the conductor, and p is the number of pole pairs of the machine.

5 Force and Torque in Electric Machines

5.1 The Role of Magnetic Energy in Electromechanical Energy Conversion

In the first chapter the energy distribution in a magnetic circuit in which one of the parts moves relative to the others was discussed. It was shown that the portion of magnetic energy stored outside the iron core in such a circuit is converted into the mechanical energy which produces the motion in the circuit. In this chapter, the exact relationship between the currents in the windings, flux linkages, magnetic energy, and the force (torque) will be derived. Here the electromechanical energy conversion is the focus of interest; hence, the portion of energy coming from the source, which is converted into the iron core losses, will not be considered. This, however, in no way influences the accuracy of the results. The physical reason why the part of the magnetic energy stored and/or converted into losses in the iron core may be neglected when the force is calculated will be elaborated in this chapter.

If there are n windings which participate in the electromechanical energy conversion in an electric machine, Eq. (1.56) can be written as

$$\sum_{j=1}^{n} i_j d\Psi_j = dW_{mg} + F dx \tag{5.1}$$

where dW_{mg} is the differential of the magnetic energy stored in the air gap of the machine, and F is the amplitude of the force which acts along the differential path dx. Utilizing the definition of the magnetic coenergy, illustrated in Fig. 1.17 in the first chapter, one can also write

$$W_{mg} + W'_{mg} = \sum_{j=1}^{n} i_j \Psi_j \tag{5.2}$$

The differential of Eq. (5.2) is

$$dW_{mg} + dW'_{mg} = \sum_{j=1}^{n} i_j d\Psi_j + \sum_{j=1}^{n} \Psi_j di_j \qquad (5.3)$$

The way the force is defined, as a function of the magnetic circuit variables, depends on the kind of applied analysis in which the *state variables* are utilized. The state variables describe the state in an electromechanical system at every time instant. The *electromagnetic state variables* give information about all electromagnetic quantities in the machine, while the *mechanical state variables* are the stator to rotor shift and the rotor angular speed. Depending on the selection of the state variables, one can distinguish three formally different approaches to the analysis of electric machines. In the first approach, the electromagnetic state variables are *the flux linkages*. The mmf drops on the reluctances and the ampereturns of the mmf sources are calculated from the known flux values, based upon the B–H curve of the magnetic material and the geometry of the magnetic circuit. In the second approach, the electromagnetic state variables are *the currents in the windings*. The fluxes in the branches of the magnetic circuits are calculated from the B–H curves of the magnetic material and the circuit geometry for given node potentials. In the third approach [6], *both the flux linkages and the currents in the windings* are the electromagnetic state variables. The B-H curve of the magnetic material is already built into the analysis strategy in this approach. Each of these three methods of analysis is valid for application in any mode of operation of any type of electromechanical energy converter. However, the choice of the method of analysis of a particular magnetic circuit at a particular mode of operation is mostly determined by the simplicity of procedure which the method offers, and by the physical clarity of the results obtained.

The magnetic energy, stored in both the iron core and the air gap of a machine is a function of all the currents, and of all the flux linkages. Only that portion of the total magnetic energy which is stored in the air gap can be converted into mechanical work because the air gap allows relative motion between the stator and the rotor. Therefore, the terms dW_{mg} and dW'_{mg} in Eqs. (5.1) and (5.3) are the total differentials of the magnetic energy and coenergy stored in the air gap, respectively.

The magnetic energy stored in the air gap can be expressed as an infinite sum of differentials of the magnetic energy along the gap periphery. Each differential of the magnetic energy in the air gap is a function of the rotor to stator coordinate, and it varies as the rotor position changes. The variation of the magnetic energy stored in the air gap, as a function of the rotor to stator shift, creates a force (torque) which can perform mechanical work, thus converting electrical energy into mechanical energy and vice versa. Mechanical work can be performed only if a force acts along a path, or a torque acts along an angle. Although the variation of the magnetic energy in the iron core creates a force, this force cannot perform useful work because the dimensions of the core are constant. Instead, the variation of the magnetic energy accumulated in iron

Force and Torque in Electric Machines

is a source of noise and vibrations.

If the flux linkages in the windings are the electromagnetic state variables, the partial derivative of the magnetic energy in Eq. (5.1) with respect to the shift x gives the force

$$F = -\left(\frac{\partial W_{mg}}{\partial x}\right)_{\Psi_j = const} \quad (5.4)$$

When a partial derivative with respect to a variable is evaluated, all other variables are assumed to be constant, and their differentials are equal to zero. In this case, the differentials of the flux linkages in all n windings are equal to zero, i.e. $d\Psi_j = 0$ for $1 < j < n$, since $\Psi_j = const$. Equation (5.4) is also valid when the electromagnetic state variables are both the flux linkages and the winding currents, since in Eq. (5.1), from which the result in Eq. (5.4) is derived, there exist no restrictions for the winding currents.

In the case of rotation, the torque can be expressed as

$$T = -\left(\frac{\partial W_{mg}}{\partial \gamma}\right)_{\Psi_j = const} \quad (5.5)$$

When the electromagnetic state variables are the currents in the windings, the partial derivative of the magnetic energy with respect to the shift x is evaluated by assuming that the currents in all n windings are constant, i.e. by setting $di_j = 0$, for $1 < j < n$. Equation (5.3) can be therefore written as

$$dW_{mg} + dW'_{mg} = \sum_{j=1}^{n} i_j d\Psi_j \quad (5.6)$$

Substituting the differential of the magnetic energy, dW_{mg}, from Eq. (5.1) into Eq. (5.6), one obtains the force equation

$$F = \left(\frac{\partial W'_{mg}}{\partial x}\right)_{i_j = const} \quad (5.7)$$

or, for rotation, the torque equation

$$T = \left(\frac{\partial W'_{mg}}{\partial \gamma}\right)_{i_j = const} \quad (5.8)$$

The torque in an electric machine is equal to the partial derivative of the accumulated magnetic coenergy, or to the negative partial derivative of the accumulated magnetic energy in the air gap with respect to the rotor shift.

It is not accidental that the torque is expressed as a function of the magnetic energy accumulated in the air gap. When analyzing other kinds of energy which are present in a machine in the electromechanical energy conversion, i.e. the rotor kinetic energy and the electrical energy of the copper losses, one can conclude that these are the functions only of mechanical or electrical quantities respectively. The rotor kinetic energy is, namely, equal to

$$W_{kin} = \frac{J\omega^2}{2} \quad (5.9)$$

where J is the rotor's moment of inertia, and ω is its angular speed. Both of these quantities are purely mechanical, defined by the rotor geometry and specific weight, and independent of the machine's electrical quantities. The energy of the copper losses is, on the other hand

$$W_{Cu} = i^2 Rt \quad (5.10)$$

This energy is a function of purely electrical quantities: the current in the winding, and the winding resistance. It does not depend on the machine's mechanical quantities. The accumulated magnetic energy is, however, a function of *both* the mechanical (geometric) and electromagnetic quantities

$$W_{mg} = \frac{Li^2}{2} \quad (5.11)$$

since the inductances and mutual inductances in a machine are dependent on the electromagnetic (number of turns, permeability μ) and geometric (magnetic circuit dimensions) quantities. The magnetic energy is the primeval electromechanical quantity, because it includes both electromagnetic and mechanical (geometric) parameters of an electromechanical energy converter.

Independently of the number of windings on the stator and rotor side of an electric machine, the total electromagnetic torque can always be represented as a superposition of the torques created between each stator and each rotor winding. Therefore, the simplest case – two windings – will be elaborated here. The obtained results can be

Force and Torque in Electric Machines

applied in the torque analysis of any kind of electric machine.

Let L_1 denote the main inductance of the first winding, L_2 the main inductance of the second winding, and L_{12} the mutual inductance between the windings, as illustrated in Fig. 5.1

The differential of the magnetic energy accumulated in the field of the first winding is equal to

$$dW_{mg1} = L_1 i_1 di_1 + L_{12} i_2 di_1 \qquad (5.12)$$

Analogously, the differential of the magnetic energy accumulated in the field of the second winding is

$$dW_{mg2} = L_2 i_2 di_2 + L_{12} i_1 di_2 \qquad (5.13)$$

making the total accumulated magnetic energy equal to

$$W_{mg} = \int dW_{mg1} + \int dW_{mg2} = \frac{i_1^2}{2} L_1 + \frac{i_2^2}{2} L_2 + i_1 i_2 L_{12} \qquad (5.14)$$

The relationships between the magnetic field quantities in the air are linear. Therefore, the magnetic coenergy is equal to the magnetic energy:

$$W'_{mg} = \frac{i_1^2}{2} L_1 + \frac{i_2^2}{2} L_2 + i_1 i_2 L_{12} \qquad (5.15)$$

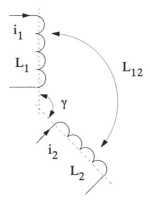

Fig. 5.1 *Illustrating the procedure for deriving the torque between two windings*

The expression for the torque between the windings is obtained by applying Eq. (5.8) to Eq. (5.15). The result is the *fundamental torque equation* for electric machine

$$T = \left(\frac{\partial W'_{mg}}{\partial \gamma}\right)_{i_j = \text{const}} = \frac{i_1^2}{2}\frac{dL_1}{d\gamma} + \frac{i_2^2}{2}\frac{dL_2}{d\gamma} + i_1 i_2 \frac{dL_{12}}{d\gamma} \qquad (5.16)$$

The torque between two windings obviously exists if at least one of the currents is different from zero. This is the **trivial** condition for the existence of torque. The torque depends on the *instantaneous* values of currents. This means that the torque at any time instant is independent of its previous value(s). The next conclusion which one can derive from the torque expression in Eq. (5.16) can be stated in the form of the **necessary** condition for existence of the electromagnetic torque: *at least one of the main- or mutual inductances must be a function of the rotor to stator shift* γ. When the main- or mutual inductance in a machine satisfies this condition, its derivative with respect to the rotor to stator angle γ is not identically equal to zero.

The main- or mutual inductance which is a function of an angle always depends on it *periodically*. Since the average value of a periodical function and its derivatives within one period is equal to zero, the average values of the derivatives of all periodically varying inductances in interval $[0, 2\pi]$ of the stator to rotor electrical angle γ must also be equal to zero. If the winding currents are constant, the average value of the torque in Eq. (5.16) in the same interval of γ is equal to zero. The average value of mechanical work, being defined as

$$W_{mech} = \int_0^{2\pi} T d\gamma \qquad (5.17)$$

in the case of constant currents in the windings, is equal to zero. In this case, the power pulsates between the mechanical and electrical part of the circuit, making the converted energy equal to zero. The pulsating mechanical power in an electromechanical circuit has the same meaning as the reactive power in an electric circuit.

The spatial dependence of the torque between two windings fed from D.C. sources is shown in Fig. 5.2 (a). When the average value of torque is equal to zero, the energy transmitted from the electrical to the mechanical part of the circuit in one half of the period is equal to the energy returned back to the electrical from the mechanical part of the circuit in the second half period. The same property would have had the torque in a hypothetical electric machine with permanent magnets on both the stator and rotor.

However, if at least one of the currents is alternating at the proper frequency, the energy flow from the electrical to the mechanical part, and vice versa, is not necessarily equal, and permanent electromechanical energy conversion can take place.

Force and Torque in Electric Machines

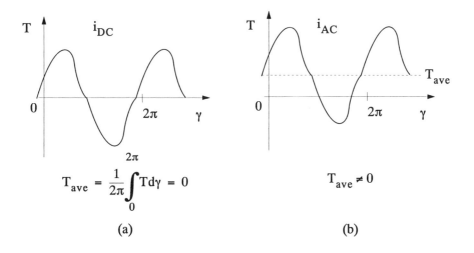

Fig. 5.2 *The torque between the D.C. (a) and A.C. (b) fed windings*

As known from calculus, a product of two periodic functions having equal frequencies and being shifted by an angle different from π/2 can have a constant term, in addition to alternating components. This can be illustrated by the product of two sinusoidal functions

$$\sin(\omega_1 t - \varphi_1)\sin(\omega_2 t - \varphi_2) = \frac{1}{2}\cos[(\omega_1 - \omega_2)t - (\varphi_1 - \varphi_2)] - \frac{1}{2}\cos[(\omega_1 + \omega_2)t - (\varphi_1 + \varphi_2)] \quad (5.18)$$

which has a constant term, different from zero, only if both the functions have the same frequency, i.e. for

$$\omega_1 = \omega_2 \quad (5.19)$$

Applying condition (5.19) to Eq. (5.18), one can express the average value of the product of the harmonic functions on the left hand side of Eq. (5.18) as

$$\int_t^{t + \frac{2\pi}{\omega_1}} \sin(\omega_1 t - \varphi_1)\sin(\omega_2 t - \varphi_2)\, dt = \cos(\varphi_1 - \varphi_2) \quad (5.20)$$

This is an important conclusion, which determines the **sufficient** condition for permanent electromechanical energy conversion: *at least one of the currents in the windings has to be alternating, and the period of the alternating current must be equal to the period of variation of the main- or mutual inductance(s)*. If this condition, expressed in (5.19), is satisfied, there exists a constant term in Eq. (5.16), which is the representative of the energy *permanently* converted from electrical to mechanical form, and vice versa.

Depending on the type of machine, the mode of operation, and the kind of source connected to the machine, the ratio between the constant torque term and its alternating (pulsating) component varies. A goal in an electric machine is to develop as high a constant torque and as low a pulsating torque component as possible. A pulsating torque in an electric machine is a source of mechanical vibrations and noise. In addition, the electric power which produces the pulsating torque thermally loads the machine. If the frequency of the pulsating torque is close to one of the natural frequencies of the rotating parts of the electric drive, serious damage can occur. The pulsating torque can be minimized in symmetric m-phase machines, in which the fundamental harmonics of the mmfs create only a constant torque. In asymmetric machines the amplitude of the pulsating torque component depends on the level of asymmetry. In an extreme asymmetry, when the positive and negative sequence mmfs have equal amplitudes, the constant torque created by the fundamental stator and rotor harmonics vanishes, and only the pulsating torque exists. There is no permanent electromechanical energy conversion in this case.

The pulsating torque components are also generated in machines which are fed from power electronics converters. Some types of converters can drive higher current harmonics, which produce pulsating torque components at the frequencies determined by the frequencies of the higher current harmonics.

The existence of a particular term in the torque equation (5.16) depends on the geometry of the machine and the placement of the windings in it. Based upon their origin, two components of torque can be distinguished in a machine: the reluctance torque, and the pure electromagnetic torque.

The **reluctance torque** is caused by the variable reluctance of the air gap, or, which is the same, by the variable main inductance of a winding. The reluctance torque created by the first winding in Fig. 5.1 is

$$T_{rel1} = \frac{i_1^2}{2}\frac{dL_1}{d\gamma} \qquad (5.21)$$

and the reluctance torque generated by the second winding is, analogously,

$$T_{rel2} = \frac{i_2^2}{2} \frac{dL_2}{d\gamma} \tag{5.22}$$

The main inductance of a winding on the stator side in an electric machine is a function of the rotor to stator angle γ when the rotor has salient poles, or teeth. The main inductance of a winding on the rotor side is a function of the rotor to stator angle γ when the stator has salient poles, or teeth.

The dependence of the main inductance of a winding on the angle γ in salient pole machines is derived in the previous chapter. The approximation of this curve by a constant and the fundamental harmonic is shown in Fig. 5.3 (c).

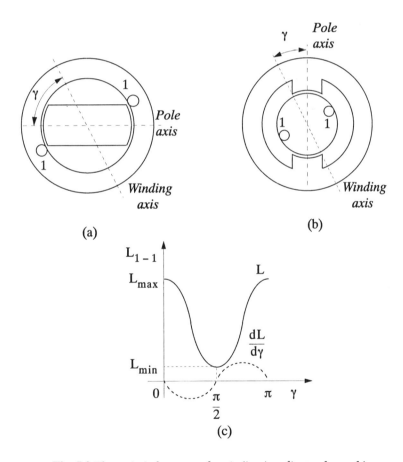

Fig. 5.3 *The main inductance of a winding in salient pole machines*

The main inductance is maximum, L_{max}, when the winding axis is aligned with the axis of the salient poles, or when the winding axis is in the d-axis of the poles. The main inductance is minimum, L_{min}, when the winding axis is perpendicular to the axis of the salient poles, or when the winding is in the q-axis of the salient poles. The period of the main inductance is π electric radians – it reaches twice its minimum and maximum values within one pole pair. The main inductance of the winding 1–1 can be approximated by the constant term and fundamental harmonic as

$$L_{1-1} = \frac{L_{max}+L_{min}}{2} + \frac{L_{max}-L_{min}}{2}\cos 2\gamma \qquad (5.23)$$

The stator to rotor angle γ in Eq. (5.23) can be replaced by the term $p\omega_m t/2$, when the winding rotates at a constant mechanical speed ω_m. With this substitution, the spatial dependence of the main inductance on the electrical angle γ, $L_{1-1}(\gamma)$, is converted into its time variation, $L_{1-1}(t)$. Such a conversion allows a comparison between the time variation of the main inductance of a winding and the time variation of the current in it.

When the winding 1–1 in Fig. 5.3 carries *constant* current, the average torque in the interval $[0,2\pi]$ is equal to zero, independently of the value of the rotor speed ω_m. This is illustrated in Fig. 5.4 (a). When the rotor is at standstill, and the angle between the stator and rotor is in the intervals $\{(-1\pm 4k)\pi/4, (1\pm 4k)\pi/4\}$ where $k = 0, 1, 2,...$, the rotor is in stable equilibrium. After being unloaded in these intervals, the rotor always returns to the stable points $\pm k\pi$, where $k = 0, 1, 2,...$. Provided that the winding current is constant, the rotor is in unstable equilibrium in all other intervals of the angle γ.

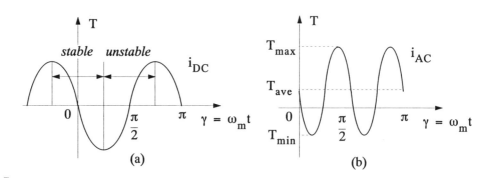

Fig. 5.4 *The reluctance torque created by D.C. (a) and A.C. current (b) in the winding in Fig. 5.3 (b). The frequency of the A.C. current and the rotor speed in case (b) satisfy Eq. (5.24).*

Force and Torque in Electric Machines

When the winding 1–1 in Fig. 5.3 is fed from a *sinusoidal* source with frequency ω, there exist rotor speeds ω_m, at which the permanent electromechanical energy conversion can take place. The constant reluctance torque component, T_{ave} in Fig. 5.4 (b), is different from zero at these rotor speeds. The condition which the speeds ω_m have to satisfy follows from Eq. (5.19)

$$\omega = \pm p\omega_m \qquad (5.24)$$

where p denotes the number of pole pairs in the machine. The main inductance of the winding has to vary *simultaneously* with the variation of the current, or, the frequency of the current has to be equal to the frequency of the main inductance variation, in order to produce a constant component of the reluctance torque. The time variation of the reluctance torque developed in this case is shown in Fig. 5.4 (b).

If the period of the current in the winding is ω, the period of the square of the current is 2ω, as is the period of variation of the main inductance. Consequently, the frequency of pulsations of the alternating component of the torque in Fig. 5.4 (b) is 4ω. The amplitude of the constant component of the torque T_{ave} is a function of the current phase shift.

As well as in machines with salient poles, the reluctance torque can be generated if teeth exist either on the stator, the rotor, or on both sides of the air gap. This torque is built by those harmonics of the winding function, the sum or difference of orders of which is an integer multiple of the number of teeth per pole pair, as shown in the previous chapter.

As an illustration, the variation of the main inductance of a winding as a function of the angle γ in a machine which has six teeth on the rotor is shown in Fig. 5.5 (b). If there are N teeth on one side of the air gap of a two pole machine, the variation of the main inductance of the winding placed on the opposite side of the gap can be approximated by the constant term and the fundamental harmonic as

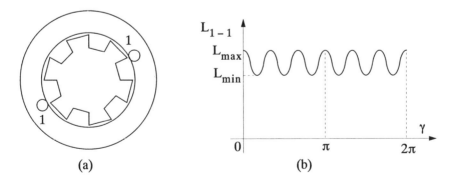

Fig. 5.5 *A machine with variable reluctance caused by the rotor teeth (a), and the main inductance of the stator winding 1–1 as a function of the rotor shift (b)*

$$L_{1-1} = \frac{L_{max}+L_{min}}{2} + \frac{L_{max}-L_{min}}{2}\cos N\gamma \qquad (5.25)$$

The angle γ in Eq. (5.25) can be replaced by $p\gamma_m$, where p is the number of pole pairs, and γ_m is the mechanical angle between the stator and the rotor. The derivative of the main inductance in Eq. (5.25) with respect to the angle γ is

$$\frac{dL_{1-1}}{d\gamma} = -\frac{L_{max}-L_{min}}{2} N\sin N\gamma \qquad (5.26)$$

When the winding is fed from a sinusoidal source, the frequency of which is ω, the constant component of the reluctance torque is generated if the frequency of change of the main inductance is equal to the frequency of the square of the source current, i.e. to 2ω:

$$N\omega_m = \pm 2\omega \qquad (5.27)$$

because the product $i^2 dL_{1-1}/d\gamma$ in this case:

$$i^2 \frac{dL_{1-1}}{d\gamma} = -I_m^2 [\sin(\omega t - \theta)]^2 \frac{L_{max}-L_{min}}{2} N\sin N\omega_m t =$$

$$= \frac{L_{max}-L_{min}}{4} I_m^2 N \{\sin N\omega_m t - \frac{1}{2}\{\sin[(N\omega_m - 2\omega)t - 2\theta] +$$

$$+ \sin[(N\omega_m + 2\omega)t + 2\theta]\}\} \qquad (5.28)$$

contains the constant term

$$\frac{L_{max}-L_{min}}{8} I_m^2 N\sin[(N\omega_m \pm 2\omega)t \pm 2\theta] = \frac{L_{max}-L_{min}}{8} I_m^2 N\sin(\pm 2\theta) \qquad (5.29)$$

only when Eq. (5.27) is satisfied.
The mechanical speeds ω_m at which a two pole machine which carries sinusoidal

Force and Torque in Electric Machines

current(s) with frequency ω develops a constant reluctance torque due to N teeth are, following Eq. (5.27),

$$\omega_m = \pm \frac{2\omega}{N} \qquad (5.30)$$

Regarding generation of the constant reluctance torque, a machine with N teeth performs as if it has N poles.

When the higher spatial harmonics in the curve of the main inductance distribution are allowed for, Eq. (5.25) becomes

$$L_{1-1} = \frac{L_{max} + L_{min}}{2} + \frac{L_{max} - L_{min}}{2} \cos N\gamma + \sum_{j=2}^{\infty} L_{(2j-1)} \cos[N(2j-1)\gamma] \qquad (5.31)$$

with $L_{(2j-1)}$ denoting the maximum value of the (2j–1)-th harmonic of the main inductance. There exist infinite values of the rotor speed, at which the constant reluctance torques appear. These torques are created by the higher harmonic terms in the main inductance curve. The rotor speeds at which constant reluctance torques appear are

$$\omega_{m,j} = \pm \frac{2\omega}{(2j-1)N}; j = 1, 2, \ldots \infty \qquad (5.32)$$

When, in addition, the winding current contains its own spectrum of higher harmonics, each of the current harmonics gives another set of infinite values of the rotor speeds at which the constant reluctance torque is generated. However, the amplitudes of the higher harmonics reluctance torques are usually small when compared to the torque created by the fundamental current and inductance harmonics. In addition, at those rotor speeds at which a higher current harmonics creates constant reluctance torque with a higher main inductance harmonic, there also exists a strong pulsating component of the reluctance torque created by the fundamental current and main inductance harmonics. As a rule, the pulsating torque component created by the fundamental terms dominates, overwhelming the small constant torque created by the higher harmonics. Therefore, only the synchronous speed in Eq. (5.30) has practical significance.

The **pure electromagnetic torque** between two windings exists only if both the currents in the windings are different from zero. The pure electromagnetic torque between the windings is:

$$T_{elmag} = i_1 i_2 \frac{dL_{12}}{d\gamma} \qquad (5.33)$$

The mutual inductance between the stator and the rotor winding is a periodic function of the rotor to stator shift γ. The period of the mutual inductance is equal to double the pole pitch, or 2π electric radians. As opposed to the main inductance of a winding which can only be positive, the mutual inductance between two windings can be either positive, or negative, depending on the rotor shift.

The approximative dependence of the mutual inductance between windings 1–1 and 2–2 in Fig. 5.6 (a) is shown in Fig. 5.6 (b). The fundamental term of the mutual inductance between the windings 1–1 and 2–2 in Fig. 5.6 (b) is

$$L_{12} = L_{max} \cos\gamma = L_{max} \cos p\omega_m t \qquad (5.34)$$

when the rotor rotates at the mechanical angular speed ω_m. Following Eq. (5.19), the pure electromagnetic torque contains a constant term when the electrical angular speed of the stator mmf ω is equal to the sum of the electrical angular speed of the rotor mmf ω_r and the rotor electric angular speed $p\omega_m$:

$$\omega = \omega_r + p\omega_m \qquad (5.35)$$

A constant torque between the stator and rotor mmfs is generated only when the relative speed between them is equal to zero. The condition in Eq. (5.35) is more general than the one expressed in Eq. (5.24), which is a special case of Eq. (5.35) for angular frequency of the rotor currents $\omega_r=0$.

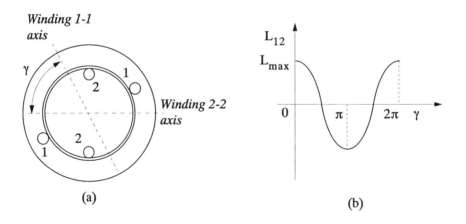

Fig. 5.6 *A machine with a cylindrical rotor and stator (a), and the dependence of the fundamental harmonic of the mutual inductance on the rotor to stator shift (b)*

5.2 Force on the Conductors in the Slots of an Electric Machine

Biot–Savart's law defines the force **F** on a current carrying conductor which is placed in the magnetic field having flux density **B** as

$$\mathbf{F} = I\mathbf{l} \times \mathbf{B} \tag{5.36}$$

where the vector of the conductor length **l** has the direction of the current I. The force on a conductor is always perpendicular to the plane defined by the vectors **l** and **B**. A simple rule to determine the direction of the force in Eq. (5.36) is illustrated in Fig. 5.7, in which the external magnetic field, the field created by the current in the conductor, and the resultant field are drawn. The flux lines created by the current in the conductor distort the external field, causing the armature reaction effect.

The strength of the own magnetic field in the centre of the current carrying conductor is equal to zero. If the current is distributed uniformly in the whole cross-sectional area of the conductor, the field strength within the conductor increases proportionally to the distance from the centre. In the analysis of torque produced by a machine, the effects of the own field within the conductor are usually neglected. Instead, it is considered that the radius of the conductor is infinitesimally small, and that the field produced by the conductor current changes stepwise when passing from one to the other side of the conductor. Therefore, the resultant flux density at the point where the conductor is placed is always equal to the external flux density. The force on the conductor is, accordingly, proportional to the value of the *external* flux density.

The current in the conductor, obviously, cannot create the force on itself. The force always tends to push the current carrying conductor out from the region with high flux density into the region with low flux density. If the conductor moves in the direction of the force which acts on it, the flux lines shorten their lengths at a maximum rate.

The conductors in an electric machine are usually placed either in slots, or between the poles in the inter-polar space. In both cases, the conductors are exposed predominantly to the leakage flux.

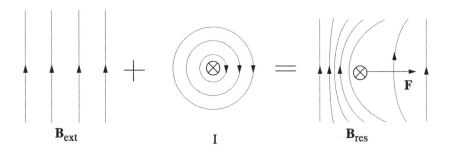

Fig. 5.7 *The force on the conductor in an external magnetic field, as a result of the tendency of the conductor to move out from a space with high flux density*

The leakage flux, by definition, does not link both the stator and rotor windings, and, therefore, cannot generate a useful torque between the stator and rotor.

The force on the conductor in the slot is illustrated in Fig. 5.8. The force created by the leakage flux is shown in Fig. 5.8 (a). This force is directed towards the bottom of the slot. When the teeth are significantly saturated, usually above 1.7 T, the component of the main flux which passes through the slot in the radial direction cannot be neglected, as shown in Fig. 5.8 (b). This flux component creates the force which pushes the conductor towards one of the sides of the slot. The resultant force, shown in Fig. 5.8 (c), is the one with which the conductor presses the slot insulation. This force is negligible when compared to the useful force between the stator and the rotor.

Linking the stator and rotor windings, the main flux passes through the air gap, teeth (poles) and yokes. The main flux on its path through the machine usually does not face the current carrying conductors. When passing through the iron, the main flux reacts on the microscopic level with the atomic currents in the domains, producing the force. This force is demonstrated on the macroscopic level as the force between the stator and the rotor.

The atomic (domain) currents, induced by external currents in the windings and represented by the intensity of magnetization vector **M**, cannot be measured. Nevertheless, the effects of the domain currents can be quantified by means of the magnetic field energy which they produce. Therefore, the expression for the force between the stator and rotor will be derived here starting from the magnetic field energy. Assume that the stator winding(s) create(s) a periodical air gap flux density distribution

$$B_s = B_s(x_s) \tag{5.37}$$

the period of which is $2\tau_p$. Here x_s denotes the stator peripheral coordinate. On the rotor side there are current carrying conductors, which produce the rotor flux density distribution in the air gap

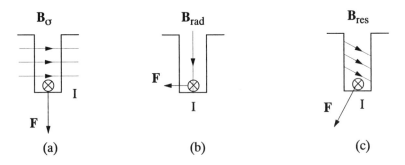

Fig. 5.8 *The force on the conductor in the slot of an electric machine created by the leakage flux density B_σ (a), by the radial component of the main flux density through the slot B_{rad} (b), and by the resultant flux density B_{res} (c)*

Force and Torque in Electric Machines

$$B_r = B_r(x_r) \tag{5.38}$$

where x_r is the rotor peripheral coordinate. The relationship between x_s and x_r is

$$x_r = x_s + g \tag{5.39}$$

with g denoting the rotor to stator peripheral shift. The magnetic energy accumulated in the double pole pitch $2\tau_p$ in the air gap which has width δ is

$$W_{mg} = \frac{1\delta}{2\mu_0} \int_0^{2\tau_p} B^2 dx_s = \frac{1\delta}{2\mu_0} \int_0^{2\tau_p} [B_s(x_s) + B_r(x_r)]^2 dx_s \tag{5.40}$$

The resultant air gap flux density is represented by the sum of the air gap flux density created by the stator currents B_s and the air gap flux density created by the rotor currents B_r. This summation of the flux densities is valid, since the magnetic medium in the air gap has constant permeability. The air gap magnetic energy W_{mg} in Eq. (5.40) can be further written as

$$W_{mg} = \frac{1\delta}{2\mu_0} \left[\int_0^{2\tau_p} B_s^2(x_s) dx_s + 2 \int_0^{2\tau_p} B_s(x_s) B_r(x_s + g) dx_s \right.$$

$$\left. + \int_g^{2\tau_p + g} B_r^2(x_s + g) d(x_s + g) \right] \tag{5.41}$$

The portion of the air gap which stores the magnetic energy given in Eq. (5.41) is limited by the stator and rotor surfaces within one pole pair. Therefore, the quantity that will be obtained by differentiating the magnetic energy with respect to the stator to rotor shift is the force between the stator and rotor per pole pair. The force between the stator and the rotor can be expressed either in terms of the stator flux density and rotor current, or in terms of the rotor flux density and the stator current. Let us use the first mode, and assume that the winding which produces the rotor flux density in the air gap B_r has only one coil with coil pitch y, as shown in Fig. 5.9.

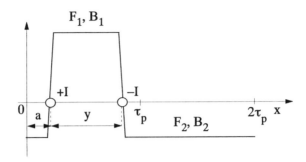

Fig. 5.9 *The mmf and flux density produced by the current in a coil with pitch y*

The rotor flux density is

$$B_r(x_s) = \begin{cases} B_1 = \dfrac{\mu_0}{\delta} \dfrac{2\tau_p - y}{2\tau_p} I_r & \text{for} \quad a - g \leq x_s \leq y + a - g \\ B_2 = -\dfrac{\mu_0}{\delta} \dfrac{y}{2\tau_p} I_r & \text{for} \quad y + a - g \leq x_s \leq 2\tau_p + a - g \end{cases} \quad (5.42)$$

The force is equal to the derivative of the magnetic energy with respect to the coordinate in the direction in which the force acts. Since the magnetic energy has to be differentiated with respect to the rotor to stator shift g, all the terms in Eq. (5.41) which are independent of the variable g do not contribute to the torque.

The first and third definite integrals in Eq. (5.41) do not depend on the value of the stator to rotor shift g. Therefore, the force can be expressed as

$$F = -\dfrac{1\delta}{\mu_0} \dfrac{\partial}{\partial g} \left[\int_0^{2\tau_p} B_s(x_s) B_r(x_s + g) \, dx_s \right] \quad (5.43)$$

or

$$F = -\dfrac{1\delta}{\mu_0} \dfrac{\partial}{\partial g} \left[B_2 \int_{-g}^{a-g} B_s(x_s) \, dx_s + B_1 \int_{a-g}^{a-g+y} B_s(x_s) \, dx_s + B_2 \int_{a-g+y}^{2\tau_p - g} B_s(x_s) \, dx_s \right]$$

$$(5.44)$$

Force and Torque in Electric Machines

Applying the rule for differentiation under the integral

$$\frac{\partial}{\partial g} \int_{a(g)}^{b(g)} f(x)\,dx = f[b(g)]\frac{d}{dg}b(g) - f[a(g)]\frac{d}{dg}a(g) \tag{5.45}$$

and utilizing the periodical character of the stator flux density

$$B_s(-g) = B_s(2\tau_p - g) \tag{5.46}$$

one can express the force from Eq. (5.44) as

$$F = -\frac{l\delta}{\mu_0}(B_1 - B_2)[B_s(a-g) - B_s(a-g+y)] \tag{5.47}$$

Substituting the values for B_1 and B_2 from Eq. (5.42) into Eq. (5.47), one obtains the expression for the force between the stator and the rotor as

$$F = lI_r[B_s(a-g+y) - B_s(a-g)] \tag{5.48}$$

Since the current path is always closed, the basic current carrying entity in a machine is the coil, not the conductor. Therefore, the force equation (5.48), which is derived from general considerations about the magnetic energy distribution in a machine, is related to a coil, not to a conductor. The force on a rotor coil is proportional to the product of the coil current and *the component of the air gap flux density created by the stator windings* at the point where the conductors of the rotor coil are currently placed. The force is generated either by the rotor currents and the stator component of the air gap flux, or by the stator currents, and the rotor component of the air gap flux. There exists no force between the stator and rotor created either by the stator component of the air gap flux and the stator currents, or by the rotor component of the air gap flux and the rotor currents.

If the rotor coil pitch y is equal to the pole pitch τ_p, the stator flux density B_s at the peripheral coordinate $a-g+\tau_p$ is equal to the negative stator flux density at the coordinate $a-g$. The total force on the coil in this case is

$$F_{tot} = 2lI_rB_s(a-g) \tag{5.49}$$

and the force *per conductor*

$$F_{cond} = lI_r B_s (a - g) \qquad (5.50)$$

The force on the rotor conductor shifted by a–g from the origin is proportional to the value of the stator component of the air gap flux density at the point where the conductor is placed, i.e. to $B_s(a–g)$. The force is unaffected by the conductor's position in the slot.

According to the previous considerations, the force expressed in Eq. (5.50) *does not act on the conductor*. The real force between the stator and the rotor, which is a consequence of the interaction between the external field and the domain currents being characterized by the intensity of magnetization vector M, can be expressed in terms of the same external flux density and the current in the conductor(s) which induces the magnetization M in the iron.

When the machine has closed slots, a significant part of the flux created by the currents in the slots passes as a leakage flux through the slot bridges, and the air gap flux density distribution cannot be derived in the manner illustrated in Fig. 5.9. The slot leakage flux causes the mmf drops in the teeth and yoke iron, so that the assumption that the slot ampereturns are equal to the air gap mmf, utilized in Fig. 5.9, is not valid here. The air gap flux in a machine with closed slots is lower than in a machine with open slots, assuming equal currents in both cases. Consequently, the force between the stator and rotor is lower if the slots are closed then if they are open, provided that all other machine parameters are identical.

5.3 Torque Produced by the Currents in the Windings – the Torque Function

By their nature, the torques in an electric machine can be either *synchronous* or *asynchronous*. A synchronous torque is created by the currents in the windings, which satisfy Eq. (5.35) at only one, *synchronous*, speed of the rotor. An asynchronous torque is created by the currents in the windings which satisfy Eq. (5.35) at *every* rotor speed. In both cases, the dependence of frequency of the currents on the rotor speed, expressed in Eq. (5.35), must provide that the stator and rotor mmf harmonics are at a standstill to each other, if a constant torque is to be generated.

The character of torques generated in a machine depends on the machine's geometry, winding distribution, and frequency of currents in the windings. The geometry of the air gap of most electric machines can be reduced to four basic forms: cylindrical rotor and stator, salient pole rotor and cylindrical stator, cylindrical stator and salient pole rotor, and doubly slotted air gap.

(a) Cylindrical stator and rotor. This is the most common form of electric machine. One can find it in almost all types of induction machine (except in shaded-pole

Force and Torque in Electric Machines

machines), in high speed synchronous machines (turbogenerators) and in some types of D.C. commutator machines. Neglecting the variable reluctance effects caused by the slots in which the stator and rotor windings are placed, only a pure electromagnetic torque is developed in a machine having cylindrical stator and rotor geometry.

Consider first a single phase winding on the stator, and a single phase winding on the rotor, as shown in Fig. 5.10 (a). The cross-section in Fig. 5.10 (a) is characteristic for single phase, two pole synchronous machines – turbogenerators. The stator winding carries sinusoidal current, whereas the rotor current is D.C. Denoting by γ the electrical angle between the axes of the stator and rotor windings, the stator to rotor mutual inductance can be expressed according to Eq. (4.40) as

$$L_{Aa} = L_{max, s, r} \cos\gamma \tag{5.51}$$

and its derivative, expressed in terms of a constant rotor angular speed ω_m, yields

$$\frac{dL_{Aa}}{d\gamma} = -L_{max, s, r}\sin\gamma = -L_{max, s, r}\sin(p\omega_m t + \delta) \tag{5.52}$$

where the argument γ of the mutual inductance function in Eq. (5.51) is replaced by

$$\gamma = \omega_m t + \delta \tag{5.53}$$

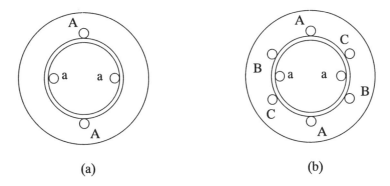

Fig. 5.10 *Cylindrical stator and rotor machines: a single phase (a), and a three phase (b) turbogenerator*

When the rotor is at standstill, the angle δ is identical to the angle γ. The stator current is

$$I_s = I_{s,\,max} \cos \omega t \qquad (5.54)$$

and the rotor current is I_r. The pure electromagnetic torque is

$$T_{em} = -L_{max,\,s,\,r} I_r I_{s,\,max} \cos \omega t \cdot \sin(p\omega_m t + \delta) \qquad (5.55)$$

or (see Appendix)

$$T_{em} = -\frac{1}{2} L_{max,\,s,\,r} I_r I_{s,\,max} [\sin \delta + \sin(\delta - 2\omega t)] \qquad (5.56)$$

when the rotor electrical angular speed $p\omega_m$ is equal to the angular frequency ω of the stator current. At all other rotor speeds there exists only an A.C. component of the torque. The torque in Eq. (5.56) has a constant component, proportional to the sine of the torque angle δ, to which a pulsating term, alternating with twice the mains frequency, is superimposed.

For the purpose of simplicity, a dimensionless quantity called *torque function* will be defined here. It depends on arguments (frequencies and phase shifts) of harmonic functions which represent the currents and inductances. When multiplied by the inductance(s) and amplitude(s) of the current(s), the torque function gives the resultant torque. The torque function has relatively simple forms when the machine carries sinusoidal currents and when the inductances are represented only by their fundamental harmonics. For the electromagnetic torque in Eq. (5.56) the torque function F_t is

$$F_t = -\frac{1}{2} [\sin \delta + \sin(\delta - 2\omega t)] \qquad (5.57)$$

and the torque in the same equation can be expressed as

$$T_{em} = L_{max,\,s,\,r} I_r I_{s,\,max} F_t \qquad (5.58)$$

When the machine is fed from rectangular current sources, or/and when the Fourier series which represents the machine's inductances has several terms, the torque function becomes complex.

Equation (5.35), which defines the rotor speeds at which a constant torque can be

Force and Torque in Electric Machines

generated, is valid both at positive and negative synchronous speeds. The constant torque, proportional to the sine of the torque angle sinδ, is developed at these two speeds in a single phase turbogenerator. Along with the constant torque, a pulsating torque component, proportional to the term sin(δ–2ωt), is produced in this machine. Twice the mains frequency 2ω appears in the term describing the pulsating torque component, because 2ω is the relative electric angular speed between the positive and negative sequence components of the air gap quantities. The speed of the positive sequence component is ω, and –ω is the speed of the negative sequence component.

If there are three symmetrically wound phases on the stator, and a single winding on the rotor, as shown in Fig. 5.10 (b), the pulsating component of the torque can be completely eliminated when certain conditions are fulfilled. This is illustrated in the following example.

The machine, the cross-section of which is shown in Fig. 5.10 (b), is a three phase synchronous generator with cylindrical rotor, or a three phase turbogenerator. The mutual inductances between the stator phases and the rotor winding in a symmetric machine are

$$L_{Aa} = L_{max, s, r} \cos\gamma \quad (5.59)$$

$$L_{Ba} = L_{max, s, r} \cos(\gamma - \frac{2\pi}{3}) \quad (5.60)$$

$$L_{Ca} = L_{max, s, r} \cos(\gamma - \frac{4\pi}{3}) \quad (5.61)$$

with $L_{max,s,r}$ denoting the maximum value of the mutual inductance between one stator phase and the rotor winding, as shown in Eq. (4.40). The symmetric stator currents can be expressed as

$$I_A = I_{s, max} \cos\omega t \quad (5.62)$$

$$I_B = I_{s, max} \cos(\omega t - \frac{2\pi}{3}) \quad (5.63)$$

$$I_C = I_{s, max} \cos(\omega t - \frac{4\pi}{3}) \quad (5.64)$$

Assuming again that the rotor rotates at a synchronous speed, and that the rotor current is D.C. and equal to I_r, one can express the pure electromagnetic torque by utilizing Eq. (5.32) as

$$T_{em} = -L_{max,\,s,\,r}I_{s,\,max}I_r\left[\sin(\omega t + \delta)\cos\omega t + \sin(\omega t + \delta - \frac{2\pi}{3})\cos(\omega t - \frac{2\pi}{3}) + \right.$$
$$\left. + \sin(\omega t + \delta - \frac{4\pi}{3})\cos(\omega t - \frac{4\pi}{3})\right] \quad (5.65)$$

After applying the trigonometric transformations, as shown in the Appendix, the torque expression from Eq. (5.65) can be simplified to

$$T_{em} = -\frac{3}{2}L_{max,\,s,\,r}I_{s,\,max}I_r\sin\delta \quad (5.66)$$

The torque developed in a symmetrically wound turbogenerator carrying symmetric sinusoidal currents is *constant*, if the rotor rotates at synchronous speed in the direction of the air gap field. If the speed of rotation of the rotor is equal to the negative synchronous speed, i.e. if the sequence of windings is opposite to the sequence of currents, the torque expression becomes

$$T_{em} = -L_{max,\,s,\,r}I_{s,\,max}I_r\left[\sin(\omega t + \delta)\cos\omega t + \sin(\omega t + \delta + \frac{2\pi}{3})\cos(\omega t - \frac{2\pi}{3}) + \right.$$
$$\left. + \sin(\omega t + \delta + \frac{4\pi}{3})\cos(\omega t - \frac{4\pi}{3})\right] \quad (5.67)$$

After simplifying this expression in the manner shown in the Appendix, one obtains

$$T_{em} = -\frac{3}{2}L_{max,\,s,\,r}I_{s,\,max}I_r\sin(\delta - 2\omega t) \quad (5.68)$$

The electromagnetic torque in this case has only a pulsating component. The frequency of pulsations of the torque is twice the mains frequency. This mode of operation occurs when the rotor of the turbogenerator rotates in a direction opposite to the direction of the rotating air gap field, where the direction of the rotating air gap field is defined by the sequence of the stator phase currents. In this case the rotor field meets the stator rotating field twice per period of rotation; hence, the torque pulsates with double the synchronous speed frequency, 2ω.

Another kind of electric machine, in which the cylindrical stator – cylindrical rotor configuration is utilized, is the induction machine. A wound rotor induction machine

Force and Torque in Electric Machines

usually has an equal number of phases on the rotor as on the stator, whereas the number of rotor phases in a squirrel cage induction machine is equal to the number of bars. The cross-section of a symmetric wound rotor induction machine is shown in Fig. 5.11 (a).

By the nature of operation of an induction machine, the frequency of the rotor currents, when added to the electrical angular frequency of rotation of the rotor, always gives the stator angular frequency ω. According to the criterion expressed in Eq. (5.19), this provides generation of torque at any speed of rotation. The torque thus produced is asynchronous. Utilizing the rotor slip s, the electrical angular frequency $p\omega_m$ of the rotor at the mechanical speed ω_m in Eq (5.35) can be written as

$$p\omega_m = (1-s)\omega \tag{5.69}$$

because the angular frequency of the rotor currents is

$$\omega_r = s\omega \tag{5.70}$$

The symmetric currents in the rotor phases a, b and c are

$$I_a = I_{r,max}\cos(s\omega t - \theta_r) \tag{5.71}$$

$$I_b = I_{r,max}\cos(s\omega t - \theta_r + \frac{2\pi}{3}) \tag{5.72}$$

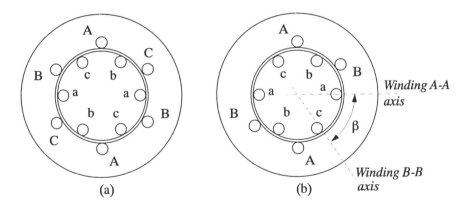

Fig. 5.11 *The cross section of a symmetric, three phase (a), and an unsymmetric, two phase (b), wound rotor induction machine*

$$I_c = I_{r,\max} \cos(s\omega t - \theta_r + \frac{4\pi}{3}) \qquad (5.73)$$

The mutual inductances between the stator (A, B, C) and rotor (a, b, c) phases are

$$L_{Aa} = L_{\max,s,r} \cos\gamma \qquad L_{Ab} = L_{\max,s,r} \cos(\gamma - \frac{2\pi}{3}) \qquad (5.74)$$

$$L_{Ac} = L_{\max,s,r} \cos(\gamma - \frac{4\pi}{3}) \qquad L_{Ba} = L_{\max,s,r} \cos(\gamma - \frac{2\pi}{3}) \qquad (5.75)$$

$$L_{Bb} = L_{\max,s,r} \cos(\gamma - \frac{4\pi}{3}); \qquad L_{Bc} = L_{\max,s,r} \cos\gamma; \qquad (5.76)$$

$$L_{Ca} = L_{\max,s,r} \cos(\gamma - \frac{4\pi}{3}); \qquad L_{Cb} = L_{\max,s,r} \cos\gamma; \qquad (5.77)$$

$$L_{Cc} = L_{\max,s,r} \cos(\gamma - \frac{2\pi}{3}) \qquad (5.78)$$

and the symmetric stator currents can be expressed as

$$I_A = I_{s,\max} \cos\omega t \qquad (5.79)$$

$$I_B = I_{s,\max} \cos(\omega t - \frac{2\pi}{3}) \qquad (5.80)$$

$$I_C = I_{s,\max} \cos(\omega t - \frac{4\pi}{3}) \qquad (5.81)$$

The pure electromagnetic torque is, according to Eq. (5.33), equal to

Force and Torque in Electric Machines

$$T_{em} = -L_{max, s, r} I_{s, max} I_{r, max} \{\cos\omega t \{\sin[(1-s)\omega t + \delta] \cdot \cos(s\omega t - \theta_r) +$$

$$+ \sin\left[(1-s)\omega t + \delta - \frac{2\pi}{3}\right] \cdot \cos(s\omega t - \theta_r + \frac{2\pi}{3}) +$$

$$+ \sin\left[(1-s)\omega t + \delta - \frac{4\pi}{3}\right] \cdot \cos(s\omega t - \theta_r + \frac{4\pi}{3}) \} +$$

$$+ \cos(\omega t - \frac{2\pi}{3}) \{\sin\left[(1-s)\omega t + \delta - \frac{2\pi}{3}\right] \cdot \cos(s\omega t - \theta_r) +$$

$$+ \sin\left[(1-s)\omega t + \delta - \frac{4\pi}{3}\right] \cdot \cos(s\omega t - \theta_r + \frac{2\pi}{3}) +$$

$$+ \sin[(1-s)\omega t + \delta] \cdot \cos(s\omega t - \theta_r + \frac{4\pi}{3}) \} +$$

$$+ \cos(\omega t - \frac{4\pi}{3}) \{\sin\left[(1-s)\omega t + \delta - \frac{4\pi}{3}\right] \cdot \cos(s\omega t - \theta_r) +$$

$$+ \sin[(1-s)\omega t + \delta] \cdot \cos(s\omega t - \theta_r + \frac{2\pi}{3}) +$$

$$+ \sin\left[(1-s)\omega t + \delta - \frac{2\pi}{3}\right] \cdot \cos(s\omega t - \theta_r + \frac{4\pi}{3}) \} \} \quad (5.82)$$

After applying the simplifying procedure shown in the Appendix, one obtains

$$T_{em} = \frac{9}{4} L_{max, s, r} I_{s, max} I_{r, max} \sin(\theta_r + \delta) \quad (5.83)$$

The torque in a symmetric induction machine which carries symmetric sinusoidal current sources is *constant*. When there is no negative sequence mmf, no pulsating component of torque exists. Provided that the amplitudes of the currents are constant, that the phase shift between the currents is equal in all the phases, and that the amplitude of the rotor to stator mutual inductance is constant, the torque is independent of the rotor speed. If these conditions are fulfilled, the stator and rotor mmfs have constant amplitudes, and the torque is proportional to the sine of the torque angle. The phase shift θ_r between the stator and rotor current per phase in this consideration is assumed constant. However, in both the voltage- and current fed induction machines, the amplitudes of the stator and rotor currents are dependent on the rotor speed. Therefore, the torque expressed in Eq. (5.83) in reality is indirectly a function of the rotor speed.

If the stator currents are rectangular, as shown in Fig. 5.12 a), and the rotor currents are sinusoidal, the pure electromagnetic torque developed contains the terms generated by the higher harmonics of the stator currents, in addition to the constant term. These components of torque are pulsating, and have an average value equal to zero. Being created by the higher harmonic mmfs which are not at standstill to the fundamental, the pulsating components of the torque do not contribute to the energy permanently converted from one form to another.

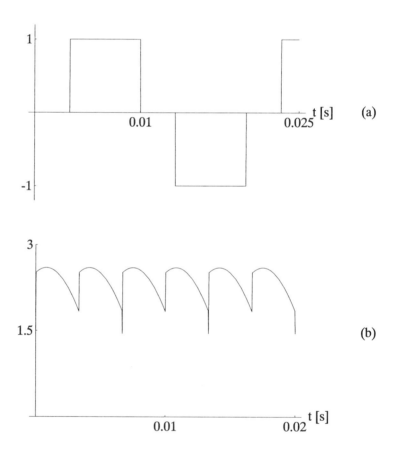

Fig. 5.12 *Stator current in one phase (a) and pure electromagnetic torque (b) in an induction machine*

Force and Torque in Electric Machines

The torque function in an induction machine in which the stator symmetric currents are rectangular, as in Fig. 5.12 (a), and the rotor currents are sinusoidal, looks as shown in Fig. 5.12 (b). The torque function in this figure was calculated by executing the statements from the Appendix. The phase shift of the rotor currents in this example is $\theta_r = \pi/12$.

When there are two phases on the stator side, which are shifted one to the other by an angle β, as shown in Fig. 5.11 (b), the torque expression becomes

$$T_{em} = -L_{max, s, r} I_{s, max} I_{r, max} \{ \cos\omega t \{ \sin[(1-s)\omega t + \delta] \cdot \cos(s\omega t - \theta_r) +$$

$$+ \sin\left[(1-s)\omega t + \delta - \frac{2\pi}{3}\right] \cdot \cos\left(s\omega t - \theta_r + \frac{2\pi}{3}\right) +$$

$$+ \sin\left[(1-s)\omega t + \delta - \frac{4\pi}{3}\right] \cdot \cos\left(s\omega t - \theta_r + \frac{4\pi}{3}\right) \} +$$

$$+ \cos(\omega t - \theta_{s,B}) \{ \sin[(1-s)\omega t + \delta + \beta] \cdot \cos(s\omega t - \theta_r) +$$

$$+ \sin\left[(1-s)\omega t - \frac{2\pi}{3} + \delta + \beta\right] \cdot \cos\left(s\omega t - \theta_r + \frac{2\pi}{3}\right) +$$

$$+ \sin\left[(1-s)\omega t - \frac{4\pi}{3} + \delta + \beta\right] \cdot \cos\left(s\omega t - \theta_r + \frac{4\pi}{3}\right) \} \}$$

(5.84)

because the stator currents in phases A and B are equal to

$$I_A = I_{s, max} \cos\omega t \tag{5.85}$$

$$I_B = I_{s, max} \cos(\omega t - \theta_{s, B}) \tag{5.86}$$

Following the procedure given in the Appendix, one can simplify Eq. (5.84) to obtain

$$T_{em} = -\frac{3}{2} L_{max, s, r} I_{s, max} I_{r, max} [\cos(\omega t + \delta) \cdot \sin(\omega t - \theta_r) +$$

$$+ \cos(\omega t - \theta_{s, B}) \cdot \sin(\beta + \delta + \omega t - \theta_r)] \tag{5.87}$$

The torque function in this case contains a constant term, to which a pulsating component is superimposed. The ratio between the amplitudes of the constant and pulsating torque is a function of both the angle between the stator currents and the angle between the stator phases A and B. In the special case of a symmetric two phase machine, one has to substitute $\theta_{s,B} = -\pi/2$ and $\beta = \pi/2$. This gives the *constant* pure electromagnetic torque

$$T_{em} = -\frac{3}{2} L_{max, s, r} I_{s, max} I_{r, max} \sin(\theta_r + \delta) \tag{5.88}$$

Two other modes of operation of the induction machine are particularly interesting: a single phase stator with a symmetrical rotor, as shown in Fig. 5.13 (a), and a symmetrical stator with a single phase rotor, shown in Fig. 5.13 (b).

The single phase stator in Fig. 5.13 (a) develops positive and negative sequence mmf components in the air gap when fed from a sinusoidal source. The symmetric rotor windings, carrying symmetric currents, develop only the positive sequence air gap mmf and flux density. The pure electromagnetic torque is equal to

$$T_{em} = -L_{max, s, r} I_{s, max} I_{r, max} \{\cos\omega t \{\sin[(1-s)\omega t + \delta] \cdot \cos(s\omega t - \theta_r) +$$

$$+ \sin\left[(1-s)\omega t + \delta - \frac{2\pi}{3}\right] \cdot \cos\left(s\omega t - \theta_r + \frac{2\pi}{3}\right) +$$

$$+ \sin\left[(1-s)\omega t + \delta - \frac{4\pi}{3}\right] \cdot \cos\left(s\omega t - \theta_r + \frac{4\pi}{3}\right) \} \}$$

$$\tag{5.89}$$

which, simplified, yields (see Appendix)

$$T_{em} = -\frac{3}{4} L_{max, s, r} I_{s, max} I_{r, max} [\sin(2\omega t + \delta - \theta_r) + \sin(\delta - \theta_r)] \tag{5.90}$$

The torque in Eq. (5.90) has two components. The stator positive sequence mmf creates a constant torque with the rotor positive sequence flux density, since they travel at the same speed and in the same direction. The stator negative sequence mmf develops a pulsating torque with the rotor positive sequence flux density. The frequency of this pulsating component is 2ω, since 2ω is the relative speed between the stator negative, and the rotor positive sequence components.

Force and Torque in Electric Machines

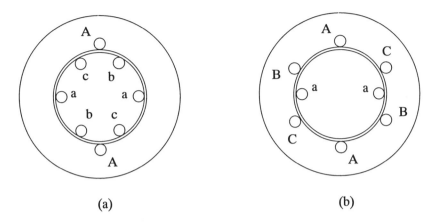

Fig. 5.13 *The cross-section of an asymmetric stator, symmetric rotor (a) and a symmetric stator, asymmetric rotor (b) induction machine*

The torque which develops the symmetric stator with the single phase rotor, shown in Fig. 5.13 (b) is

$$T_{em} = -L_{max,s,r}I_{s,max}I_{r,max}[\cos\omega t \, \sin[(1-s)\omega t + \delta] \cdot \cos(s\omega t - \theta_r) +$$

$$+ \cos\left(\omega t - \frac{2\pi}{3}\right) \sin\left[(1-s)\omega t + \delta - \frac{2\pi}{3}\right] \cdot \cos(s\omega t - \theta_r) +$$

$$+ \cos\left(\omega t - \frac{4\pi}{3}\right) \sin\left[(1-s)\omega t + \delta - \frac{4\pi}{3}\right] \cdot \cos(s\omega t - \theta_r) \,]$$

(5.91)

or

$$T_{em} = -\frac{3}{4}L_{max,s,r}I_{s,max}I_{r,max}[\sin(2s\omega t + \delta - \theta_r) - \sin(\delta - \theta_r)] \quad (5.92)$$

The torque expression in Eq. (5.92) contains a constant term which is proportional to the sine of the difference between the torque angle and rotor phase shift, and the term pulsating with frequency $2s\omega$. The symmetric currents in the stator symmetric windings create only the positive sequence mmf. The alternating current in the single phase rotor windings creates both the positive and the negative sequence mmf. The pulsating torque in Eq. (5.92) is a consequence of the interaction between the rotor negative sequence mmf and the stator positive sequence mmf in the air gap.

At slip s, the rotor electrical speed relative to the stator is $(1-s)\omega$. The speed of the rotor positive sequence relative to the rotor is $s\omega$. The speed of the rotor positive sequence relative to the stator is obtained when its speed relative to the rotor $s\omega$ is added to the rotor speed $(1-s)\omega$ relative to the stator. The result is

$$s\omega + (1-s)\omega = \omega \qquad (5.93)$$

or, the speed of the rotor positive sequence relative to the stator positive sequence is equal to zero, as expected.

The speed of the rotor negative sequence *relative to the rotor* at slip s is $-s\omega$. The speed of the rotor negative sequence *relative to the stator* is

$$-s\omega + (1-s)\omega = (1-2s)\omega \qquad (5.94)$$

The speed of the rotor negative sequence relative to the stator positive sequence is

$$\omega - (1-2s)\omega = 2s\omega \qquad (5.95)$$

This means that the angular frequency of the pulsating torque component in Eq. (5.92) is equal to $2s\omega$. When the rotor slip is equal to s = 0.5, the rotor negative sequence is at standstill relative to the stator.

(b) Cylindrical stator, salient pole rotor. This configuration is characteristic of salient pole synchronous generators, or hydrogenerators. The cross-section of a two pole machine with rotor salient poles is shown in Fig. 5.14. The stator windings carry sinusoidal currents:

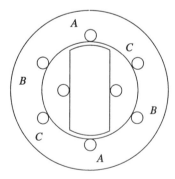

Fig. 5.14 *Cross-section of a synchronous machine with two salient rotor poles*

Force and Torque in Electric Machines

$$I_A = I_{s,\,max} \cos \omega t \tag{5.96}$$

$$I_B = I_{s,\,max} \cos \left(\omega t - \frac{2\pi}{3}\right) \tag{5.97}$$

$$I_C = I_{s,\,max} \cos \left(\omega t - \frac{4\pi}{3}\right) \tag{5.98}$$

and the rotor current is D.C. Besides the pure electromagnetic torque, the stator currents develop a reluctance torque, since the main and mutual inductances of the stator windings are dependent on the rotor angle.

The mutual inductance between the stator and rotor windings is defined in Eq. (4.42). The pure electromagnetic torque is according to Eq. (5.66) equal to

$$T_{em} = -\frac{3}{4} (L_{max,\,s,\,r} + L_{min,\,s,\,r}) I_{s,\,max} I_r \sin \delta \tag{5.99}$$

where $L_{max,s,r}$ is the maximum, and $L_{min,s,r}$ the minimum value of the mutual inductance between the stator and rotor windings.

The reluctance torque caused by the rotor saliency is created by the variation of the main inductance of each stator winding. The dependence of the main inductance of a winding on the rotor angle in a salient pole machine is described in Eq. (4.32). The reluctance torque created by the variation of the main inductances of the stator windings is therefore

$$T_{rel} = -(L_{max,\,s,\,s} - L_{min,\,s,\,s}) I_{s,\,max}^2 \{ \sin 2(\omega t + \delta)(\cos \omega t)^2 +$$

$$+ \sin\left[2(\omega t + \delta) - \frac{2\pi}{3}\right]\left[\cos\left(\omega t - \frac{2\pi}{3}\right)\right]^2 +$$

$$+ \sin\left[2(\omega t + \delta) - \frac{4\pi}{3}\right]\left[\cos\left(\omega t - \frac{4\pi}{3}\right)\right]^2 \} \tag{5.100}$$

or, as shown in the Appendix,

$$T_{rel} = -\frac{3}{4} (L_{max,\,s,\,s} - L_{min,\,s,\,s}) I_{s,\,max}^2 \sin 2\delta \tag{5.101}$$

with $L_{max,s,s}$ denoting the maximum, and $L_{min,s,r}$ the minimum value of the main inductance of each stator winding.

The third component of torque in a salient pole synchronous machine is created by variation of the main inductance between two stator phases due to the rotor saliency. The dependence of the mutual inductances between the stator phases on the rotor shift is given in Eqs. (4.44) – (4.46). The torque created by the variation of the stator mutual inductances is

$$T_{em} = -(L_{max,s,s} - L_{min,s,s}) I_{s,max}^2 \times$$

$$\times \{ \sin\left[2(\omega t + \delta + \frac{\pi}{6})\right] \cos\omega t \left[\cos(\omega t - \frac{2\pi}{3})\right] +$$

$$+ \sin[2(\omega t + \delta)]\left[\cos(\omega t - \frac{2\pi}{3})\right]\left[\cos(\omega t - \frac{4\pi}{3})\right] +$$

$$+ \sin\left[2(\omega t + \delta - \frac{\pi}{6})\right]\left[\cos(\omega t - \frac{4\pi}{3})\right] \cos\omega t \} \quad (5.102)$$

or

$$T_{em} = -\frac{3}{4}(L_{max,s,s} - L_{min,s,s}) I_{s,max}^2 \sin 2\delta \quad (5.103)$$

The electromagnetic torque produced by the variation of the mutual inductance between the stator windings is equal to the reluctance torque caused by variation of the main inductances of the stator windings.

The total torque is equal to the sum of the three components of the torque, given in Eqs (5.99), (5.101), and (5.103)

$$T_{em} = -\frac{3}{4}(L_{max,s,r} + L_{min,s,r}) I_{s,max} I_r \sin\delta -$$

$$-\frac{3}{2}(L_{max,s,s} - L_{min,s,s}) I_{s,max}^2 \sin 2\delta \quad (5.104)$$

When the stator currents in a salient pole synchronous machine are rectangular, as shown in Fig. 5.12 (a), and shifted by $2\pi/3$ radians to each other, the torque function for the pure electromagnetic torque in this example looks as shown in Fig. 5.15 (a). The torque function in the same machine for the reluctance torque is shown in Fig. 5.15 (b).

Force and Torque in Electric Machines 269

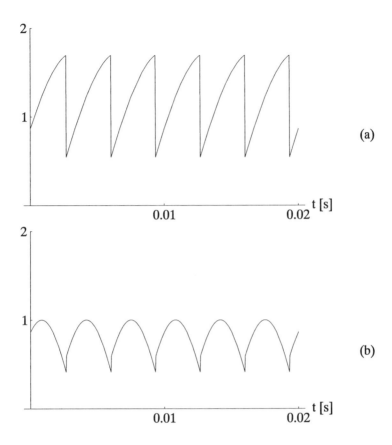

Fig. 5.15 *The torque functions for the pure electromagnetic (a) and reluctance torque (b) at the phase shift of the stator currents $\theta = 2\pi/5$*

Both torque functions in Fig. 5.15 are generated by executing the statements from the Appendix. The pulsating torque components in both the reluctance and the pure electromagnetic torque are created by the higher harmonics components in the stator currents.

(c) Salient pole stator, cylindrical rotor. This configuration is common in classical D.C. machines. The poles of a D.C. machine carry field winding. The cylindrical rotor has slots with the commutator (armature) winding in them, as shown in Fig. 5.16 (a). Neglecting the rotor slotting, the stator winding has constant main inductance. The main inductance of each rotor coil is a function of the double stator to rotor shift. The stator (field) current in a D.C. machine, I_{field}, is constant. The rotor current is alternating, and has a waveform similar to the one shown in Fig. 5.16 (b).

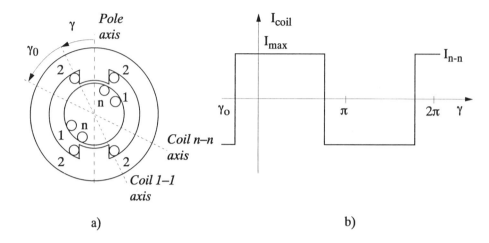

Fig. 5.16 *The cross-section of an electric machine with salient stator poles and a cylindrical rotor (a) and the current in the rotor coil n–n, shifted by γ_0 radians relative to the coil 1–1 (b)*

In a two pole machine the main inductance of the n-th rotor coil, which is shifted by γ_0 radians relative to the first rotor coil, can be described by the constant term and fundamental harmonic as

$$L_n = \frac{L_{max} + L_{min}}{2} + \frac{L_{max} - L_{min}}{2} \cos[2(\gamma + \gamma_0)] \qquad (5.105)$$

whereas the fundamental term of the mutual inductance between the stator winding and the rotor n-th coil is

$$L_{1,n} = L_{12} \cos(\gamma + \gamma_0) \qquad (5.106)$$

The frequency of the rotor currents is proportional to the rotor speed, and the frequency of the stator current is zero. According to the nature of torques in electric machines, a D.C. machine develops *synchronous* torque at every rotor speed. Therefore, a D.C. machine can be considered a synchronous machine with variable frequency of the rotor (armature) currents.

The A.C. currents in the rotor (armature) coils of a D.C. commutator machine contain very rich spectra of odd harmonics. However, only the fundamental harmonic of the rotor coil currents, which is $4/\pi$ times greater than the D.C. current value I_{max}, creates a constant pure electromagnetic torque with the stator field current I_{field}. Higher harmonics in the rotor coil currents result in pulsating torques, the average values of which are equal to zero.

Force and Torque in Electric Machines

If the spatial shift between two adjacent rotor coils is $2\pi/k$, where k is the number of coils, and if the fundamental component of current in the i-th armature coil is equal to

$$I_i = I_{max} \sin(\gamma - \frac{i-1}{k} 2\pi) \tag{5.107}$$

the pure electromagnetic torque created by the fundamental components of the coil currents and the field current is

$$T = -\frac{4}{\pi} L_{12} I_{field} I_{max} \sum_{i=1}^{k} \sin(\gamma - \frac{i-1}{k} 2\pi) \sin(\gamma - \frac{i-1}{k} 2\pi) \tag{5.108}$$

or, as shown in the Appendix

$$T = -\frac{4}{\pi} L_{12} I_{field} I_{max} \frac{k}{2} \tag{5.109}$$

The pure electromagnetic torque in a D.C. commutator machine, created by the fundamental terms of the rotor coil currents, is *constant*. This is illustrated in Fig. 5.17 (a), where the absolute value of the torque function for the case k=9 in Eq. (5.109) is drawn. If all harmonics of the rotor coil currents are allowed for, i.e. if the current is rectangular as in Fig. 5.16 (b), the dependence of the pure electromagnetic torque on the rotor shift γ looks as shown in Fig. 5.17 (b). The torque ripple, which is generated by the higher harmonics in the rotor currents, is superimposed on the constant torque generated by the fundamental current harmonic. The more coils there are on the rotor, the lower the amplitude and the higher the frequency of the torque ripple.

When only the fundamental harmonic of the rotor currents is considered, the torque function is constant. According to Eq. (5.109), the constant term in the torque function for k=9 in this example is equal to

$$\frac{4}{\pi} \frac{9}{2} = 5.73 \tag{5.110}$$

as shown in Fig. 5.17. The reluctance torque created by each rotor coil is proportional to the sine of double the rotor to stator shift. Since at any rotor position there is an equal number of rotor coils on both sides of the pole axis, the sum of all reluctance torques in a commutator machine is always equal to zero.

The *universal commutator motors* are fed from sinusoidal sources, as mentioned in the second chapter. The armature and the field currents are sinusoidal,

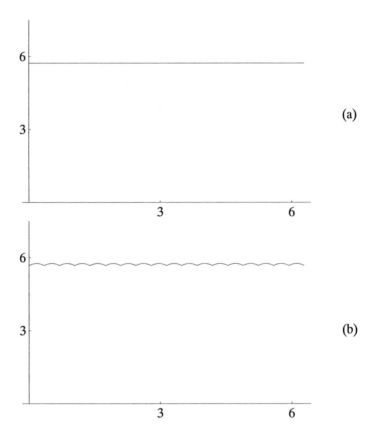

Fig. 5.17 *The torque functions in a D.C. machine considering only the fundamental harmonic of the rotor current (a), and its full spectrum (b). The variable on the x-axis is the rotor to stator shift in radians.*

$$i_a = I_{max} \sin \omega t \qquad (5.111)$$

$$i_f = I_{field} \sin(\omega t - \theta) \qquad (5.112)$$

with θ denoting the phase shift between the field and armature current. The expression for torque created by the fundamental spatial components of the mmfs is obtained by replacing the constant values of the armature and field current in Eq. (5.109) by their instantaneous values, defined in Eqs. (5.111) and (5.112),

$$T = -\frac{4}{\pi}L_{12}I_{field}I_{max}\frac{k}{2}\left[\frac{\cos\theta}{2} - \frac{\cos(2\omega t - \theta)}{2}\right] \qquad (5.113)$$

The torque in a universal commutator motor pulsates with frequency 2ω, where ω is the frequency of the applied voltage.

The speed of rotation of an induction or synchronous motor is firmly related to the frequency of the applied voltages or currents. The frequency of the field and armature current in a universal commutator motor, on the contrary, has no direct influence on its speed of rotation. The average value of torque is proportional to the cosine of the phase shift between the field and armature current, $\cos\theta$. For this reason universal commutator motors are always manufactured as *series motors*, i.e. the field current is equal to the armature current, and $\cos\theta = 1$.

If the field winding is connected parallel to the armature winding, the phase shift between the field and armature current is close to $90°$, due to the almost pure inductive character of the field winding. The constant torque in this case is negligible, and only the alternating torque component exists.

(d) Doubly slotted air gap. Most types of electric machines have both stator and rotor slotted. Very often the effects of double slotting are secondary when compared to those caused by interaction of the windings in a smooth air gap, or in a salient pole geometry. However, there are cases when the interaction between the mmf and slot harmonics cannot be neglected, the most pronounced of them being the generation of higher harmonic torques during starting or reversal of a squirrel cage induction machine.

The main- and mutual inductances in a machine which has slots on both the stator and rotor sides have been defined in the previous chapter. The orders of harmonics of inductances are functions either of the number of stator slots, or of the number of rotor slots, or of the orders of harmonics of the winding functions which produce them. Therefore, one can expect a very rich spectrum of speeds at which the torque dips created by the higher harmonics of the winding functions and numbers of teeth appear.

The cross-sectional view of a machine which has slots and windings on both sides of the air gap is shown in Fig. 5.18. Such a machine generates both pure electromagnetic and reluctance torque.

The higher harmonic components of the main- and mutual inductances create torques with the stator and/or rotor currents at various rotor speeds. The variation of the following inductances causes higher harmonic torques:

(a) Main inductance of one stator winding;
(b) Mutual inductance between two stator windings;
(c) Mutual inductance between a stator and a rotor winding;
(d) Mutual inductance between two rotor windings;
(e) Main inductance of one rotor winding.

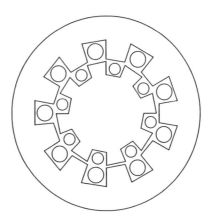

Fig. 5.18 *A doubly slotted electric machine with windings on both sides of the air gap*

The variation of the main inductance of a stator winding which creates the torque in case (a) is described in Eq. (4.70). Substituting

$$x_0 = \frac{D}{2}\gamma_0; \quad \gamma_0 = p\omega_m t \qquad (5.114)$$

one can express the argument of the cosine function in Eq. (4.70) in terms of the rotor mechanical angular speed ω_m as

$$kN_r p\omega_m t \qquad (5.115)$$

The reluctance torque created by the variation of the inductance $L_{s,r,k}$ in Eq. (4.70) is

$$T = \frac{1}{2}i_1^2 \frac{dL_{s,r,k}}{d\gamma_0} \qquad (5.116)$$

If the frequency of the stator current i_1 is ω_1, the torque in Eq. (5.116) has a constant term when the rotor mechanical angular speed ω_m is equal to

$$\omega_m = \frac{2\omega_1}{kN_r p}; \quad k = 1, 3, 5, \ldots \qquad (5.117)$$

Force and Torque in Electric Machines

The mutual inductance between two stator windings $L_{s,s,r,k}$ which varies as a function of the rotor slotting depends on the rotor shift as shown in Eq. (4.70). The pure electromagnetic torque produced by two stator currents i_1 and i_2 which have frequencies ω_1 and ω_2 respectively is

$$T = i_1 i_2 \frac{dL_{s,s,r,k}}{d\gamma_0} \qquad (5.118)$$

The torque in Eq (5.118) contains a constant term when the rotor mechanical angular speed ω_m is equal to

$$\omega_m = \frac{\omega_1 \pm \omega_2}{kN_r p}; \quad k = 1, 3, 5, \ldots \qquad (5.119)$$

Eq (5.119) is also valid when one of the frequencies ω_1 or ω_2 is equal to zero, i.e. when that current is D.C.

The torque between one stator and one rotor winding is caused either by the variation of the mutual inductance $L_{1,2,0}$ in Eq. (4.76), or $L_{1,2,s}$ in Eq. (4.78), or $L_{1,2,r}$ in Eq. (4.81).

The k-th harmonic $L_{1,2,0,k}$ of the inductance $L_{1,2,0}$ generates the pure electromagnetic torque

$$T = i_1 i_2 \frac{dL_{1,2,0,k}}{d\gamma_0} \qquad (5.120)$$

along with the stator current i_1 and rotor current i_2 when the rotor mechanical angular speed ω_m is equal to

$$\omega_m = \frac{\omega_1 \pm \omega_2}{kp}; \quad k = 1, 3, 5, \ldots \qquad (5.121)$$

The k-th harmonic $L_{1,2,s,k}$ of the inductance $L_{1,2,s}$ generates the pure electromagnetic torque

$$T = i_1 i_2 \frac{dL_{1,2,s,k}}{d\gamma_0} \qquad (5.122)$$

along with the stator current i_1 and rotor current i_2 when the rotor mechanical angular speed ω_m is equal to

$$\omega_m = \frac{\omega_1 \pm \omega_2}{kp}; \quad k = N_s \pm 1, N_s \pm 3, ..., 3N_s \pm 1, ... \qquad (5.123)$$

The k-th harmonic $L_{1,2,r,k}$ of the inductance $L_{1,2,r}$ generates the pure electromagnetic torque

$$T = i_1 i_2 \frac{dL_{1,2,r,k}}{d\gamma_0} \qquad (5.124)$$

along with the stator current i_1 and rotor current i_2 when the rotor mechanical angular speed ω_m is equal to

$$\omega_m = \frac{\omega_1 \pm \omega_2}{kp}; \quad k = N_r \pm 1, N_r \pm 3, ..., 3N_r \pm 1, ... \qquad (5.125)$$

The mutual inductance between two stator windings $L_{r,r,s,k}$ which varies as a function of the stator slotting depends on the rotor shift analogously as the mutual inductance $L_{s,s,r,k}$ shown in Eq. (4.66). However, the number of rotor teeth in this equation has to be replaced by the number of stator teeth, in order to obtain the mutual inductance between two rotor windings. The pure electromagnetic torque produced by two rotor currents i_1 and i_2 which have frequencies ω_1 and ω_2 respectively is, therefore,

$$T = i_1 i_2 \frac{dL_{r,r,s,k}}{d\gamma_0} \qquad (5.126)$$

The torque in Eq. (5.126) contains a constant term when the rotor mechanical angular speed ω_m is equal to

$$\omega_m = \frac{\omega_1 \pm \omega_2}{kN_s p}; \quad k = 1, 3, 5, ... \qquad (5.127)$$

Force and Torque in Electric Machines 277

The reluctance torque created by the variation of the rotor inductance $L_{r,s,k}$ in Eq. (4.63) is

$$T = \frac{1}{2}i_1^2 \frac{dL_{r,s,k}}{d\gamma_0} \qquad (5.128)$$

The torque in Eq. (5.128) will have a constant term when the rotor mechanical angular speed ω_m is equal to

$$\omega_m = \frac{2\omega_1}{kN_s p}; \quad k = 1, 3, 5, \ldots \qquad (5.129)$$

5.4 Electromagnetic Torque as a Function of the Air Gap Quantities

The forces and torques in various types of electric machines were derived in the previous sections by utilizing the machines' equivalent circuit parameters: their inductances and currents in their windings. In this section, the electromagnetic torque as a function of the machine's air gap quantities: the current sheet, magnetomotive force and the air gap flux density will be expressed.

Let B_s denote the fundamental term of the component of the air gap flux density created by the stator currents, and A_r the fundamental term of the current sheet produced by the rotor currents

$$B_s = B_{s,\max} \sin\left(\frac{\pi}{\tau_p}x - \alpha_s\right) \qquad A_r = A_{r,\max} \sin\left(\frac{\pi}{\tau_p}x - \alpha_r\right) \qquad (5.130)$$

The component B_r of the air gap flux density created by the rotor currents is

$$B_r = \frac{\mu_0}{\delta} F_r = \frac{\mu_0}{\delta} \int A_r dx =$$

$$= -\frac{\mu_0 \tau_p A_{r,\max}}{\delta \pi} \cos\left(\frac{\pi}{\tau_p}x - \alpha_r\right) = -B_{r,\max} \cos\left(\frac{\pi}{\tau_p}x - \alpha_r\right) \qquad (5.131)$$

The magnetic energy stored in the air gap is, according to Eq. (5.40)

$$W_{mg} = \frac{1\delta}{\mu_0} \int_0^{\tau_p} \left[B_{s,max} \sin\left(\frac{\pi}{\tau_p}x - \alpha_s\right) - B_{r,max} \cos\left(\frac{\pi}{\tau_p}x - \alpha_r\right) \right]^2 dx \qquad (5.132)$$

After integration, Eq. (5.132) becomes

$$W_{mg} = \frac{1\delta}{\mu_0} \left[\frac{\tau_p}{2} (B_{s,max}^2 + B_{r,max}^2) - B_{s,max} B_{r,max} \tau_p \sin(\alpha_r - \alpha_s) \right] \qquad (5.133)$$

The derivative of the magnetic energy with respect to the shift $\alpha_r - \alpha_s$ between the stator component of the air gap flux density B_s and the rotor current sheet A_r gives the torque between the stator and rotor

$$\frac{\partial W_{mg}}{\partial (\alpha_r - \alpha_s)} = -\frac{1\delta}{\mu_0} B_{s,max} B_{r,max} \tau_p \cos(\alpha_r - \alpha_s) \qquad (5.134)$$

or, after substitutions from Eq. (5.131),

$$T = -\frac{\partial W_{mg}}{\partial (\alpha_r - \alpha_s)} = V B_{s,max} A_{r,max} \cos(\alpha_r - \alpha_s) \qquad (5.135)$$

with V denoting the rotor volume. Since the volume of the machine's active part is proportional to the rotor volume, an important conclusion can be derived from Eq. (5.135): the torque developed in an electric machine is proportional to the machine's volume, and its magnetic ($B_{s,max}$) and current ($A_{r,max}$) loading. Keeping the values of $B_{s,max}$ and $A_{r,max}$ constant, which is always true if the machines are built in the same fashion, the torque increases proportionally to the machine's volume, or to the third power of the machine's linear dimensions. Another conclusion, which can be derived from Eq. (5.135), is that for the same machine's volume more torque can be obtained by increasing its specific loadings B and A. Since the air gap flux density is limited by the B–H characteristics of the magnetic material to not more than 1T, the only way to increase the torque in this case is to increase the current sheet. A higher current sheet means more copper losses, which, however, can be taken away from the conductors by utilizing special cooling techniques. This can be illustrated by an example of large synchronous machines, the conductors of which are directly cooled by air, water, or hydrogen. The current sheet in the air gap of such a machine is much higher than in a conventional, indirectly cooled, small or medium machine.

In some kinds of electric machines the air gap flux density created either by the stator or rotor currents is almost rectangularly distributed along the rotor peripheral coordinate, as illustrated in Fig. 5.19 (b). Besides the fundamental, such a rectangular distribution of the flux density contains a rich spectrum of higher spatial harmonics. The useful torque, i.e. the one whose average value is different from zero, is created only by those harmonics which have equal periods. Although the peak value of the sinusoidal flux density distribution in Fig. 5.19 (a) and the maximum value of the rectangular flux density distribution in Fig. 5.19 (b) are equal, the machine with rectangular flux density distribution shown in Fig. 5.19 (b) gives $4/\pi$ more torque. In the latter case, the higher torque is developed because the fundamental term of the rectangular flux density distribution in Fig. 5.19 (b) has an amplitude which is $4/\pi$ times greater than the amplitude B_{max} of the sinusoidal flux density.

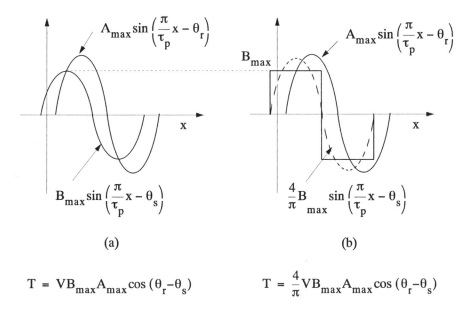

Fig. 5.19 *Air gap quantities in a machine having sinusoidal (a) and rectangular (b) air gap flux density. The maximum values of the flux densities B_{max} are equal in both machines.*

6 Steady State Performance of Induction Machines

The induction motor, particularly with a squirrel cage rotor, is the most widely used source of mechanical power fed from an A.C. power system. Its robustness, reliability, low manufacturing cost, and low sensitivity to disturbances during operation make the squirrel cage motor the first choice when selecting a motor for a particular application.

The stator and rotor fundamental components of the air gap quantities in an induction machine develop electromagnetic torque at all speeds of rotation except at synchronous speed. An induction machine has a self starting capability, so it does not need an auxiliary machine to start it up. Its rotor can be temporarily excessively overloaded, without real danger of damage caused either by too high a current or a torque. An induction machine does not have a separate field winding. Therefore, no additional source to supply the field current is necessary. When an induction machine is loaded, the fundamental component of its air gap flux is not distorted by the load current due to armature reaction. In an induction machine the mechanical load does not introduce a fictitious voltage drop as a consequence of the loss of the air gap flux.

The squirrel cage induction machine has the most robust rotor of all electric machines. In addition, the squirrel cage machine is easy to maintain, because it does not have slip rings or a commutator as its current terminals. The manufacturing technology of induction machines is mature and it does not demand special materials or production procedures. Therefore, the unit price per Nm of an induction machine is the lowest among all types of electric machines.

6.1 Construction and Principles of Operation

The air gap on the stator side of an induction machine is slotted. A three- or two phase winding is placed into the slots. As a rule, the three phase winding is symmetric. When connected to a three phase source, the three phase winding produces a rotating air gap flux density, as elaborated in the second and third chapters. If an induction machine carries a two phase stator winding, the phases are often electrically perpendicular to each other. In order to obtain a rotating air gap field, the currents in the two phases must have equal amplitudes and must be shifted 90° to each other. If the currents are not perpendicular to each other, and/or when their amplitudes are not equal, an elliptical air gap field in a symmetrical two phase machine is generated. The stator windings in a two phase induction machine usually have different numbers of turns. An elliptical air gap mmf in a two phase induction machine is produced either by adding an external impedance in series with one of the phases or by manufacturing one phase in the form of a copper ring. The latter construction is applied in shaded pole induction machines.

The rotor of an induction machine is cylindrical and slotted. The slots carry the rotor winding, which can be either squirrel cage, or wound. The squirrel cage winding described in the second chapter contains as many bars as there are rotor slots, and two rings, one on each side of the rotor. The squirrel cage of large and medium size induction machines is made of copper, or of brass combined with copper. The rotor slots in such machines are never closed. The squirrel cage of small machines is often made of die-cast aluminum, in order to reduce costs and simplify manufacturing. The rotor slots in these machines are almost always closed to withstand the high pressure of liquid aluminum during die casting.

A wound rotor electric machine carries a regular, three phase symmetrical winding on the rotor side. The ends of the three phase winding are connected to the slip rings, against which the graphite brushes on the stator side are pressed. The brushes along with the slip rings assure the contact of the rotor windings to an external circuit during rotation.

When connected to a symmetrical A.C. source, the stator winding in an induction machine creates the air gap flux in which the positive sequence component running at the synchronous speed of the fundamental dominates. The rotating air gap flux induces voltages in conductors of the rotor winding. When the rotor winding is closed, the induced voltages drive currents which, along with the air gap flux, create the electromagnetic torque between the stator and rotor, as described in the fifth chapter. The electromagnetic torque accelerates the rotor, which starts to follow the positive sequence of the fundamental component of the air gap flux. As the rotor speed increases, the relative speed between the positive sequence of the fundamental component of the air gap flux and the rotor conductors decreases. Therefore, the frequency and the amplitude of the induced voltage in the rotor winding decreases. The lower induced voltages drive fewer currents through the rotor windings and, consequently, produce less torque. When the rotor speed becomes equal to the speed of the fundamental component of the air gap flux, the fundamental component of the air gap flux does not induce voltage in the rotor conductors, and the electromagnetic torque at this operating point is equal to zero.

6.2 Effects of the Fundamental Spatial and Time Air Gap Flux Harmonics in an Induction Machine

As any other electric machine, an induction machine is built to convert electrical energy to mechanical energy, and vice versa, at the frequency of the fundamental time and spatial harmonics of the fluxes, mmfs, currents, voltages, etc. However, the slotting on both sides of the air gap, along with the discrete winding distributions, are the sources of the higher spatial harmonics of the flux density in the air gap of an induction machine. Each of these harmonics creates its own torque and losses. The higher harmonic torques are particularly emphasized during the start up of an induction machine. The amplitudes of the higher harmonic torques can be large enough to cause the rotor to stall at speeds which are much lower than the rated speed. Because of their importance, the higher harmonic torques will be analyzed separately from the asynchronous torque created by the fundamental time and spatial components of the stator and rotor currents and fluxes.

6.2.1 Equivalent Circuit for the Fundamental Spatial Harmonic

A symmetrical induction machine has an m_1-phase symmetrical stator winding fed from a symmetrical source, and an m_2-phase symmetrical rotor winding. At steady states, when the rotor speed is constant, the amplitudes of the currents in the stator phases are equal. The amplitudes of the rotor currents at steady states are also equal. Therefore, the state in a machine can be completely described by only one stator and only one rotor phase current. These two currents can be visualized to flow through the stator and rotor branches of the machine's equivalent circuit, introduced in the fourth chapter. The parameters of the equivalent circuit of an induction machine are usually referred to one stator phase, as shown in Fig. 4.24 (a). The equivalent rotor current I'_r in this figure is fictitious, because its amplitude is referred to the parameters of one stator phase winding, and its frequency is equal to the stator frequency. Nevertheless, the electric (losses) and magnetic (torque) effects of the current I'_r are equal to the ones produced by the actual rotor phase current I_r, the frequency of which varies as a function of the rotor speed.

The fundamental component of the rotor current I_r is driven by the fundamental component of the voltage E_r induced in the rotor phase winding which links the air gap flux $\Phi_{g,Max,1}$. The induced voltage E_r is given in Eq. (4.132) as

$$E_r = 4.44 \Phi_{g,Max,1} f_r w_r f_{w,1,r} \qquad (6.1)$$

where f_r is the frequency of the fundamental time harmonic of the rotor voltage, w_r is the number of turns in each rotor phase and $f_{w,1,r}$ is the rotor winding factor for the

Steady State Performance of Induction Machines

fundamental spatial component. By utilizing the normalized relative speed between the rotor and air gap mmf, or slip s as defined in the second chapter, one can express the frequency of the fundamental rotor time harmonic f_r as a function of the fundamental stator time harmonic frequency f_s

$$f_r = sf_s \tag{6.2}$$

Denoting by $E_{r,0}$ the induced voltage in a rotor phase at standstill, i.e. for a slip equal to one

$$E_{r,0} = 4.44 \Phi_{g,Max,1} f_s w_r f_{w,1,r} \tag{6.3}$$

one can express the induced voltage E_r as

$$E_r = sE_{r,0} \tag{6.4}$$

The amplitude of the rotor induced voltage is proportional to the slip.

Since the rotor frequency is proportional to the slip, all rotor reactances are also functions of the slip. Denoting by $X_{\sigma,r}$ the rotor leakage reactance per phase calculated at stator frequency f_s, one can express the rotor leakage reactance per phase $X_{\sigma,r,s}$ at slip s as

$$X_{\sigma,r,s} = sX_{\sigma,r} \tag{6.5}$$

The R.M.S. value I_r of the rotor current is equal to the quotient between the applied voltage and the impedance per phase

$$I_r = \frac{E_r}{\sqrt{X^2_{\sigma,r,s} + R^2_r}} = \frac{sE_{r,0}}{\sqrt{(sX_{\sigma,r})^2 + R^2_r}} = \frac{E_{r,0}}{\sqrt{X^2_{\sigma,r} + \left(\frac{R_r}{s}\right)^2}} \tag{6.6}$$

with R_r denoting the rotor resistance per phase. A simultaneous variation of the frequency of the induced voltage in the rotor phase and the rotor reactance have the same effect on the rotor phase current I_r as if the rotor resistance were fictitiously divided by the slip, keeping the rotor frequency constant.

The equivalent resistance per phase R_r/s can be represented as the sum of the rotor resistance per phase R_r and a fictitious resistance $R_r(1-s)/s$

$$\frac{R_r}{s} = R_r + R_r \frac{1-s}{s} \tag{6.7}$$

The rotor induced voltage per phase reflected to one stator phase E'_r is equal to the stator induced voltage per phase E_s, because

$$E'_r = \frac{w_s f_{w,s}}{w_r f_{w,r}} E_r = E_s \tag{6.8}$$

Allowing for the previous considerations, one can draw an equivalent circuit per stator phase of an induction machine for the fundamental spatial harmonic as shown in Fig. 6.1. The applied voltage V in this figure is the stator phase voltage and the current I_s is the stator phase current. The current I_s has two components: the no load current I_0 and the reflected rotor current I'_r. The no load current I_0 has two components: the magnetizing current I_m, and the current I_{Fe}, which carries the power from the source to cover the iron core losses. The magnetizing current I_m builds the air gap flux, the measure of which is the main reactance X_m. The air gap flux in an induction machine builds a current which cannot be localized to a particular winding, as can be done in a commutator, or a synchronous machine. The magnetizing current in an induction machine is just a component of the total stator current which flows from the power system to the machine. The current I_{Fe} dissipates, on the fictitious resistor R_{Fe}, the losses which, multiplied by the number of stator phases, give the iron core losses P_{Fe}. The resistor R_{Fe} is connected in parallel to the main reactance. The voltage drop across the main reactance and resistor R_{Fe} is equal to E_s because the iron core losses P_{Fe} are proportional to the square of the main flux magnitude, which is proportional to the amplitude of the induced voltage E_s.

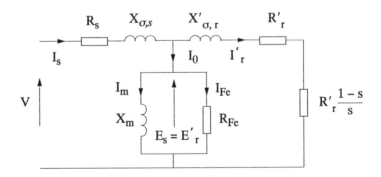

Fig. 6.1 *Equivalent circuit per one stator phase for the fundamental spatial harmonic of an induction machine*

6.2.2 Distribution of Power and Torque–Slip Characteristic

The expression for the electromagnetic torque developed in a symmetrical induction machine having three stator and three rotor phases, and being fed from symmetrical three phase sinusoidal sources, is derived in the fifth chapter and given in Eq. (5.77). This equation is derived by differentiating the magnetic energy stored in the air gap of the machine with respect to the rotor shift. By multiplying Eq. (5.77) by the electrical speed of rotation of the air gap field ω_s, i.e. by the synchronous speed, one obtains the *air gap power* P_{ag} defined as

$$P_{ag} = \omega_s T_{em} = \frac{9}{4} \omega_s L_{max,s,r} I_{s,max} I_{r,max} \sin\theta_r \qquad (6.9)$$

with θ_r denoting the angle between the stator and rotor currents in corresponding phases, and, therefore, also in the equivalent circuit. The air gap power is the power which is transferred from the stator to the rotor of an induction machine through its air gap. The induction machine, the electromagnetic torque of which is given in Eq. (5.77), has three stator and three rotor phases. It can be easily proved for any number m_1 of the stator and m_2 of the rotor phases that the air gap power is equal to

$$P_{ag} = \omega_s m_2 \frac{m_1}{2} L_{max,s,r} \frac{I_{s,max}}{\sqrt{2}} \frac{I_{r,max}}{\sqrt{2}} \sin\theta_r \qquad (6.10)$$

or, after substituting the expression for the main inductance and the corresponding main reactance X_m in a symmetrical m_1 phase machine, derived in the fourth chapter

$$P_{ag} = m_2 X_m I_s I_r \sin\theta_r \qquad (6.11)$$

where I_s and I_r are the R.M.S. values of the stator and rotor currents, respectively. Let us assume, for simplicity, an equal number of stator and rotor phases, an equal number of turns in the phases and an equal winding distribution in the stator and rotor. The reflected rotor current I'_r in this case is equal to the rotor phase current I_r. The no load current I_0 can be further written as

$$I_0 = I_s - I_r \qquad (6.12)$$

Utilizing the notation in Fig. 6.1 and the relationship between the stator and rotor current, given in Eq. (6.12), one can write

$$(I_s - I_r e^{-j\theta_r}) jX_m = I_r e^{-j\theta_r} (\frac{R_r}{s} + jX_r) \qquad (6.13)$$

from which one can write for the R.M.S. values of the currents I_s and I_r

$$I_s X_m = I_r \sqrt{(\frac{R_r}{s})^2 + (X_r + X_m)^2} \qquad (6.14)$$

The sine of the angle θ_r between the currents I_s and I_r is

$$\sin\theta_r = \frac{\frac{R_r}{s}}{\sqrt{(\frac{R_r}{s})^2 + (X_r + X_m)^2}} \qquad (6.15)$$

Substituting Eqs. (6.14) and (6.15) into (6.11), one obtains for the air gap power

$$P_{ag} = m_2 I_r^2 \frac{R_r}{s} = m_1 I_r'^2 \frac{R_r'}{s} \qquad (6.16)$$

because the electric power does not change when reflecting quantities from one winding to another.

The expression (6.16) relates the power which passes through the air gap of an induction machine to the parameters of the equivalent circuit in Fig. 6.1. If the equivalent resistance per phase R_r/s is substituted by the sum of the actual resistance per phase R_r and the fictitious resistance $R_r(1-s)/s$, as shown in Eq. (6.7), the air gap power P_{ag} in Eq. (6.16) can be written as

$$P_{ag} = m_1 I_r'^2 R_r' + m_1 I_r'^2 R_r' \frac{(1-s)}{s} \qquad (6.17)$$

The air gap power P_{ag} has two components: the current losses in the rotor circuit $P_{r,loss}$

$$P_{r,loss} = m_1 I_r'^2 R_r' \qquad (6.18)$$

and the *mechanical power* P_{mech}

$$P_{mech} = m_1 I'^2_r R'_r \frac{(1-s)}{s} \quad (6.19)$$

Knowing the air gap power, one can express the rotor current losses by utilizing Eqs. (6.16) and (6.18)

$$P_{r,loss} = s P_{ag} \quad (6.20)$$

The mechanical power as a function of the air gap power is obtained by substituting Eq. (6.16) into (6.19)

$$P_{mech} = (1-s) P_{ag} \quad (6.21)$$

The relationship between the power components which appear in an induction motor can be visualized as shown in Fig. 6.2.
The input electrical power which the motor takes from the source is equal to

$$P_1 = m_1 U I_s \cos\varphi \quad (6.22)$$

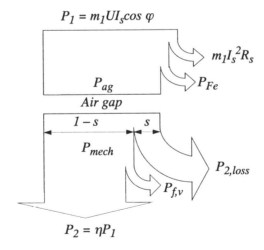

Fig. 6.2 *The distribution of power in an induction motor*

with φ denoting the angle between the applied phase voltage and stator current I_s. A portion of the input power P_1 is converted into the stator winding copper losses, and its other portion covers the stator iron core losses P_{Fe}. The rest of the input power is supplied to the rotor as the air gap power P_{ag}. The air gap power is divided into the rotor current losses and the mechanical power, as a function of the rotor slip s. When the friction and windage losses $P_{f,v}$ are subtracted from the mechanical power P_{mech}, the power on the shaft P_2 is obtained. The ratio between the output power on the shaft and input electrical power of an electric motor is called the *efficiency*

$$\eta = \frac{P_2}{P_1} \qquad (6.23)$$

The properties of an electric machine as an electromechanical energy converter are described by its *external characteristic*. The external characteristic of an electric motor is its torque–slip curve at a constant applied voltage for voltage fed machines or at a constant current, for current fed machines. The external characteristic of an electric generator is the dependence of its terminal voltage on the load current at a constant speed.

Voltage fed machine: The torque–slip characteristic of a voltage fed induction motor can be obtained from Eq. (6.19) divided by the angular speed of rotation ω

$$T = \frac{P_{mech}}{\omega} = m_1 I_r'^2 R_r' \frac{(1-s)}{s\omega} = m_1 I_r'^2 R_r' \frac{1}{s\omega_s} \qquad (6.24)$$

with ω_s denoting the synchronous angular speed of rotation.

Introducing the following impedances associated with the equivalent circuit in Fig. 6.1:

$$Z_s = R_s + jX_{\sigma,s}; \qquad Z_0 = j\frac{R_{Fe}X_m}{R_{Fe}+jX_m}; \qquad Z_r' = \frac{R_r'}{s} + jX_{\sigma,r}'; \qquad (6.25)$$

one can express the stator current I_s as

$$I_s = I_0 + I_r' = \frac{E_s}{Z_0} + \frac{E_s}{Z_r'} \qquad (6.26)$$

The applied voltage V in Fig. 6.1 is now

$$V = I_1 Z_1 + E_s = (I_0 + I'_r) Z_1 + E_s = E_s \left[1 + \frac{Z_1}{Z_0} + \frac{Z_1}{Z'_r} \right] \quad (6.27)$$

The magnetizing reactance X_m of an induction machine is usually at least an order of magnitude greater than the stator leakage reactance $X_{\sigma,s}$. The ratio between the two is a function of the number of poles and the air gap width. The more poles the induction machine has, the smaller the area of one pole, and therefore, the higher the magnetizing current and the lower the magnetizing reactance. The magnetizing reactance of high speed machines can reach values which are two orders of magnitude greater than the stator leakage reactance. In low speed machines, however, the difference between the two is smaller. Therefore, the magnitude of the ratio between the impedances Z_1 and Z_0 in Eq. (6.27) is usually very small, as is its phase shift. Introducing a complex constant Σ, defined as

$$\Sigma = 1 + \frac{Z_1}{Z_0} \quad (6.28)$$

one can write

$$V = E_s \left(\Sigma + \frac{Z_1}{Z'_r} \right) \quad (6.29)$$

from which

$$\frac{E_s}{V} = \frac{Z'_r}{Z'_r \Sigma + Z_1} \quad (6.30)$$

The dependence of the amplitude and the phase of the ratio between the induced and applied voltage for the induction machine, the equivalent circuit data of which are given in the Appendix, is shown in Fig. 6.3 (a). As shown in Fig. 6.3 (a), the induced voltage E_s at start up falls to approximately one half of its rated value. Since the main flux is proportional to the induced voltage, it is also reduced to about 50% of its rated value. The influence of the main reactance on the starting current is negligible. The starting current can be considered dependent only on the stator and rotor leakage reactances and resistances.

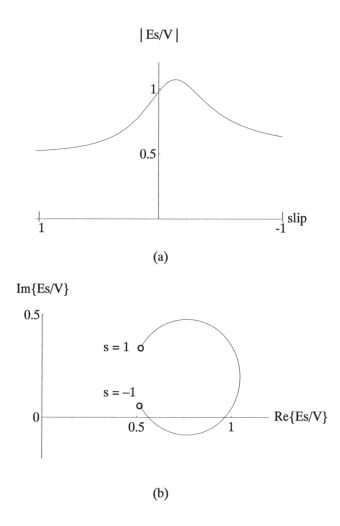

Fig. 6.3 *The magnitude (a) and phase (b) of the ratio between the induced and applied voltage of an induction machine*

According to the voltage distribution in the machine's equivalent circuit in Fig. 6.1, at start up the induced voltage E_s is proportional to the voltage drop across the rotor reactance. The value of the rotor reactance is close to the value of the stator reactance; hence, the short-circuit current makes the voltage drop across the rotor reactance which is equal to approximately one half of the applied voltage.

The ratio between the induced and applied voltages, with the rotor slip as a parameter, is shown in a polar diagram in Fig. 6.3 (b). One can see in this figure a characteristic circular form as a function of the slip. This property of the machine's impedance is the basis for the circular diagram of an induction machine. In the past, the circular diagram was a frequently used tool when determining the electromagnetic properties of a machine. However, with the advent of modern computers, more accurate methods of machine analysis became available, so that a circular diagram of an induction machine is only of historic importance today.

The rotor current is equal to

$$I'_r = \frac{E_s}{Z'_r} = \frac{V}{Z'_r \Sigma + Z_1} = \frac{V}{(R_s + jX_{\sigma,s}) + (\frac{R'_r}{s} + jX'_{\sigma,r})\Sigma} \quad (6.31)$$

The variation of the absolute value of the rotor current as a function of the rotor slip in the range $-1 < s < 2$ is shown in Fig. 6.4. At slip 1 a voltage fed squirrel cage induction machine draws a stator current from a source which is 4–7 times greater than the rated current. The magnitude of the reduced rotor current at zero speed is almost equal to the magnitude of the stator current, since the magnetizing current at this operating point is negligible when compared to each of them. As the rotor accelerates, the rotor current decreases and falls to zero at synchronism. For negative slip values the rotor current again increases, quickly reaching its blocked rotor value.

The stator phase current is obtained by utilizing Eq. (6.26). The variation of the absolute value and the real value of the stator current in the slip range from 2 to –1 is shown in Fig. 6.5 (a) and (b).

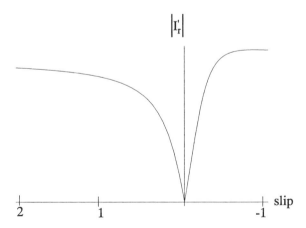

Fig. 6.4 *Absolute value of the rotor current reduced to a stator phase as a function of the slip*

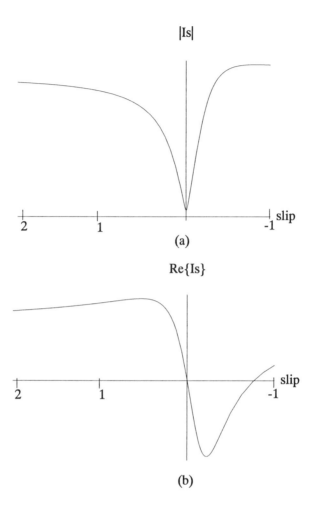

Fig. 6.5 *The absolute value (a), and real component (b) of the stator phase current as a function of the rotor slip*

Except at a low slip, the absolute value of the stator current is almost equal to the absolute value of the rotor current, as shown in Fig. 6.4. At low slip values the rotor current becomes very small, whereas the stator current almost equals the no load current I_0. At synchronism, when the slip is equal to zero, the rotor impedance for the fundamental becomes infinitely high. Therefore, the rotor current is equal to zero at this point, and the stator draws the no load current.

The real component $Re\{I_s\}$ of the stator current is positive in the slip range from

Steady State Performance of Induction Machines

+∞ to 0. A positive real component of the stator current is in phase with the applied voltage, which means that the machine *draws* the electrical power from the source. For negative slip values around zero the real component of the stator current becomes negative as a consequence of the phase shift of 180° between the vector of the applied voltage and the vector of the stator current. The machine operates as a generator in this slip range, and *delivers* the electric power to the source. As the slip decreases, the real component of the stator current again becomes positive, as shown in Fig. 6.5 (b). Although the slip is negative, the electric power generated in the machine is not delivered to the source because it is less than the generator losses. Besides the converted mechanical power coming from the shaft, the generator in this slip range takes additional power from the source in order to cover its own copper and iron core losses.

The negative slip at which the real component of the stator current becomes positive, or at which the generator stops to deliver the electric power to the power system, can be simply estimated by neglecting the iron core and mechanical losses in the machine. In this case the mechanical power from the shaft is equal to the sum of the stator and rotor copper losses, or

$$I'^2_r R'_r \frac{1-s_1}{s_1} + I'^2_r R'_r + I_s^2 R_s = 0 \qquad (6.32)$$

Neglecting the no load component of the stator current, one obtains from Eq. (6.32) for the slip s_1 at which the mechanical power on the shaft is equal to the losses in the generator

$$s_1 = - \frac{R'_r}{R_s} \qquad (6.33)$$

For a slip below s_1 the machine takes the electrical power from the terminals and mechanical power from the shaft and converts both into losses in the stator and rotor windings.

Considering that the phase shift of the complex coefficient Σ defined in Eq. (6.28) is equal to zero, one can express the square of the rotor current as

$$I'^2_r = \frac{V^2}{(R_s + \frac{R'_r}{s}\Sigma)^2 + (X_{\sigma,s} + X'_{\sigma,r}\Sigma)^2} \qquad (6.34)$$

Substituting for the square of the rotor current from Eq. (6.34) into the torque equation (6.24), one obtains the expression for the torque of a voltage fed induction machine

$$T = m_1 \frac{V^2}{\omega_s \left[\left(R_s + \frac{R'_r}{s} \Sigma \right)^2 + (X_{\sigma,s} + X'_{\sigma,r} \Sigma)^2 \right]} \frac{R'_r}{s} \qquad (6.35)$$

The qualitative shape of the torque–slip characteristic of the machine, the equivalent circuit data of which are given in the Appendix, is shown in Fig. 6.6.

As any other electric machine, the induction machine can operate as a *motor*, *generator*, or a *brake*. In the motor mode of operation, the input electric power from the machine's terminals is converted into mechanical power on the shaft and the machine's losses. In the generator mode of operation, the input mechanical power on the shaft is converted into electrical power on the machine's terminals and the machine's losses. When an electric machine operates as a brake, both the input mechanical power from the shaft and the electrical power from the machine's terminals are converted into losses in the machine, or in the machine and external resistors. The operating mode of an induction machine is determined by its slip.

An induction machine acts as a brake for slips smaller than $-R_r'/R_s$, and greater than one. When the slip is greater than one, the rotor rotates in the direction opposite to that of the air gap field. One possibility for turning the rotor in a direction opposite to that of the air gap field is to exchange the terminals of any two phases of a three phase machine.

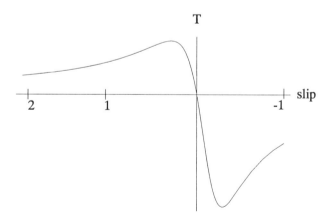

Fig. 6.6 *The torque of an induction machine as a function of the rotor slip*

Steady State Performance of Induction Machines

Another way to obtain a slip greater than one is to add a huge resistance to the rotor circuit of a wound rotor induction machine to the shaft of which a source of external mechanical power is connected. The induction machine in this case must be connected to oppose the motion of the external mechanical source.

An induction machine acts as a motor in the slip interval between one and zero. The usual values of the rated slip of an induction motor are a few percent. The rotor rated speed should be as close to the synchronous speed as possible, in order to decrease the losses in the rotor windings, and, therefore, to better utilize the material from which the motor is made. The torque–slip characteristic of an induction motor is almost linear for slip values between zero and the rated slip.

An induction machine operates as a generator for a negative slip between R_r'/R_s and zero. In this slip range the rotor speed is greater than the speed of the air gap field. In the generator mode of operation an induction machine takes the mechanical power from the shaft and converts it to electrical power. When an induction generator is connected to a power system, it delivers active power to it but draws reactive power from the power system. Since the field current of an induction generator, i.e. the current which builds the air gap flux, is just a component of the total stator current, it has to be supplied to the generator through its stator terminals. Therefore, when an induction generator feeds its own consumers independently of a power system, a reactive component of the stator current has to be supplied to it in order to build the air gap flux. This is usually performed by connecting capacitors parallel to the generator terminals.

The torque of an induction machine is proportional to the square of the applied voltage. It is different from zero at all speeds except at synchronous, when the fundamental harmonic of the air gap flux does not induce voltage in the rotor winding. The torque T_{st} which the machine develops at standstill is called the *starting torque*. It is obtained from Eq. (6.35) by substituting for s = 1

$$T_{st} = m_1 \frac{V^2}{\omega_s \left[(R_s + R_r'\Sigma)^2 + (X_{\sigma,s} + X_{\sigma,r}'\Sigma)^2 \right]} R_r' = \frac{m_1 I_{st}^2 R_r'}{\omega_s} \quad (6.36)$$

where I_{st} is the motor starting current. The starting torque of an induction motor is proportional to the rotor resistance, as long as the pull out slip is less than one. The property of the starting torque of an induction machine to be proportional to the rotor resistance is used when the starting torque of an already manufactured squirrel cage machine is too low. In this case, the rings of the squirrel cage machine are machined so that their cross-sectional area is decreased. This way the ring resistance and, therefore, the complete rotor resistance is increased. The price for such a modification is an increased rated slip of the machine, and, therefore, its deteriorated efficiency.

Another means of obtaining high values of starting torque in a squirrel cage machine is to utilize the skin effect in the rotor conductors. As shown in the first chapter, an A.C. current in a conductor which is placed into a slot of an electric machine is shifted towards the slot opening. Therefore, the conductor's resistance is increased with respect to its D.C. value, whereas its leakage reactance is decreased. The change of the rotor bar's electric parameters is as strong as the rotor conductors are high, and as the rotor frequency is high. The skin effect is observable in machines with a rotor conductor height which is greater than the critical height. Since the frequency of the rotor current at standstill is equal to the stator frequency, and at the rated point only a few Hz, the skin effect is strongest during the motor's start up, and, at the rated slip, it is negligible.

Typical shapes of rotor slots with conductors in which the skin effect occurs are shown in Fig. 6.7.

The shapes of the rotor slots shown in Fig. 6.7 (a) and (b) are the most common in squirrel cage induction machines, especially in medium and large size ones. The trapezoidal form in Fig. 6.7 (b) has better thermal properties than the rectangular one in Fig. 6.7 (a), because here the heat transfer from the bar to the teeth on its left and right sides is extremely good. Good contact between the bar and slot sides is ensured by the centrifugal force which acts on the bar. The bars in the rotor slot in Fig. 6.7 (c) are often made of different material: the lower bars of copper, and the upper ones of brass. At standstill, the rotor current flows mostly through the upper bars. The resistance of the upper bars is high, since the specific resistance of brass is high, and the cross-section of the upper bars is small. At the rated operating point the current flows mostly through the lower copper bars. The slot forms in Fig. 6.7 (d) and (e) are used in machines in which the rotor yoke flux density is high, and the rotor bars have yet to be higher than the allowed teeth height. This can sometimes happen in high speed machines. The effective bar height in Fig. 6.7 (e), which is a decisive parameter for the skin effect, is equal to the sum of the heights of two parts of the bar.

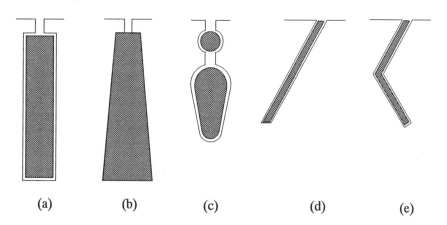

Fig. 6.7 *The rotor slot shapes for utilizing the skin effect*

Steady State Performance of Induction Machines

The increase of the rotor resistance due to the skin effect has two advantageous consequences: a higher starting torque, and a lower starting current. This is illustrated in Fig. 6.8, in which the solid line represents the torque–slip characteristics of a machine in which the skin effect is negligible, and the dashed line represents the torque–slip characteristics in a machine in the rotor of which the skin effect occurs. The starting torque of a machine in which the skin effect is utilized is denoted in Fig. 6.8 by $T_{st,skin}$, and of a machine with a negligible skin effect by T_{st}.

At slip one, the frequency of the rotor currents is equal to the frequency of the stator currents. Therefore, the rotor resistance is fictitiously increased, and inductance is fictitiously decreased relative to their D.C. values. The corresponding pull-out slip, being proportional to the ratio between the rotor resistance and inductance, is, therefore, relatively high on starting, which means that the maximum torque is shifted towards the left in Fig. 6.8. Since the pull-out slip is supposedly less than one, the starting torque must accordingly be relatively large. As the rotor speed increases, the frequency in it decreases and the skin effect becomes less influential. The pull out slip decreases which means that the corresponding starting torque also decreases. The dashed line in Fig. 6.8 should in this sense be considered an envelope of the motor torque–slip curves, each of them being defined at one rotor speed by its current resistance and inductance values.

The grade of utilization of skin effect depends on the rotor bar geometry and cage construction. This is illustrated in Fig. 6.9, in which the torque–slip characteristics of induction machines with various rotor bar geometries are shown for the same current–slip characteristic. The lowest starting torque has a motor with a wound rotor, in which the influence of the skin effect is not observable. A slight deviation from the torque–slip characteristic of a wound rotor motor can be seen in a squirrel cage motor with round rotor bars. If the rotor bars' geometry is similar to that shown in Fig. 6.7 (a), (b), (d), or (e), the starting torque in a such machine is significantly higher than in a machine with round bars.

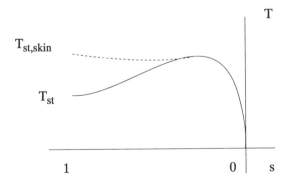

Fig. 6.8 *Torque–slip characteristic of a machine without skin effect (solid line), and of a machine in which the skin effect in the rotor conductors is utilized (dashed line)*

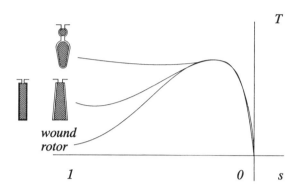

Fig. 6.9 *The torque–slip characteristics of machines with various rotor slot shapes. The starting current is equal in all three cases.*

The highest starting torque is developed in a machine with a double cage, the external bars of which are made of brass, and the internal bars of copper.

The torque–slip curve of an induction machine described in Eq. (6.35) has two extremes: the maximum torque, when the machine operates as a motor, and the minimum torque, when the machine operates as a generator. The extreme torques are obtained by differentiating Eq. (6.35) with respect to the slip, and solving the equation obtained for the slip. In this way the so-called *pull out* slip is obtained

$$s_p = \pm \frac{R'_r \Sigma}{\sqrt{R_s^2 + (X_{\sigma,s} + X'_{\sigma,r}\Sigma)^2}} \qquad (6.37)$$

The positive sign of the pull out slip in Eq. (6.37) is related to the motor maximum (pull out) torque, and the negative sign is related to the generator minimum torque. The generator pull out slip lies between zero and $-R_r'/R_s$, which is the limit slip for the generator mode of operation.

Extreme values of torque are obtained by substituting the values of the pull out slip from Eq. (6.37) into the torque equation (6.35)

$$T_m = m_1 \frac{V^2}{2\omega_s \Sigma \left[\sqrt{R_s^2 + (X_{\sigma,s} + X'_{\sigma,r}\Sigma)^2} \pm R_s\right]} \qquad (6.38)$$

The maximum torque of an induction motor is inversely proportional to its rotor leakage reactance $X'_{\sigma,r}$ and independent of the rotor resistance. If the maximum torque of an already manufactured squirrel cage induction motor has to be increased, its rotor slot openings have to be made wider. This way, its rotor leakage reactance is decreased, and the maximum torque is increased. However, the price for such an action is an increase in the starting current.

The positive sign of the stator resistance in Eq. (6.38) is related to the motor mode of operation, and the negative sign stands for the generator mode. The absolute value of the motor maximum torque is less than the absolute value of the generator minimum torque. This can be explained by the property of an induction machine having the torque proportional to the square of the applied voltage and, therefore, approximately, to the square of the air gap flux. As shown in Fig. 6.3 (a), the induced voltage for the negative pull out slip is greater than the induced voltage for the positive pull out slip, if the applied voltage is kept constant. Therefore, the absolute value of the minimum generator torque must be greater than the absolute value of the maximum motor torque.

Introducing the coefficient β, defined as

$$\beta = \frac{2R_s}{R'_r \Sigma} \qquad (6.39)$$

and dividing Eq. (6.35) by Eq. (6.37), one obtains the *Kloss* equation

$$\frac{T}{T_m} = \frac{2 + \beta s_p}{\frac{s}{s_p} + \frac{s_p}{s} + \beta s_p} \qquad (6.40)$$

The Kloss equation allows comparison between the performance of induction machines which have various values of rated power (torque) and synchronous speed, because both the speed and torque in a Kloss equation are expressed in their relative values.

Current fed machine: The torque–slip characteristic of a constant current fed induction machine is different from the torque–slip characteristic of a constant voltage fed machine. In a voltage fed machine, the magnetizing current creates an air gap flux the amplitude of which is firmly fixed to the amplitude of the applied voltage. In a current fed induction machine, however, the total current is not defined by the rotor speed, but by the source. The stator impedance has no influence on the current amplitude in a current fed machine – it only determines the values of terminal voltages. The current which is pumped from the source into the machine becomes divided between the magnetizing and rotor branches in the machine's equivalent circuit. For low slip values, the rotor impedance is very high, so that the stator current is mostly magnetizing. A high magnetizing current causes a high air gap flux and, therefore, high iron core losses. For this reason the motor current at low slip values has to be limited.

When the rotor current is expressed as a function of the stator current, as shown in Eq. (6.14), and substituted into Eq. (6.16), the torque–slip equation of a current fed induction machine is obtained

$$T = m_1 \frac{I_s^2 X_m^2}{\omega_s} \frac{\frac{R'_r}{s}}{(\frac{R'_r}{s})^2 + (X'_r + X_m)^2} \qquad (6.41)$$

The torque–slip characteristic of the machine, the parameters of which are used in this section, is shown in Fig. 6.10.

At rotor speeds close to the synchronous, almost complete stator current builds the main flux. Therefore, the torque–slip characteristic in this slip range is extremely stiff, because the rotor speed hardly changes as the torque on the shaft increases. For higher values of slip, the fictitious rotor resistance R_r'/s decreases, and a part of the stator current flows through the branch of the equivalent circuit which represents the rotor.

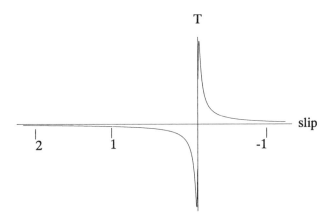

Fig. 6.10 *The torque–slip characteristic of a current fed induction machine*

Steady State Performance of Induction Machines

The pull out slip is obtained by differentiating Eq. (6.41) with respect to the slip. The expression obtained is set equal to zero, and then solved for the slip. This gives

$$s_p = \frac{R'_r}{X'_r + X_m} \quad (6.42)$$

The main inductance X_m in the denominator of expression (6.42) is the reason why the pull out slip of a current fed machine is significantly smaller than the pull out slip of a voltage fed machine. The lower the pull out slip, the stiffer the torque–slip characteristic of an induction machine.

The maximum torque of a current fed machine is obtained by substituting the value of the pull out slip from Eq. (6.42) into the torque–slip equation (6.41)

$$T_m = m_1 \frac{I_s^2 X_m^2}{2\omega_s (X_m + X'_{\sigma,r})} \quad (6.43)$$

Neglecting the rotor leakage reactance, the maximum torque is proportional to the value of the main reactance. According to the results of the no load characteristic computation performed in the fourth chapter, the main reactance is not constant: at pull out slip the main reactance is due to a saturation lower than at slip which is greater than pull out slip.

Dividing Eq. (6.41) by (6.43), one again obtains the Kloss equation, but this time without the correction factor β for influence on the stator impedance

$$\frac{T}{T_m} = \frac{2}{\frac{s}{s_p} + \frac{s_p}{s}} \quad (6.44)$$

The pull out slip s_p used in Eq. (6.44) is evaluated in Eq. (6.42).

Although the induction machine develops a torque at every speed, except at synchronous, its capability to operate stably at a given speed depends on the character of mechanical torque on its shaft. Every load has its own torque–slip characteristic. The intersection of the torque–slip characteristic of the load and of the machine determines the operating point. The criterion to define whether the machine operates stably at a given operating point can be derived based upon the following analysis: suppose that the machine rotates at a constant speed at a point P. At a given time instant, a disturbance in the system source–machine–load appears. This disturbance can occur in any part of the system: a change of the amplitude or frequency of the applied voltage, a short circuit in the machine's winding, or a sudden change of the mechanical load. The operating point P is then said to be stable if the motor returns to it after the disturbance disappears.

A general equation for an electric machine states that the developed (electromagnetic) torque T_m is equal to the sum of the load torque T_m and accelerating/decelerating torque $J\, d\omega/dt$

$$T_m = T_1 + J\frac{d\omega}{dt} \qquad (6.45)$$

The accelerating/decelerating torque which is proportional to the inertia J on the shaft, exists as long as the variation of the rotor angular speed is different from zero. If the torque–slip curves are linearized in the vicinity of the observed operating point, one can write

$$\frac{\Delta T_m}{\Delta \omega} = \tan\mu \qquad \frac{\Delta T_1}{\Delta \omega} = \tan\lambda \qquad (6.46)$$

with angles μ and λ given in Fig. 6.11. The small-signal analysis of Eq. (6.45) gives

$$(T_m + \Delta T_m) - (T_1 + \Delta T_1) = J\frac{d\omega}{dt} + J\frac{d\Delta\omega}{dt} \qquad (6.47)$$

Substituting Eq. (6.45) into (6.47) one obtains

$$\Delta T_m - \Delta T_1 = J\frac{d\Delta\omega}{dt} \qquad (6.48)$$

or, by utilizing Eq. (6.46)

$$\frac{\tan\mu - \tan\lambda}{J}dt = \frac{d\Delta\omega}{\Delta\omega} \qquad (6.49)$$

The solution of the differential equation (6.49) is

Steady State Performance of Induction Machines

$$e^{\frac{\tan\mu - \tan\lambda}{J}t} = C\Delta\omega \tag{6.50}$$

The constant of integration C in Eq. (6.50) is determined from the initial condition that at time instant t = 0 when the disturbance is removed from the system, the difference between the angular speed after and before the disturbance is equal to $\Delta\omega_0$.

Replacing this condition in Eq. (6.50), and solving it for $\Delta\omega$, one obtains

$$\Delta\omega = \Delta\omega_0 e^{\frac{\tan\mu - \tan\lambda}{J}t} \tag{6.51}$$

The motor speed at a stable operating point after the disturbance has disappeared returns to its value before the disturbance appeared. This condition is fulfilled if the argument of the exponential function in Eq. (6.51) is negative, i.e. for

$$\tan\mu < \tan\lambda \tag{6.52}$$

The slope of the torque–slip curve of the motor is smaller than the slope of the torque–slip curve of the load at a stable operating point. Applying this criterion, points B, C, and D in Fig. 6.11 are stable, and A is unstable. Point D in Fig. 6.11 is a regular operating point, characteristic for various types of loads. The load torque–slip characteristic which has intersection with the motor curve at point C is typical for centrifugal load – fans, pumps, etc. The same (centrifugal) nature has the torque–speed characteristic of the load which intersects the torque–speed characteristic of the motor at point B.

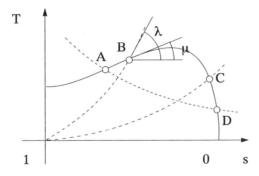

Fig. 6.11 *Illustrating the definition of stability*

6.3 Effects of Higher Spatial Harmonics of Flux Density in the Air Gap of an Induction Machine

Besides the fundamental term, the air gap flux density in an induction machine contains higher spatial harmonics. Being a function of the air gap mmf, saturation and air gap permeance, the air gap flux density contains the harmonics coming from each of these components. The higher spatial harmonics travel at speeds which are functions of the order of the spatial harmonic, as well as of the order of the time harmonics of the current which produces the particular harmonic. The speeds of higher spatial harmonics in a three phase machine are given in Table 2 in the second chapter. Limiting the analysis to the effects of the fundamental time harmonic of the stator current only, the air gap flux density harmonics which have synchronous speed given in the first column in Table 2 are selected. Each of these harmonics creates its own *asynchronous* torque, which is superimposed on the torque of the fundamental. The torque–slip characteristic of an induction machine in which, besides the fundamental harmonic torque T_1, the fifth and seventh spatial harmonics create their own torques T_5 and T_7, is shown in Fig. 6.12.

The synchronous speed of the fifth harmonic of the air gap flux density is $-\omega_s/5$, where ω_s is the synchronous speed of the fundamental. The negative sign of the synchronous speed of the fifth harmonic means that it rotates in a direction opposite to the direction of the fundamental. Analogous conclusions can be derived for the seventh and all other higher spatial harmonics.

When an induction motor runs at or near its rated speed, the operating point for all higher spatial harmonic torques is deep in their generator range. Therefore, the higher spatial harmonics act as an additional load for the fundamental, decreasing its efficiency. Induction machines are thus designed so as to minimize the amplitudes of the higher spatial harmonics of the air gap flux density.

The amplitudes of the higher harmonics of the air gap flux density decrease as the order of harmonics increase, because their winding factors decrease. This is, however, not true for the slot harmonics. It has been shown in the fourth chapter that the winding factor for the slot harmonics is equal to the winding factor for the fundamental. Therefore, the higher harmonic torques created by the air gap flux density, the order of which is equal to the order of the slot harmonics, are not negligible. These harmonics of the air gap flux density create asynchronous torques, the synchronous speeds of which are

$$\omega = \frac{\omega_s}{iN_s \pm p}; \quad i = 1, 3, 5, \ldots \tag{6.53}$$

where N_s is the number of stator slots.

Steady State Performance of Induction Machines 305

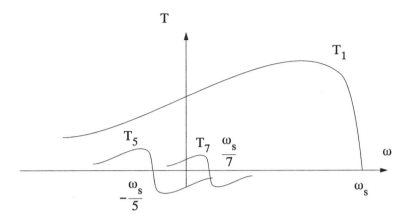

Fig. 6.12 *The fundamental, fifth and seventh harmonic components of asynchronous torque in an induction machine*

Besides asynchronous, *synchronous* torques can also be generated in an induction machine. A synchronous torque has an average value different from zero at only one, the synchronous, speed of rotation. At all other speeds the average value of a synchronous torque is equal to zero. Therefore, a synchronous torque is represented in a torque–slip diagram as a line perpendicular to the x-axis, positioned at the synchronous speed of the harmonic which causes the torque. This is illustrated in Fig. 6.13

The air gap of an induction machine is rich in the harmonics of the mmf and the flux density created by the stator and rotor currents and slotting. The synchronous speeds of the harmonics generated by the stator currents are constant as long as the stator frequency is constant. Since the frequency of the rotor currents is a function of the rotor speed, the synchronous speeds of the rotor-created harmonics are dependent on the rotor speed. Therefore, there exist rotor speeds at which a particular stator harmonic is in synchronism with a particular rotor harmonic. The two harmonics create a synchronous torque at such a speed. The rotor speeds at which a synchronous torque can occur have been calculated at the end of Section 5.3. Here five mechanisms which generate torques between the stator and the rotor in a doubly slotted machine have been given. The equations derived in Section 5.3 will be elaborated here in the sense that the angular frequency of the rotor currents will be expressed in terms of the rotor slip.

A synchronous torque in an induction machine is generated by:

(a) The stator currents of the angular frequency ω_1, due to variation of the self inductance of a stator phase, or the mutual inductance between two stator phases. The mechanical speeds ω_m of the rotor at which the synchronous torques appear in this case are:

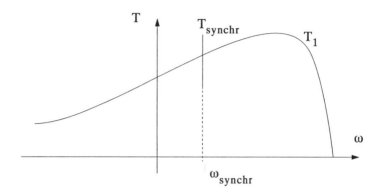

Fig. 6.13 *The fundamental (T_1) and a synchronous (T_{synchr}) torque in an induction machine*

$$\omega_m = \frac{2\omega_1}{kN_r p}; \quad k = 1, 3, 5, \ldots \qquad (6.54)$$

as given in Eqs. (5.111) and (5.113).

(b) The stator currents of frequency ω_1 and the rotor currents of frequency ω_2, due to variation of the mutual inductance between the stator and rotor phases. Substituting for $\omega_2 = s\omega_1$ into Eqs. (5.115), (5.117), and (5.119), where the slip s is defined as

$$s = \frac{\omega_1 - p\omega_m}{\omega_1} \qquad (6.55)$$

one obtains the speeds at which the synchronous torque can appear

$$\omega_m = 2\frac{\omega_1}{p(k+1)}; \quad k = 1, 3, 5, \ldots \qquad (6.56)$$

and

$$\omega_m = 0 \qquad (6.57)$$

An induction machine develops a synchronous torque at standstill, when $\omega_m = 0$. This torque creates starting problems in an induction machine, especially when the number of stator slots is equal to the number of rotor slots.

(c) The rotor currents of frequency $\omega_2 = s\omega_1$, due to variation of the self inductance of one rotor phase, or the mutual inductance between two rotor phases. In this case, the mechanical speeds ω_m of the rotor, at which the synchronous torques appear, are:

$$\omega_m = \frac{2\omega_1}{p(kN_s + 2)}; \quad k = 1, 3, 5, \ldots \quad (6.58)$$

6.4 Self Excitation of an Induction Machine

When an induction machine is connected to an A.C. source, a component of its stator current is used to build the air gap flux, independently of whether the machine operates as a motor, brake, or a generator. If an induction machine has to operate as a generator feeding its own consumers, a reactive energy in the circuit has to be provided. This energy builds the main magnetic flux. Other than in the machine's magnetic field, the reactive energy can be stored in capacitors. The reactive energy needed to excite the induction generator oscillates between the electric field of the capacitors and the magnetic field of the generator. A necessary condition which allows reactive energy to oscillate between the two fields is that a part of, or the complete magnetic circuit of, the machine is saturated. Therefore, the main reactance is not constant, and the generator's open circuit characteristic can be represented in an I–V diagram as a curve. When the machine is not saturated, its main reactance is constant, and its open circuit characteristic is a line. The operating point of an induction generator excited by capacitors is defined by the intersection of the line representing the capacitor's reactance and a nonlinear curve standing for the machine's I–V characteristic. At the operating point, the frequency of the stator currents, as well as the rotor slip, are defined by the parameters of the machine's equivalent circuit, load on the shaft, and the capacitance of the capacitors [13] - [16]. Since the stator frequency is variable, the relative frequency λ can be introduced, defined as

$$\lambda = \frac{f}{f_{ref}} \quad (6.59)$$

where f is the actual stator frequency, and f_{ref} is the reference frequency at which the reactances in the equivalent circuit in Fig. 6.14 (b) are calculated.

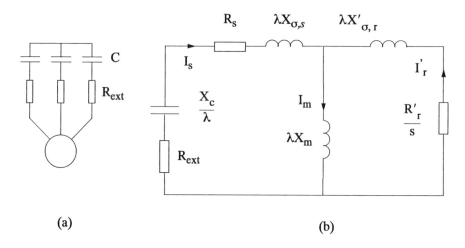

Fig. 6.14 *The induction machine self excited by capacitors (a), and its equivalent circuit (b). The branch containing the equivalent iron core resistance is neglected.*

The slip s in Fig. 6.14 (b) is defined as

$$s = \frac{\omega - p\omega_m}{\omega} \qquad (6.60)$$

where $\omega = 2\pi f$ and p is the number of the machine's pole pairs. The induction machine can be self excited with capacitors only if its speed lies between the lower boundary speed n_l and the upper boundary speed n_u, defined as

$$n_l = \lambda_l n_{ref,s}(1 - s_l) \qquad n_u = \lambda_u n_{ref,s}(1 - s_u) \qquad (6.61)$$

where

$$n_{ref,s} = \frac{60 f_{ref}}{p} \qquad (6.62)$$

is the mechanical synchronous speed at the reference frequency. At the lower boundary speed, the capacitor is in resonance with the main inductance. The rotor current is therefore equal to zero at this operating point, and the machine does not develop torque. At the upper boundary speed, the capacitor is in resonance with the rotor branch. The magnetizing current through the main reactance at this point is equal to zero, and the developed torque is again equal to zero.

The imaginary part of the impedance of a circuit which is in series resonance is equal to zero. From this condition one can express the relative stator frequency and slip at lower and upper boundary speed as

$$\lambda_u = \sqrt{\frac{X_c}{X_{\sigma,s} + X'_{\sigma,r}}} \qquad s_u = -\frac{R'_r}{R_s}$$

$$\lambda_l = \sqrt{\frac{X_c}{X_{\sigma,s} + X_{m,M}}} \qquad s_l = 0$$

(6.63)

where $X_{m,M}$ is the maximum (unsaturated) value of the main inductance. The slip of a self excited induction generator varies between $-R'_r/R_s$ and zero. The same slip values have been defined earlier as the operating range of an induction generator connected to a power system. The dependences of the braking torque T, rotor slip s, relative frequency λ, and main inductance X_m on the rotor speed are shown in Fig. 6.15.

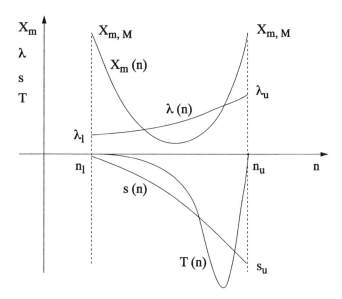

Fig. 6.15 *The variation of the induction machine parameters during capacitor braking*

The slip, relative frequency and main inductance of a self excited induction generator vary between two bounds in such a way that at each operating point the following conditions are satisfied, see [13] - [16]:

(a) The mechanical power on the shaft is equal to the copper losses in the machine and in the external stator resistance;

(b) The reactive power in the machine and load is equal to the reactive power of the capacitors;

(c) The equivalent circuit in Fig. 6.14 (b) is in permanent resonance. This means that the Thevenin impedance between any two points obtained by opening the equivalent circuit is equal to zero.

The external characteristic of the induction generator, i.e. the dependence of its terminal voltage on the load current at constant speed of rotation, can be obtained from equations derived from any of the three previously defined conditions. This can be illustrated with an example of active power distribution in the generator. From Fig. 6.16 (a) one can relate the air gap power P_{ag} to the stator current losses P_{Cus} and rotor current losses P_{Cur} as

$$P_{Cur} = sP_{ag} \qquad P_{Cus} = -P_{ag} \qquad (6.64)$$

from which one can express the slip as

$$s = -\frac{P_{Cur}}{P_{Cus}} = -\frac{I_r'^2 R_r'}{I_s^2 R_s} \qquad (6.65)$$

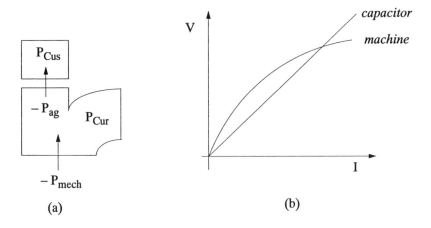

(a)   (b)

Fig. 6.16 *The distribution of power in a self excited induction machine (a), and the V–I characteristic of the system (b)*

From the machine's equivalent circuit in Fig. 6.14 (b) one can write the voltage equation

$$\frac{I'_r}{s}\sqrt{R'^2_r + s^2\lambda^2 X'^2_{\sigma,r}} = I_s\sqrt{R^2_1 + (\lambda X_{\sigma,s} - \frac{X_c}{\lambda})^2} \quad (6.66)$$

where R_1 is the total resistance in the stator circuit

$$R_1 = R_s + R_{ext} \quad (6.67)$$

Raising Eq. (6.66) to the power of two, and substituting in the equation obtained the square of the reflected rotor current from Eq. (6.65), one obtains the relationship between the stator current, slip and stator relative frequency. Assuming that neither the slip, nor the stator current is equal to zero, one may divide the equation obtained by sI_s^2. This way, one obtains a biquadratic equation for the stator relative frequency, the coefficients of which are functions of the rotor slip

$$\lambda^4 s(sR_1 X'^2_{\sigma,r} + R'_r X^2_{\sigma,s}) + \lambda^2 R'_r \left[s(R_1^2 - 2X_{\sigma,s}X_c) + R_1 R'_r \right] + sR'_r X_c^2 =$$

$$= 0$$

$$(6.68)$$

Denoting coefficients of Eq. (6.68) by

$$A = s(sR_1 X'^2_{\sigma,r} + R'_r X^2_{\sigma,s}) \quad B = R'_r \left[s(R_1^2 - 2X_{\sigma,s}X_c) + R_1 R'_r \right]$$

$$C = sR'_r X_c^2 \quad (6.69)$$

one can express the solution of Eq. (6.68) as

$$\lambda = \sqrt{\frac{-B + \sqrt{B^2 - 4AC}}{2A}} \quad (6.70)$$

The results of the previous calculation are illustrated in Fig. 6.17, in which the function on the left side of Eq. (6.68) is drawn. The solution of Eq. (6.68) lies on the intersection of the surface in Fig. 6.17 with the plane z = 0.

The next step in the computation of the external characteristic of a self excited induction generator is to define the value of the main reactance for a given pair of values of the slip and a reduced frequency. Expressing the sum of the currents in the equivalent circuit in Fig. 6.14 (b) in algebraic form as

$$I_s^2 = I_m^2 + I'^2_r + 2I_m I'_r \sin\theta_r \tag{6.71}$$

where the sine of the angle θ_r is defined as

$$\sin\theta_r = -\frac{\lambda X'_{\sigma,r}}{Z'_r} \tag{6.72}$$

and introducing the impedance Z'_r in Eq. (6.25), one can relate the main reactance X_m to the other parameters of the equivalent circuit and the capacitance as

$$\lambda^2 \frac{X_m^2}{Z_s^2} = 1 + \lambda^2 \frac{X_m^2}{Z'^2_r} + 2\lambda^2 \frac{X'^2_r X_m^2}{Z'^2_r} \tag{6.73}$$

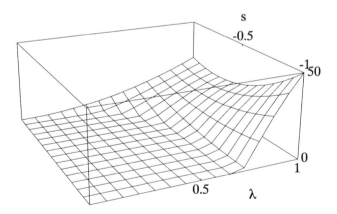

Fig. 6.17 *Illustrating the relationship between the relative frequency λ and the rotor slip in an induction generator*

Steady State Performance of Induction Machines 313

where

$$Z_s = R_s + R_{ext} + j(\lambda X_{\sigma,s} - \frac{1}{\lambda X_c}) \qquad (6.74)$$

In deriving Eq. (6.73) the voltage equations

$$I_s Z_s = I_m \lambda X_m = I'_r Z'_r \qquad (6.75)$$

were used to eliminate the machine's currents. The main reactance as a function of λ and s can be expressed from Eq. (6.73) as

$$X_m = \frac{Z_s^2 X'_{\sigma,r}}{Z'^2_r - Z_s^2} + \sqrt{\left(\frac{Z_s^2 X'_{\sigma,r}}{Z'^2_r - Z_s^2}\right)^2 + \frac{Z_s^2 Z'^2_r}{\lambda(Z'^2_r - Z_s^2)}} \qquad (6.76)$$

When the dependence of the main reactance on the stator relative frequency λ and the rotor slip s is determined, the magnetizing current from the machine's no load characteristic is evaluated for given values of X_m. The product of the magnetizing current and main reactance gives the induced voltage, from which the stator and rotor currents are calculated. Then the terminal voltage is evaluated as a product of the stator phase current and the external impedance per phase. The torque on the shaft is

$$T = \frac{P_{ag}}{\omega_s} = -\frac{3 I_s^2 R_1 p}{2 \pi f_{ref} \lambda} \qquad (6.77)$$

The self excitation of an induction machine with capacitors is a process which can exist only when the magnetizing curve of the iron core is nonlinear. In order to excite the machine with capacitors, the residual magnetic flux in the machine must exist, along with the source of mechanical power on its shaft. If the machine enters a self excitation mode directly from normal operation on a power system, its residual flux is identical to the main flux. The machine can be reconnected from the power systems to the capacitors very quickly, or, which is often the case, the capacitors are connected along with the machine to the power system. In both cases, the residual magnetic flux linked by the machine's windings remains almost completely saved. The residual flux in the self excited generator induces voltages in its stator windings. These voltages drive the currents through the capacitors. The voltage drops across the capacitors support the currents, and, after a certain time period, the machine reaches a new operating point, being defined by the parameters of the equivalent circuit in Fig. 6.14 (b).

Another possibility for self exciting an induction machine with capacitors is to start to rotate it with another motor mechanically coupled to its shaft. If the capacitors, along with its electrical load, are connected to the stator terminals, the residual flux in the rotor iron induces voltages in the stator phases which drive the currents through the load impedance and capacitors. After a certain time, the voltage on the stator terminals reaches its stationary value.

Independently of the mode with which the machine has to enter the self excitation, firmly determined relationships between the total resistance and the capacitance of the capacitors have to exist, in order to initiate the self excitation procedure at all. This can be illustrated by the simple example of an extremely large load resistance (open circuit), for which the machine certainly cannot be excited. In order to obtain the maximum value of the load resistance with which a given machine can be excited with given capacitors, the characteristic of the system to be in permanent resonance will be utilized. Replacing the parallel combination of the stator and magnetizing branch by impedance Z_p, as shown in Fig. 6.18,

$$Z_p = \frac{j\lambda X_m \, Z_1}{j\lambda X_m + Z_1} = \frac{-\lambda X_m (\lambda X_1 - \frac{X_c}{\lambda}) + j\lambda X_m R_1}{R_1 + j\left[\lambda (X_m + X_1) - \frac{X_c}{\lambda}\right]} \qquad (6.78)$$

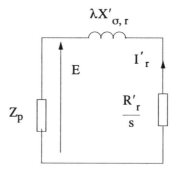

Fig. 6.18 *Illustrating the resonance principle in a self excited induction machine*

one can express the induced voltage per phase E as

$$E = I'_r Z_p = -I'_r \left(\frac{R'_r}{s} + j\lambda X'_r \right) \tag{6.79}$$

The total reactance per stator phase in Eqs. (6.78) – (6.79) is denoted by X_1, and the total resistance per stator phase by R_1. Since the impedances are complex numbers, one can express the relationships between their real and imaginary components, which are valid at any load point,

$$\text{Re}\{Z_p\} = -\text{Re}\{Z'_r\} \qquad \text{Im}\{Z_p\} = -\text{Im}\{Z'_r\} \tag{6.80}$$

Negative signs in Eq. (6.80) are caused by a negative rotor slip, and by the different character of the machine's reactances and the capacitance of the capacitor.

Two algebraic expressions in Eq. (6.80) define the same relationships between the slip, relative frequency and main inductance as the relationships derived by exploiting the principle of active power distribution. When the relative frequency λ is eliminated from Eq. (6.80), one obtains the following relationship between the slip and main reactance:

$$s^2 X_c R_1 (X_m + X'_{\sigma,r})^2 + s R'_r \left[X_c X_m^2 + R_1^2 (X_m + X'_{\sigma,r}) \right] + R'^2_r R_1 (X_m + X_1) = 0 \tag{6.81}$$

The main reactance is maximum on the boundaries of the interval in which a self excited induction generator can operate, because the magnetizing current at these operating points is equal to zero. Between the upper and lower boundaries, the main reactance significantly decreases relative to its maximum value.

The solution of Eq. (6.81) is

$$s_{1,2} = -\frac{R'_r \left[X_c X_m^2 + R_1^2 (X_m + X'_{\sigma,r}) \right]}{2 X_c R_1 (X_m + X'_{\sigma,r})^2} \pm$$

$$\pm \sqrt{\frac{R'^2_r \left[X_c X_m^2 + R_1^2 (X_m + X'_{\sigma,r}) \right]^2}{4 X_c^2 R_1^2 (X_m + X'_{\sigma,r})^4} - \frac{R'^2_r (X_m + X_1)}{X_c (X_m + X'_{\sigma,r})^2}} \tag{6.82}$$

The polynomial under the square root in Eq. (6.82) is a function of the main reactance X_m. Therefore, it is denoted by $P(X_m)$. The polynomial $P(X_m)$ must be positive in order to obtain a real solution for the slip. Its value varies between zero and $P(X_{m,M})$, where $P(X_{m,M})$ is the value of the polynomial at the lower and upper bounds, when the main reactance X_m has its maximum $X_{m,M}$. A necessary condition which an induction generator in a self excited mode has to satisfy in order to convert energy between the lower and upper speed bounds is that the slip between these bounds is a real number. This way, it is ensured that the slip on the bounds is also a real number. Therefore, the necessary condition for self excitation can be expressed as $P(X_{m,M}) \geq 0$, or

$$\frac{\left[X_c X_{m,M}^2 + R_1^2 (X_{m,M} + X'_{\sigma,r})\right]^2}{4 X_c R_1^2 (X_m + X'_{\sigma,r})^2} - X_{m,M} - X_1 \geq 0 \tag{6.83}$$

Inequality (6.83) describes an area in the (X_c, R_1) coordinate system in which the condition for the self excitation of an induction machine with capacitors is fulfilled. This area is limited by the curves the equations of which are obtained when the left side in Eq. (6.83) is set equal to zero. Introducing coefficients a_1, b_1, and c_1, defined as

$$a_1 = (X_{m,M} + X'_{\sigma,r})^2$$

$$b_1 = 2(X_{m,M} + X'_{\sigma,r})\left[X_{m,M}^2 + 2X_{m,M}(X_1 + X'_{\sigma,r}) + 2X_1 X'_{\sigma,r}\right] \tag{6.84}$$

$$c_1 = X_{m,M}^4$$

one can write the equations of square parabolas which limit the area in the (X_c, R_1) coordinate system on which the condition for self excitation of an induction machine with capacitors is fulfilled as

$$R_{1_{1,2}} = k_{1,2} \sqrt{X_c} \tag{6.85}$$

where

$$k_{1,2} = \sqrt{\frac{b_1 \pm \sqrt{b_1^2 - 4a_1 c_1}}{2a_1}} \tag{6.86}$$

Steady State Performance of Induction Machines

The two parabolas described by Eqs. (6.85) divide the first quadrant of the (X_c, R_1) coordinate system (only positive X_c and R_1 are possible!) into three areas. Only in two of these areas can an induction machine be self excited by capacitors, which can be checked by examining the sign of the polynomial $P(X_m)$. This criterion gives the areas of possible self excitation, as shown in Fig. 6.19 (a). Since one of the areas is not limited from above, i.e. for $R_1 \to \infty$, when the machine is disconnected from the capacitors, an additional physical criterion has to be applied in order to define the area in which the self excitation is possible. This criterion is derived from the function $\lambda(X_m)$, which is obtained from Eq. (6.80) when slip is eliminated from it. The result is

$$\lambda^4 (X_m + X_1) [X_1 (X_m + X'_{\sigma,r}) + X_m X'_{\sigma,r}] + \lambda^2 \cdot$$
$$\{ (X_m + X'_{\sigma,r}) [R_1^2 - X_c (X_m + 2X_1)] - X_c X_m X'_{\sigma,r} \} + X_c^2 (X_m + X'_{\sigma,r}) = 0$$

(6.87)

The term which multiplies the square of λ in Eq. (6.87) has to be negative if the solution of Eq. (6.87), i.e. the stator relative frequency, is to be a real number. This condition has to be satisfied at every speed of rotation at which the machine is excited; hence, one can write for the maximum value of the main reactance

$$(X_{m,M} + X'_{\sigma,r}) [R_1^2 - X_c (X_{m,M} + 2X_1)] - X_c X_{m,M} X'_{\sigma,r} < 0 \qquad (6.88)$$

From inequality (6.88) one can relate the total stator resistance to the capacitive reactance as

$$R_1^2 < \frac{X_{m,M}^2 + 2X_{m,M}(X_1 + X'_{\sigma,r}) + 2X_1 X'_{\sigma,r}}{X_{m,M} + X'_{\sigma,r}} X_c \qquad (6.89)$$

or, by substituting coefficients introduced in Eq. (6.84),

$$R_1 < \sqrt{X_c} \sqrt{\frac{b_1}{2a_1}} \qquad (6.90)$$

This condition is illustrated in Fig. 6.19 b).

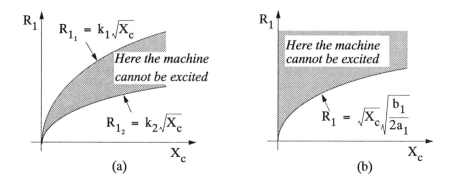

Fig. 6.19 *Illustrating the conditions for self excitation of an induction generator, expressed in Eqs. (6.85) and (6.91)*

Combining conditions illustrated in Fig. 6.19 (a) and (b), one obtains that if an induction machine has to operate in a self excited mode, the following relationship between the total stator resistance and capacitor reactance has to be satisfied:

$$R_1 < k_2 \sqrt{X_c} \qquad (6.91)$$

with k_2 being defined in Eq. (6.86).

6.5 Single Phase Induction Machine

It was assumed in the previous sections that the induction machines which were analyzed were connected to symmetrical m-phase sources, with m usually being equal to three. Three phase induction machines are used mostly for industrial applications. However, induction machines are often used in domestic settings, where, as a rule, only single phase sources are available. Here it should be noted that the term *single phase machine* describes, in fact, the type of source to which the machine is connected, and which in this case has only one phase. The machine itself very often has *two* phases: either the main and an auxiliary phase, or a wound pole and short-circuited rings in a shaded pole single phase machine.

When the single winding of an electric machine is connected to a single phase A.C. source, it generates the air gap flux density, the fundamental harmonic of which has two components, as shown in the second chapter. The positive sequence component of the fundamental harmonic of the air gap flux density rotates in a positive direction at synchronous speed; the negative sequence component has the same amplitude as the

Steady State Performance of Induction Machines

positive sequence, but rotates in the opposite direction.

If a winding is placed in such a field, each component of the flux density induces a voltage in it. If the winding is closed, as in a squirrel cage machine, the induced voltages can drive currents. The positive sequence current generates with the positive sequence component of the air gap flux an asynchronous torque, the average value of which is different from zero at all speeds of rotation, except at synchronous speed. The negative sequence component of the current creates, with the air gap flux, an asynchronous torque, the average value of which is equal to zero only at a negative synchronous speed of the rotor.

Following this explanation, a single phase induction machine can be represented as a combination of two symmetrical, three phase machines which are electrically and mechanically coupled, as shown in Fig. 6.20. In order to obtain equal electromagnetic conditions in a real and in two fictitious symmetrical machines, each phase of such a symmetrical machine has w/3 turns, where w is the number of turns of the single phase machine. The first symmetrical machine produces an air gap field which rotates in the positive direction and its impedance per phase is $Z_{pos}/2$. The second symmetrical machine produces an air gap field which rotates in the negative direction and its impedance per phase is $Z_{neg}/2$. The rotors of equivalent machines are firmly coupled and rotate in a positive direction.

The rotor slip is defined as its normalized speed relative to the positive synchronous speed

$$S_{pos} = \frac{\omega_s - \omega}{\omega_s} = s \qquad (6.92)$$

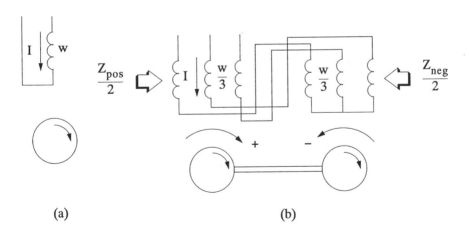

Fig. 6.20 *A single phase induction machine (a), and its representation as two mechanically and electrically coupled, symmetrical three phase machines (b)*

The rotor slip relative to the negative synchronous speed is

$$s_{neg} = \frac{-\omega_s - \omega}{-\omega_s} = 2 - s \qquad (6.93)$$

The machine operates as a brake for the negative sequence component if for the positive sequence component it is operating as a motor or a generator. Analogously, the machine operates as a brake for the positive sequence component if it is, for the negative sequence component, in the motor or generator mode of operation.

Each component of the air gap flux density creates an asynchronous torque with the rotor squirrel cage. The amplitude of the positive and negative sequence components of the asynchronous torque, as well as the currents in the motor, can be calculated from the motor's equivalent circuit. The equivalent circuit is derived from the scheme in Fig. 4.21. Combining the equivalent circuit of a single winding shown in this figure with the equivalent circuit of a symmetrical m-phase machine in Fig. 6.1, one obtains the equivalent circuit of a single phase induction machine as shown in Fig. 6.21.

The impedance $Z_{pos}/2$ in Fig. 6.21 is a function of the rotor slip s relative to the positive sequence component of the air gap flux density; the impedance $Z_{neg}/2$ is a function of the rotor slip (2–s) relative to the negative sequence component of the air gap flux density. The voltage drop V_{pos} is called the *positive sequence voltage*, whereas the voltage V_{neg} is the *negative sequence voltage*.

The applied voltage V is equal to the sum of voltages V_{pos} and V_{neg}. Since the same stator current I_s flows through both the impedance Z_{pos} and Z_{neg}, and since these two impedances are equal only at the rotor standstill, when s = 1, the voltages V_{pos} and V_{neg} are equal only at the rotor standstill. The electromagnetic torque produced by each component of the air gap flux density is evaluated by using Eq. (6.35), in which either s, or 2–s has to be substituted for the slip, depending on whether the torque for the positive or negative sequence component is evaluated. At slip s=1, the positive and negative sequence components of torque are equal. Since their directions of rotation are opposite to each other, the total torque is equal to zero and the machine cannot start. At other speeds of rotation, either the positive or negative sequence component of torque prevails. The torques developed in a single phase induction machine are given in Fig. 6.22.

The torque–slip characteristic of a single phase induction machine is qualitatively different from that of a three phase induction machine in two details: the starting torque of a single phase machine is equal to zero, and its no load speed is below the synchronous speed. The no load speed of a symmetrical induction machine is equal to the synchronous speed, because at synchronism the fundamental harmonic does not develop torque in a symmetrical machine.

The torque–slip characteristics in Fig. 6.22 are derived by assuming that the voltage drop across both the positive and negative sequence impedance is independent of the

Steady State Performance of Induction Machines

rotor slip. Therefore, the resultant torque–slip characteristic could be represented here as a sum of the positive and negative sequence torques, each of them being generated independently of the other. In reality, however, the amplitude of the torque is additionally modulated by the rotor slip. Since the motor current is common to both the positive and negative sequence components, the voltage drops V_{pos} and V_{neg} are dependent on the slip.

The voltage drop across the impedances Z_{pos} and Z_{neg} at standstill is equal to V/2 ; at any other slip one can write

$$V_{pos} = V \frac{Z_{pos}}{Z_{pos} + Z_{neg}} = V c_{V, pos}; \quad c_{V, pos} = \frac{Z_{pos}}{Z_{pos} + Z_{neg}} \quad (6.94)$$

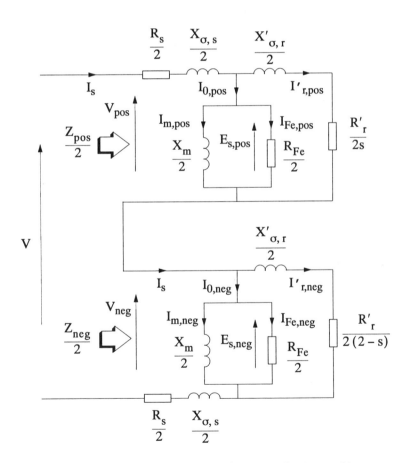

Fig. 6.21 *Equivalent circuit of a single phase induction machine*

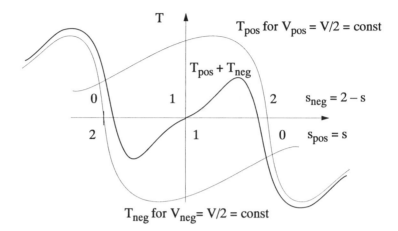

Fig. 6.22 *Torques produced by the positive sequence air gap flux density (T_{pos}), negative sequence air gap flux density (T_{neg}), and the total torque $T_{pos} + T_{neg}$. The voltage drops V_{pos} and V_{neg} across the branches representing the positive and negative sequence quantities are considered constant in this analysis.*

and

$$V_{neg} = V \frac{Z_{neg}}{Z_{pos} + Z_{neg}} = V c_{V,neg}; \qquad c_{V,neg} = \frac{Z_{neg}}{Z_{pos} + Z_{neg}} \qquad (6.95)$$

Since the electromagnetic torque of an induction machine is proportional to the square of the applied voltage, the curves denoted by T_{pos} and T_{neg} in Fig. 6.22 have to be multiplied by the squares of factors $c_{V,pos}$ and $c_{V,neg}$ respectively, in order to obtain the real torque for a given speed. This is illustrated in Fig. 6.23, in which the first and fourth quadrant of the torque–slip coordinate system from Fig. 6.22 have been redrawn, along with torques obtained by multiplying the sequence component torques by the squares of factors $c_{V,pos}$ and $c_{V,neg}$. This multiplication influences the torque values only when the machine is running; at standstill, the total developed torque is zero.

The zero starting torque is the most serious drawback of a single phase induction machine. It is usually overridden by adding one more phase to the stator. The additional stator phase can be wound similarly to the main phase, in which case the former is called the *auxiliary phase*, or it can be made in the form of a closed copper ring, as is found in *shaded pole* machines. In both cases, a more or less elliptical field is produced in the air gap. In a motor with an auxiliary stator phase, the elliptical field is generated by shifting the mmf produced by the auxiliary phase relative to the mmf

Steady State Performance of Induction Machines

produced by the main phase. In order to do this, the auxiliary phase winding is geometrically shifted to the main phase by an electrical angle between 90° and 110°. In addition, the current in the auxiliary phase is shifted to the current in the main phase. This can be done because the impedance of the auxiliary phase has a different magnitude and angle than the main phase impedance.

Unequal stator impedances are obtained either by manufacturing the auxiliary winding with a higher resistance, or, more commonly, by adding an external impedance in series with the auxiliary phase. The external impedance is almost exclusively a capacitor. Therefore, such single phase motors are called *capacitor motors*. Depending on whether the capacitor in the auxiliary phase is used to help produce an elliptical field only at starting, or permanently, the capacitor motors are manufactured either as *capacitor start*, or *capacitor run* motors (sometimes called *permanent split capacitor* motors).

The circuit connection of a single phase induction motor with an auxiliary phase is shown in Fig. 6.24. The number of turns in the main and auxiliary phases, as well as their winding factors, are usually different. Only when an induction machine has to operate in both directions of rotation with equal torque, have the main and auxiliary phases to be manufactured to be electrically identical.

In order to simplify the analysis, the parameters of the auxiliary phase are usually reflected to the parameters of the main phase. Introducing the factor of reflection a, defined as

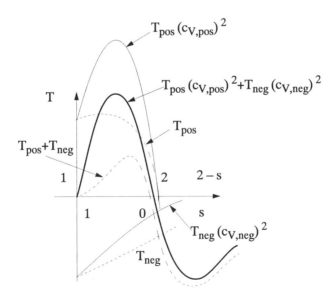

Fig. 6.23 *The torque components in a single phase induction machine*

$$a = \frac{w_a f_{w,a}}{w_m f_{w,m}} \tag{6.96}$$

where $f_{w,a}$ and $f_{w,m}$ are the winding factors of the auxiliary and main phase, respectively, one can express the parameters of the auxiliary phase reflected to the main phase as

$$R'_a = \frac{R_a}{a^2} \qquad X'_{\sigma,s,a} = \frac{X_{\sigma,s,a}}{a^2} \qquad Z'_{ext} = \frac{Z_{ext}}{a^2} \tag{6.97}$$

The voltage and current in the auxiliary phase are reflected following the relationships derived in the fourth chapter

$$V'_a = \frac{V_a}{a} = \frac{V}{a} \qquad I'_a = aI_a \tag{6.98}$$

If, in addition, the difference between the electromagnetic parameters of the stator phases is assigned to the external impedance, a fictitious two phase motor with symmetrical windings is obtained. This way, the asymmetry of the motor is converted into changed parameters of the external impedance. The circuit connection of a fictitious symmetrical two phase motor, which is equivalent to the real, asymmetrical motor, is shown in Fig. 6.25 [12]. Here the impedance Z''_{ext} is defined as

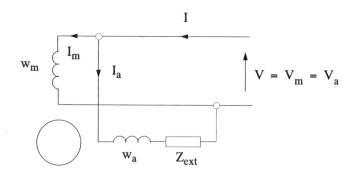

Fig. 6.24 *The circuit connection of a single phase induction machine with an auxiliary phase*

Steady State Performance of Induction Machines 325

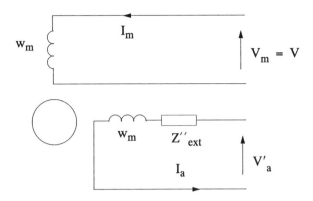

Fig. 6.25 *The circuit connection of a fictitious two phase, symmetrical motor which is equivalent to an asymmetric single phase motor*

$$Z''_{ext} = \frac{Z_{ext}}{a^2} + R'_a - R_a + j(X'_{\sigma,s,a} - X_{\sigma,s,a}) \quad (6.99)$$

The current I_m in the main phase produces the positive and negative sequence mmfs in the air gap. Instead of using the phase current I_m, two fictitious currents can be introduced, one of them producing the positive sequence mmf, and the other the negative sequence mmf

$$I_m = I_{m,pos} + I_{m,neg} \quad (6.100)$$

The current $I_{m,pos}$ is called the positive sequence current in the main phase, and $I_{m,neg}$ is the negative sequence current in the main phase. The current in the auxiliary phase can be then expressed as

$$I_a = jI_{m,pos} - jI_{m,neg} \quad (6.101)$$

The advantage of the representation of the stator currents introduced here is that the positive and negative sequence components of the currents, voltages, torques, losses, etc., can be separated in each phase.

The voltage equations of a single phase induction machine can be expressed in terms of the positive and negative sequence components of currents and impedances as

– the main phase voltage:

$$V = V_m = V_{m,\,pos} + V_{m,\,neg} \tag{6.102}$$

- the auxiliary phase voltage:

$$V'_a = V'_{a,\,pos} + V'_{a,\,neg} \tag{6.103}$$

By utilizing the positive Z_{pos} and negative Z_{neg} sequence components of the impedances introduced in Fig. 6.21, one can express the voltage in the main phase as

$$V_m = Z_{pos} I_{m,\,pos} + Z_{neg} I_{m,\,neg} \tag{6.104}$$

and in the auxiliary phase as

$$V'_a = (Z_{pos} + Z''_{ext}) I'_{a,\,pos} + (Z_{neg} + Z''_{ext}) I'_{a,\,neg} \tag{6.105}$$

Introducing the ratio γ between the negative and positive sequence components of the stator main phase current

$$\gamma = \frac{I_{m,\,neg}}{I_{m,\,pos}} = -\frac{I'_{a,\,neg}}{I'_{a,\,pos}} \tag{6.106}$$

and expressing the positive and negative sequence components of the currents as functions of the applied voltages, one obtains

$$V_m = Z_{pos} I_{m,\,pos} + Z_{neg} I_{m,\,neg} = Z_{pos} I_{m,\,pos} + \gamma Z_{neg} I_{m,\,pos} \tag{6.107}$$

The following form of the complex number γ was used in these equations

$$\gamma = \frac{Z''_{ext} + (1 + \frac{j}{a}) Z_{pos}}{Z''_{ext} + (1 - \frac{j}{a}) Z_{neg}} \tag{6.108}$$

Based upon the previous equations, one can draw the equivalent circuit of the main phase of a single phase induction machine with an external impedance as shown in

Fig. 6.26, and the equivalent circuit of the auxiliary phase in Fig. 6.27.

Utilizing the equivalent circuits of the single phase induction machine developed in this section, the torque–slip characteristics of a capacitor motor were calculated for the motor and capacitor parameters given in the Appendix. The computation was carried out for two values of an external capacitor: the start capacitor of 250 µF, and the run capacitor of 25 µF. The results of the computations are presented in Fig. 6.29.

The torque–slip characteristics in Fig. 6.29 (a) illustrate the difference between the motor performance for two different values of the capacitor. The larger capacitor helps develop a larger starting torque, which is a guarantee that the machine will start. However, large capacitors can be manufactured only as electrolytic capacitors. The losses of A.C. current in the dielectric of electrolytic capacitors are usually large, which prevents them from being permanently connected in series with the auxiliary phase.

Another drawback of an electrolytic capacitor is its high price compared to the price of a paper capacitor. Therefore, when the motor has successfully started, the run capacitor is connected instead of the start capacitor. The dielectric in the run capacitor is paper. The capacitance of the run capacitor is therefore significantly smaller than the capacitance of the start capacitor, but its dielectric losses to A.C. current are very small. Therefore, it may be permanently connected in series with the auxiliary phase.

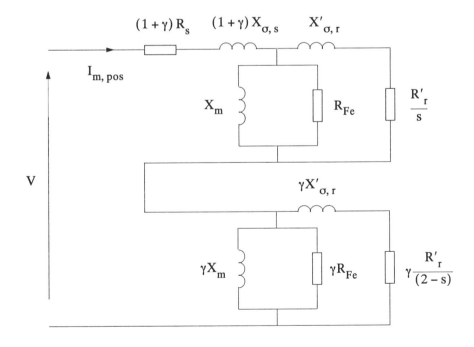

Fig. 6.26 *Equivalent circuit of the main phase of a single phase motor with external impedance*

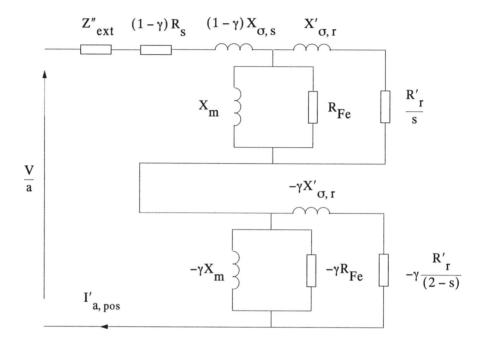

Fig. 6.27 *Equivalent circuit of the auxiliary phase of a single phase motor with external impedance*

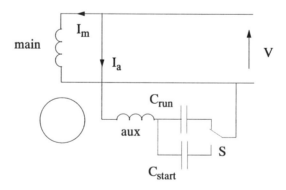

Fig. 6.28 *Circuit connection of a single phase, capacitor-start, capacitor-run motor*

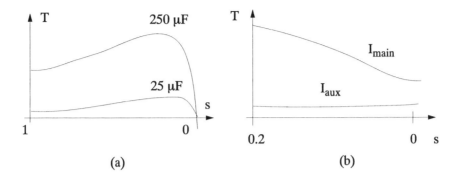

Fig. 6.29 *Torque–slip characteristic of a single phase induction machine for various values of the capacitor (a), and the main and auxiliary phase currents for 25 μF (b)*

The circuit connection of a single phase induction machine with a starting and running capacitor is shown in Fig. 6.28, and its torque–slip characteristics in Fig. 6.29 (a). As seen in Fig. 6.29 (a), the starting torque of a motor with 25 μF in the auxiliary phase is significantly lower than the one created by 250 μF.

After the motor has successfully started, the switch S in Fig. 6.28 disconnects the starting capacitor and connects the running capacitor. The switch *S* can be either a centrifugal switch, activated at a certain speed, or a voltage activated electromagnet connected in parallel with the auxiliary phase.

The currents in the main and auxiliary phase of the motor are shown in Fig. 6.29 (b). The dependence of the main phase current on the rotor slip is similar to that of the stator current in a symmetrical machine. The stator phase current of a symmetrical machine is minimum at zero slip, when the torque is equal to zero.

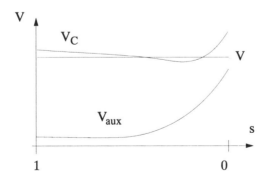

Fig. 6.30 *The voltage across the capacitor VC, across the auxiliary phase V_{aux} and the applied voltage V as functions of the rotor slip with a connected capacitor of 25 μF*

The main phase current of a single phase capacitor machine is minimum at the slip at which the sum of the positive and negative sequence torques is equal to zero. This slip is always greater than zero; hence, the minimum of the main phase current is shifted towards the positive slips. The current in the auxiliary phase, dominated by the capacitor reactance, increases steadily as the motor speeds up.

The voltage drops in the auxiliary phase circuit are dependent on the rotor slip. Since the voltage drop on a capacitor is shifted 180° to the voltage drop across an inductor, the amplitudes of the voltage drops in the auxiliary phase circuit can be higher than the applied voltage, keeping the vector sum of the voltages across the auxiliary phase and the capacitor equal to the applied voltage. Therefore, the voltage across the capacitor can sometimes be extremely high, which is a sign that the auxiliary circuit is close to one of its resonant frequencies. The resonance frequencies in the auxiliary phase circuit at various rotor slips are evaluated by setting the imaginary part of the auxiliary phase impedance equal to zero. This is illustrated in Fig. 6.31, in which the imaginary part of the auxiliary phase impedance is drawn as a function of the rotor slip and stator frequency. The imaginary part of the auxiliary phase is equal to zero on the boundary line obtained as a cross-section between the surfaces Im{Z_{aux}} and Z=0.

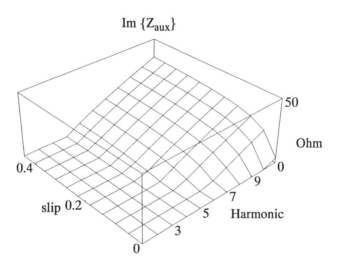

Fig. 6.31 *Imaginary part of the auxiliary phase impedance as a function of the rotor slip and stator frequency, expressed as a multiple (harmonic) of the frequency of the applied voltage*

6.6 Capacitor Braking of Single Phase Induction Machines

An asymmetrical induction machine, just like a symmetrical one, can be excited by capacitor(s) and can produce a braking torque in a certain interval of rotor speeds. Unlike a symmetrical machine, in which only a positive sequence air gap mmf exists which produces a positive sequence braking torque, in an asymmetric machine there also exists a negative sequence air gap mmf, which produces additional negative sequence torque. The air gap field in an asymmetric, capacitor braked induction machine is elliptical. Its amplitude pulsates at a frequency defined by the motor, load and capacitor parameters. A pulsating amplitude of the air gap mmf produces a pulsating braking torque. The average value of this torque can be found in a similar way as for the symmetrical connection, as shown in Section 6.4.

Various connections of the stator winding of a three phase induction machine which enable single phase capacitor braking are shown in Fig. 6.32. The analysis of the circuits shown in this figure will be carried out by utilizing a modified equivalent circuit of the single phase induction motor shown in Fig. 6.21.

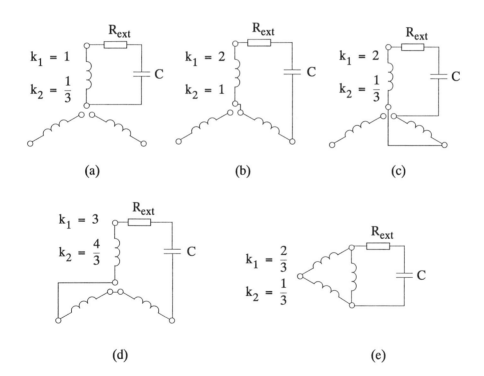

Fig. 6.32 *Stator phase connections of a three phase induction machine for a single phase capacitor braking mode*

This circuit is modified to allow for the change of the stator and rotor parameters as a function of the type of connection from Fig. 6.32. Therefore, two factors – k_1 for the stator, and k_2 for the rotor – reflect the phase connection. These factors multiply the circuit parameters calculated for a symmetrical, three phase connection of the motor. The equivalent circuit of a single phase connected three phase induction machine in the capacitor braking mode is shown in Fig. 6.33.

The branch representing the iron core losses is not included in the analysis, because these losses have negligible influence on the physics of single phase capacitor braking. The dependence of the relative frequency λ on the rotor slip and the value of the main inductance X_m can be found here in the same manner as for a symmetrical, three phase capacitor braking induction machine. It was shown in Section 6.4 that there exist three independent ways to obtain these curves.

Utilizing the method of natural frequencies, one can write for the circuit in Fig. 6.33

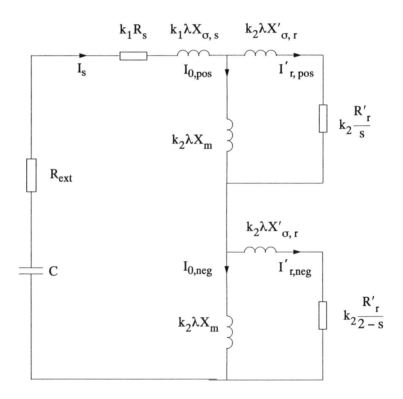

Fig. 6.33 *The equivalent circuit of a single phase connected three phase induction machine in the capacitor braking mode of operation*

$$Z_1 + Z_{pos} + Z_{neg} = 0 \tag{6.109}$$

where

$$Z_1 = k_1 R_s + R_{ext} + j(k_1 \lambda X_{\sigma,s} - \frac{X_c}{\lambda}) \tag{6.110}$$

$$Z_{pos} = jk_2 \lambda X_m \frac{\frac{R'_r}{s} + j\lambda X'_{\sigma,r}}{\frac{R'_r}{s} + j\lambda (X_m + X'_{\sigma,r})} \tag{6.111}$$

$$Z_{neg} = jk_2 \lambda X_m \frac{\frac{R'_r}{2-s} + j\lambda X'_{\sigma,r}}{\frac{R'_r}{2-s} + j\lambda (X_m + X'_{\sigma,r})} \tag{6.112}$$

After substitution of Eqs. (6.110), (6.111), and (6.112) into (6.109), one obtains two algebraic equations which relate the slip s, relative frequency λ, and main inductance X_m as variables. The first of these equations can be written as a condition that the real part of the sum in Eq. (6.109) is equal to zero, and the second equation is analogous for the imaginary part of that sum. If one expresses the term $s(2-s)$ from the first equation and substitutes it into the second one, a biquadratic equation in λ is obtained:

$$a\lambda^4 + \lambda^2 (bX_c + cR_1 + dR_1^2) + dX_c^2 = 0 \tag{6.113}$$

where

$$R_1 = k_1 R_s + R_{ext} \tag{6.114}$$

and

$$a = -\left[k_2 X_m^2 + X_m(k_1 X_{\sigma,s} + 2k_2 X'_{\sigma,r}) + k_1 X_{\sigma,s} X'_{\sigma,r}\right] \times$$
$$\times \left[X_m(k_1 X_{\sigma,s} + 2k_2 X'_{\sigma,r}) + k_1 X_{\sigma,s} X'_{\sigma,r}\right] \qquad (6.115)$$

$$b = (X_m + X'_{\sigma,r})\left[k_2 X_m^2 + 2X_m(k_1 X_{\sigma,s} + 2k_2 X'_{\sigma,r}) + 2k_1 X_{\sigma,s} X'_{\sigma,r}\right] \qquad (6.116)$$

$$c = -k_2 R'_r X_m^2 \qquad (6.117)$$

$$d = -(X_m + X'_{\sigma,r})^2 \qquad (6.118)$$

Eq. (6.113) has four solutions: two negative, and two positive roots for every value of the main inductance and slip. Since the relative frequency can only be a positive number, there exist only two physically correct solutions of this equation. Both of them are valid, while for each value of the main inductance there exist two values of the rotor slip and two values of the relative frequency. This can be checked in Fig. 6.15. The physically valid solutions $\lambda_{I,II}$ of Eq. (6.113) are

$$\lambda_{I,II} = \sqrt{\frac{-(bX_c + cR_1 + dR_1^2) \pm \sqrt{(bX_c + cR_1 + dR_1^2)^2 - 4adX_c^2}}{2a}} \qquad (6.119)$$

Eq. (6.119) relates relative stator frequency λ to the value of the main inductance X_m. The next step is to find the appropriate values of the rotor slip s. The slip is obtained as a solution of the real or imaginary part of Eq. (6.109). After that, the solution steps are identical to those for a symmetrical three phase capacitor braking.

The boundary values for the rotor slip and stator relative frequency are obtained by applying the condition for resonance along various paths within the equivalent circuit in Fig. 6.33. The resonance loop in the motor on the lower speed limit is shown in Fig. 6.34 (a), and on the upper limit in Fig. 6.34 (b). The lower boundary slip s_l and relative frequency λ_l are

$$\lambda_l = \sqrt{\frac{X_c}{k_1 X_{\sigma,s} + k_2(X_{m,M} + X'_{\sigma,r})}} \qquad s_l = 0 \qquad (6.120)$$

Steady State Performance of Induction Machines

whereas the upper boundary slip s_u and relative frequency λ_u can be expressed as

$$\lambda_u = \sqrt{\frac{X_c}{k_1 X_{\sigma,s} + 2k_2 X'_{\sigma,r}}} \qquad s_u = 1 - \sqrt{1 + 2\frac{k_2 R'_r}{R_s}} \qquad (6.121)$$

A single phase induction machine can be braked with capacitors within the previously defined slip and relative frequency limits only if certain relationships between the capacitance C, external resistance R_{ext} and motor equivalent circuit parameters is established. The conditions which these quantities have to satisfy can be determined by using Eq. (6.119). The relative frequency λ, evaluated in this equation, can be a real number only if the expression under the inner root

$$S \equiv (bX_c + cR_1 + dR_1^2)^2 - 4adX_c^2 \qquad (6.122)$$

is greater than zero, assuming that the main inductance is equal to $X_{m,M}$, in order to satisfy this condition on the upper and lower boundary. With the auxiliary function F, defined as

$$F = R_1(c + dR_1) \qquad (6.123)$$

one can rewrite the condition above as

$$(bX_c + F)^2 - 4adX_c^2 > 0 \qquad (6.124)$$

The roots of the curve which in the X_c,F coordinate system limits the region defined in expression (6.124) are

$$F_I = -X_c(b + 2\sqrt{ad}) \qquad F_{II} = -X_c(b + -2\sqrt{ad}) \qquad (6.125)$$

Expression (6.124) is satisfied for all F's, except for $F_I < F < F_{II}$. Since the coefficients c and d defined in Eqs. (6.117) and (6.118) are negative, and the total stator resistance R_1 is a positive number, condition (6.124) is satisfied for $F < F_I$ and $F_{II} < F < 0$, or, expressed in terms of the total stator resistance R_1:

$$R_1(c + dR_1) < -X_c(b + 2\sqrt{ad}) \qquad (6.126)$$

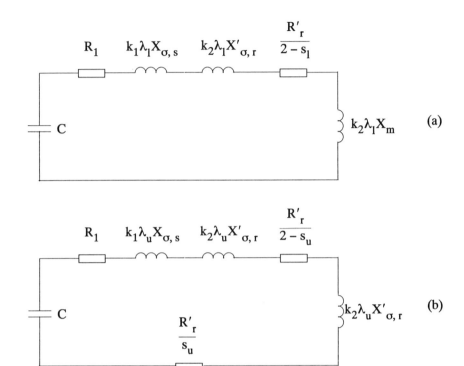

Fig. 6.34 *The resonant circuit of a single phase capacitor braked induction machine on the lower (a) and upper (b) resonant frequency*

and

$$R_1(c + dR_1) + X_c(b + 2\sqrt{ad}) < 0 \qquad (6.127)$$

Conditions (6.126) and (6.127) are simultaneously satisfied if the total stator resistance R_1 lies within the following interval:

$$k_1 R_s < R_1 < \frac{-c - \sqrt{c^2 - 4dX_c(b - 2\sqrt{ad})}}{2d} \qquad (6.128)$$

6.7 Speed Control of Induction Machines

6.7.1 Wound Rotor Machines

Although the manufacturing expenses of wound rotor induction machines are higher than those of squirrel cage machines, wound rotor machines were very often used in the past. The first reason for this is the possibility of controlling the starting current of the wound rotor machine by connecting external resistors in series with its rotor windings. The starting current of a squirrel cage machine can be between 4 and 7 times greater than the rated current, thus causing too large a voltage drop on the internal impedance of a weak power source, and, therefore, a decreased applied voltage across the motor's terminals. If the motor is loaded with a high torque at start up, such a reduced supply voltage can cause too low a starting torque, which is proportional to the square of the applied voltage, and the motor can stall at zero speed.

Another reason for using a wound rotor induction motor is the ease of its speed variation. The rotor speed of a wound rotor induction machine is, among other things, a function of the rotor resistance. Without changing the frequency and amplitude of the applied stator voltage, it is possible to vary the speed of the wound rotor induction machine at a given torque only by varying the resistance in the rotor circuit.

When a resistance is added to the rotor of an induction machine at a certain operating point, the induced voltage in the rotor phase has to cover the voltage drop on the fictitious resistance representing the load on the shaft, and, in addition, the voltage drop on the external resistor. The energy dissipated on the external resistor is supplied by increasing the rotor induced voltage. This can be done only by increasing the rotor slip. The higher the rotor slip, the lower the speed of rotation. Therefore, the additional resistance in the rotor circuit causes the rotor speed to decrease. This can be proved by using Eqs. (6.37) for the pull out slip, and (6.38) for the maximum torque. The maximum torque of an induction machine is independent of the rotor resistance; the pull out slip is proportional to the rotor resistance. By varying the rotor resistance, the pull out slip is shifted towards a higher slip, as illustrated in Fig. 6.35. In this figure, the torque–slip characteristics of a machine with two different values of rotor resistance are given. The pull out slips which correspond to various rotor resistances are denoted by $s_{p,1}$ and $s_{p,2}$ in this figure.

Along with the pull out slip, the slip for a given torque increases as the rotor resistance increases. The rate of change of the rotor slip depends on the shape of the torque–slip curve of the load. This is illustrated in the following example, in which the rotor slip as a function of the rotor resistance for a constant load torque is evaluated. Rewriting Eq. (6.24)

$$T = m_1 I_r'^2 R_r' \frac{1}{s\omega_s} \qquad (6.129)$$

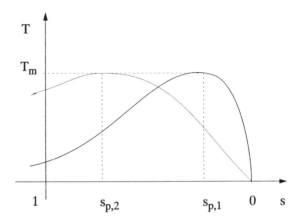

Fig. 6.35 *Torque control of a wound rotor induction machine by varying its rotor resistance*

and substituting for the rotor current, one obtains

$$T = \frac{R'_r m_1}{s\omega_s} \frac{E'^2_r}{\left(\frac{R'_r}{s}\right)^2 + X'^2_{\sigma,r}} \qquad (6.130)$$

The rotor resistance per phase is of the same order of magnitude as the rotor leakage reactance. However, when the rotor resistance per phase is divided by the slip which is in the range of several percent, one obtains a much bigger resistance than the rotor leakage reactance, or

$$\frac{R'_r}{s} \gg X'_{\sigma,r} \qquad (6.131)$$

Eq. (6.107) is also valid for high slip values which can be obtained only by adding resistance to the rotor. The ratio between the total rotor resistance and the slip in this case is again high, so that one may neglect the influence of the rotor leakage reactance. With this assumption one may write Eq. (6.130) as

$$T = \frac{m_1 E'^2_r}{\omega_s \dfrac{R'_r}{s}} = c\frac{s}{R'_r} \qquad (6.132)$$

Considering a constant load torque, one can express the total rotor resistance $R'_{new} = R'_{add} + R'$ with which the slip is s_{new}, as

$$R'_{new} = \frac{s_{new}}{s} R'_r \qquad (6.133)$$

or

$$R_{new} = \frac{s_{new}}{s} R_r \qquad (6.134)$$

from which the additional resistance per rotor phase can be expressed as

$$R_{add} = \frac{s_{new} - s}{s} R_r \qquad (6.135)$$

Since the rotor electrical losses are proportional to the slip, speed control by varying the rotor resistance is not economical. It creates additional thermal problems because of the need for cooling the external rotor resistance. A contemporary way to control the speed of a wound rotor induction motor is to save the energy which is in resistance control dissipated into losses in the external resistor. This can be done either by returning this energy to the motor through its shaft, or by sending it back to the source. The former procedure demands an additional motor on the shaft, connected into a so-called *Kraemer*, or constant power drive. The latter mode, i.e. the one in which the energy from the rotor circuit is sent directly to the power system, is known as a *Scherbius*, or a constant torque drive.

The circuit connection of the Kraemer drive is shown in Fig. 6.36.

The part sP_{ag} of the energy transferred through the air gap of the wound rotor induction machine in a Kraemer drive is not converted into losses, as in a regular induction machine, but brought back to the shaft. This is performed by rectifying the rotor voltages in the rectifier in Fig. 6.36, and then by converting the so obtained D.C. voltage into A.C. voltage with a frequency $(1-s)f_s$, where f_s is the stator frequency. Neglecting the losses in the power electronic components of the circuit, as well as in both the induction and synchronous motors, the power sP_{ag} is then returned back to the shaft of the induction motor.

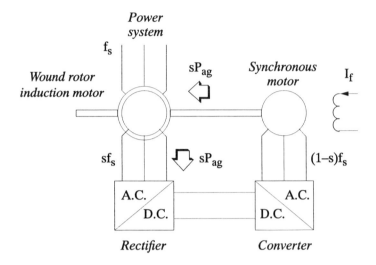

Fig. 6.36 *Kraemer drive*

Since there is no power loss in the external rotor resistors, a Kraemer drive is also known as a constant power drive. The torque–slip characteristic of the wound rotor induction motor in the Kraemer drive is shown in Fig. 6.37 (a).

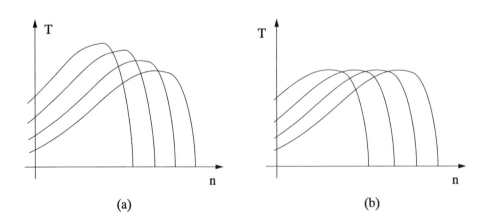

Fig. 6.37 *Torque–slip characteristics of a wound rotor induction motor in Kraemer drive (a), and in Scherbius drive (b)*

Steady State Performance of Induction Machines

The circuit connection of a wound rotor induction motor in a Scherbius, or subsynchronous, drive is given in Fig. 6.38. The energy sP_{ag} in the Scherbius drive is not returned to the motor through the shaft, but to the power system. Since the induced rotor voltages at low slip values are lower than the stator voltage, a transformer usually has to be added between the converter in the rotor circuit and the power system. Neglecting losses in the rotor circuit, the energy sP_{ag} flows through the rectifier, converter, transformer, and then back to the power system. The torque–slip characteristic of the wound rotor induction motor in the Scherbius drive is shown in Fig. 6.37 (b).

In both families of torque–slip characteristics in Fig. 6.37 one can see that the motor speed at which the developed electromagnetic torque is equal to zero is variable.

In a regular, symmetrical induction machine the torque is equal to zero only at synchronous speed, when the rotor current is equal to zero. In both Kraemer and Scherbius drives, however, the zero rotor current is obtained at a speed different from synchronous by adding a voltage source to the rotor circuit. In a Kraemer drive the role of the external voltage source in the rotor circuit is played by the induced voltage in the motor connected to the shaft of the wound rotor.

In the Scherbius drive the external voltage in the rotor circuit is the induced voltage on the secondary of the transformer, the primary of which is connected to the power system. Since the rotor circuit in both cases is connected to the external voltage via a rectifier, the equivalent circuit of a wound rotor induction machine in the Kraemer and Scherbius drives looks as shown in Fig. 6.39.

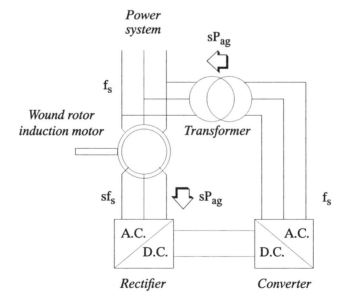

Fig. 6.38 *Scherbius drive*

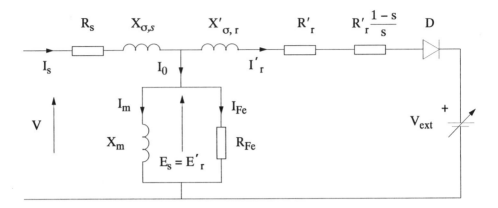

Fig. 6.39 *Equivalent circuit of a wound rotor induction motor in Kraemer and Scherbius drives*

The variable voltage source V_{ext} in the rotor circuit represents the induced voltage on the D.C. terminals of the converter. The product of the rotor current and voltage V_{ext} stands for the power returned to the motor in the Kraemer drive, or to the power system in the Scherbius drive. The diode D represents the rectifier, which enables the connection between the A.C. and D.C. part of the rotor circuit.

The electrical power fictitiously dissipated on the resistor $R'_r(1-s)/s$ is equivalent to the mechanical load on the shaft. At a given slip, the rotor current is determined by the induced voltage in the rotor phase and by the amplitude of the external voltage. The rotor current, and, therefore, the torque on the shaft, can be set to zero at any rotor speed only by varying the external voltage.

6.7.2 Squirrel Cage Machines

The synchronous speed n_s of the fundamental spatial and time component of the air gap mmf is equal to

$$n_s = \frac{60f}{p} \tag{6.136}$$

where f is the frequency of the applied voltage and p is the number of pole pairs. The synchronous speed, and therefore any other speed on the torque–slip characteristic of a squirrel cage induction motor, can be varied either by changing the number of poles of the motor, or by varying the frequency of the applied voltage.

Pole changing is performed in two ways: the machine is either manufactured with several stator windings (usually not more than three), or the coils of a single stator winding are connected in different ways, in order to obtain various numbers of pole pairs.

The simplest way to discretely control the speed of a squirrel cage machine is to build several windings into the stator slots, one for each speed. Since the number of poles of a squirrel cage is always equal to the number of poles of the stator winding, one squirrel cage rotor can be used for any number of stator poles. However, one has to make sure that the electrical parameters of the squirrel cage are within a reasonable range for each number of poles. A drawback of using one winding for each speed of rotation is poor utilization of the stator slot area by the conductors of each winding since room has to be left for the conductors of other windings. Squirrel cage induction machines with more than one stator winding are utilized in elevator drives, where the numbers of poles of two stator windings are usually 4/16, or 6/24. The low polarity (high speed) is used for the run connection of an elevator, and the high polarity (low speed) for its braking.

The speed of a squirrel cage induction machine can be controlled gradually only by one winding, if it allows a so-called *consequent poles* connection. This connection is illustrated in Fig. 6.40, in which the air gap mmfs created by a six pole winding connected to have six poles (a) and four poles (b) is shown.

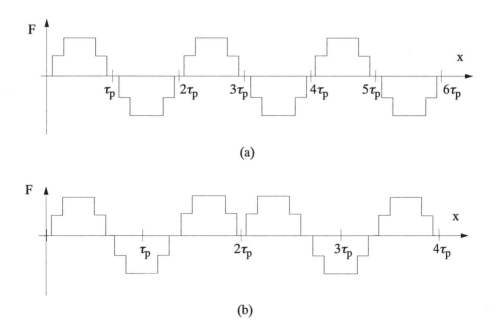

Fig. 6.40 *Air gap mmf created by a six pole winding connected to have six poles (a), and four poles (b)*

The coil groups under each pole of a regular six pole winding produce the symmetrical mmf shown in Fig. 6.40 (a), if connected in the sequence +, −, +, −, +, −. Each sign in this scheme represents the mmf polarity under one pole in the air gap. If the coil groups under the poles are connected in the sequence +, −, +, +, −, +, the air gap mmf has asymmetric poles, as shown in Fig. 6.40 (b). As one can see in Fig. 6.40 (b), the lower polarity winding generates higher spatial harmonics of the air gap mmf, the amplitudes of which are higher than the higher harmonic mmf amplitudes in the original, high polarity winding. Therefore, the windings in the machine the speed of which is varied by applying the consequent poles principle, are usually manufactured with asymmetric coils. The level of asymmetry of the coils is adjusted to minimize the higher spatial harmonics at both polarities.

A special case of the consequent pole connection, when the winding has 2/4 poles, is known as *Dahlander's* connection. Squirrel cage machines in this connection can develop torque–slip characteristics which meet either constant torque, constant power, or centrifugal torque needs of the load.

Speed control by means of **frequency variation** on the stator side is used in both voltage- and current source fed machines. The power source is mostly a six step, or a PWM inverter. A six step inverter creates a very rich higher harmonics spectrum of currents in the stator windings. The higher harmonics of the stator current create their own torques, which act as a mechanical load for the torque created by the fundamental. Each harmonic of the stator current dissipates its own I^2R losses on the stator winding resistances. The ampereturns of each stator current harmonic induce corresponding harmonics in the rotor winding, each of which again dissipates I^2R losses. For all these reasons, the rated power of an induction machine fed from a six step inverter has to be reduced relative to its rated power when fed from sinusoidal sources. The rated power of the machine fed from a six step inverter is usually 5% – 15% lower than when it is fed from a sinusoidal source. In this reduction factor the loss of the rated power due to increased losses in the iron in the machine fed from a six step inverter is also taken into account.

When the speed of an induction machine fed from a voltage source is controlled so that its overload capability is preserved, the main flux in the machine has to be constant over a wide range of applied frequency. A constant main flux means that the maximum torque of an induction machine is also constant, because the maximum torque is proportional to the square of the main flux. The fundamental component of the main flux can be written by means of Eq. (4.132) as

$$\Phi_{g, Max, 1} = \frac{E_{1,1}}{4.44 f w f_{w,1}} \qquad (6.137)$$

If the main flux has to be held constant when the frequency is varied, the voltage $E_{1,1}$ induced by the fundamental spatial and time harmonics of the main flux also has

Steady State Performance of Induction Machines

to be kept constant. The induced voltage in an induction machine can be controlled only indirectly by changing the magnitude of the applied voltage. Therefore, the overload capability of an induction machine is constant if both the applied voltage and frequency vary in such a manner that their ratio remains constant.

Previous considerations are illustrated in Fig. 6.41, in which the ratio between the induced and applied voltages in an induction machine as a function of the slip and frequency of the applied voltage is shown. At low slip values, the induced voltage is proportional to the applied voltage over a wide range of frequency of the applied voltage. However, when the stator frequency becomes very low, the induced voltage starts to decrease relative to the applied voltage. Therefore, the applied voltage at low frequencies must be increased in order to compensate for the loss of the induced voltage. The drop of the induced voltage relative to the applied voltage is caused by a disproportional increase of the stator resistance relative to the main reactance. Since the stator resistance is independent of the frequency of the applied voltage, the voltage drop on it increases relative to the induced voltage, which is proportional to the frequency.

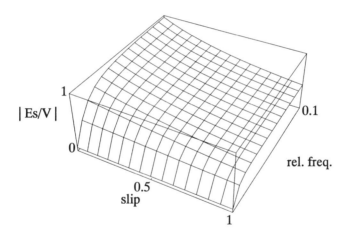

Fig. 6.41 *Ratio between the induced and applied voltage in an induction machine as a function of the slip and the frequency of the applied voltage*

7 Steady State Performance of Commutator Machines

A D.C. commutator motor was the very first electric machine which successfully converted electrical energy into mechanical work. This happened in 1838 in St. Petersburg, where inventor Jacobi used what is today called a D.C. commutator machine to run a boat over the Neva river. After constructional improvements by the Belgian inventor Gramme who introduced segments into the commutator, and after the invention of the cylindrical stator, the commutator motor reached the shape which has more or less remained unchanged until today.

The greatest advantage of a commutator machine over other types of machines is the ease of its speed control. This can be performed either by varying the armature voltage or the field current. As shown in the second chapter, a commutator machine is, in fact, a synchronous machine, the synchronous speed of which is always equal to the actual speed of the rotor. Unlike a synchronous machine, a commutator machine cannot lose synchronism, whatever the load on its shaft.

7.1 Principles of Operation of a D.C. Commutator Machine

The armature (rotor) conductors of a commutator machine rotate in the air gap field created by the armature, commutating pole, compensating, and field windings. The induced voltage in a rotor conductor is proportional to the flux density at the point on the air gap periphery where the conductor is placed at a given time instant. The principle of inducing and rectifying voltage in a commutator machine is illustrated in Fig. 7.1, in which a simple machine with only two commutator sectors is shown. Assuming that the flux Φ produces the flux density B similar to the one shown in Fig. 3.15, the induced voltage in series connection of two conductors in Fig. 7.1 looks as shown in Fig. 7.2.

The induced voltage before the commutator is alternating, because the conductors pass periodically through positive and negative flux densities. The commutator, along with the brushes, is a mechanical converter.

Steady State Performance of Commutator Machines

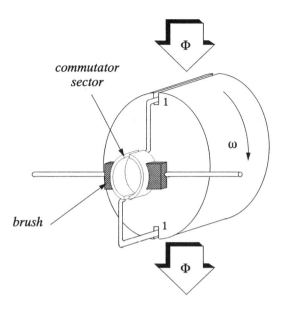

Fig. 7.1 *Illustrating the action of a commutator*

The A.C. current in the rotor winding becomes D.C. when exiting the brushes on the stator side. Similarly, the D.C. current, when entering from the stator through the brushes to the commutator sectors, exits on the rotor side as an A.C. current, the frequency of which is proportional to the rotor's speed of rotation.

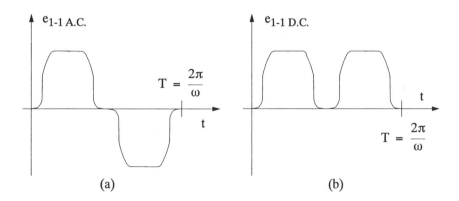

Fig. 7.2 *Induced voltage on the rotor side before the commutator (a) and on the stator side after the brushes (b) of the commutator machine in Fig. 7.1*

The rectified voltage on the brushes in Fig. 7.2 (b) has a constant term (D.C. component) and alternating terms. The amplitudes of the alternating components of the induced voltage are relatively high here compared to the fundamental. Denoting by V_{max} the maximum value of the induced voltage in Fig. 7.2 (b), and by V_{min} its minimum value, one can define the *ripple factor* f_p as

$$f_p = \frac{V_{max} - V_{min}}{V_{max} + V_{min}} 100 \tag{7.1}$$

Since the induced voltage in Fig. 7.2 (b) varies between zero and maximum, the ripple factor in this case is equal to 100%. In an ideal case, when the voltage has no alternating components, the ripple factor would be equal to zero. The ripple factor can be decreased by increasing the number of slots, along with the number of commutator sectors. This is illustrated in Fig. 7.3, in which the induced voltage in a D.C. commutator machine with two coils placed in four rotor slots is shown.

Both ends of each coil are fixed to the commutator segments. Since there are four coil ends, the number of commutator segments is also four. The coil axes are shifted by 90° electrically, which means that the induced voltage in the coil 2 – 2 is 90° delayed relative to the induced voltage in the coil 1 – 1, as shown in Fig. 7.3. The sum of the two voltages, also shown in this figure, has ripple factor equal to 33%. One can imagine that the additional coils, connected in series and placed in slots which are shifted relative to each other, increase the total induced voltage, simultaneously decreasing the pulsating factor. The D.C. induced voltage contaminated with a small A.C. component can do no harm in a regular D.C. motor or generator. In special applications, however, the A.C. component must be minimized. This is the case in D.C. tachogenerators, which are manufactured by utilizing permanent magnets instead of a field coil.

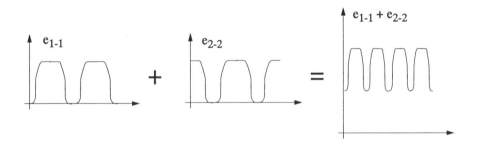

Fig. 7.3 *Induced voltage in a D.C. commutator machine with four commutator sectors*

The output signal from the tachogenerator is the induced voltage, the average value of which is proportional to the rotor speed. When the induced voltage contains A.C. components, the speed signal is erroneous, and has to be filtered.

Neglecting the A.C. component, the induced voltage in a D.C. commutator machine at constant speed and main flux is constant. The applied voltage is also constant, as well as the armature current. The relationships between the voltages and currents are algebraic. For the motor mode of operation one can write:

$$V = E + I_a R_a \qquad (7.2)$$

where V is the applied voltage, E is the induced voltage, I_a is the armature current, and R_a is the armature resistance. When a D.C. commutator machine operates as a motor, the applied voltage is greater than the induced voltage, and the armature current flows from the source to the motor. When a D.C. commutator machine operates as a generator, the induced voltage is greater than the applied voltage, and the current flows from the generator to the source. The voltage equation for the generator can be written as

$$E = V + I_a R_a \qquad (7.3)$$

At ideal no load, the induced voltage is equal to the applied voltage, and the armature current is equal to zero

$$E = V \qquad (7.4)$$

A commutator machine can operate at ideal no load only if it is run by another motor which covers its mechanical and electrical losses.

7.2 Induced Voltage and Electromagnetic Torque

When the armature conductors rotate at speed n [rpm], the induced voltage in each conductor is proportional to its length l, speed v, and flux density B at the current position of the conductor in the air gap. Denoting by e(t) the instantaneous value of the induced voltage per conductor, one can write

$$e(t) = B(x) l v \qquad (7.5)$$

The instantaneous value of the induced voltage per conductor is a function of the rotor position. When more conductors are connected in series, the instantaneous value of voltage no longer makes sense. Instead, the *average value* e of the induced voltage is utilized, being defined as

$$e = B_{av} l v \qquad (7.6)$$

where B_{av} is the average value of the flux density under one pole.

The rotor (armature) winding, which consists of conductors grouped into parallel circuits, was analyzed in the second chapter. Depending on whether the armature winding is a lap or wave type, the number of parallel circuits a can be equal either to 2pm, or 2m, where m is the multiplicity of the winding, and p is the number of pole pairs. If there are z conductors in total in the armature winding, the number of conductors per parallel circuit is z/a. The induced voltage E in a parallel circuit is equal to the induced voltage in the machine, i.e.

$$E = \frac{z}{a} e = \frac{z}{a} B_{av} l v \qquad (7.7)$$

Substituting for the conductor speed v

$$v = \omega \frac{D}{2} = \frac{n}{30} p \tau_p \qquad (7.8)$$

one obtains for the induced voltage

$$E = \left(\frac{pz}{30a}\right) \Phi_m n = c_e \Phi_m n \qquad (7.9)$$

The induced voltage in Eq. (7.9) is expressed, among other variables, as a function of the main flux in the machine. The main flux Φ_m is created by the field current, and is equal to

$$\Phi_m = B_{av} l \tau_p \qquad (7.10)$$

The induced voltage is proportional to the main flux and the speed of rotation. The inverse relationship, which is very often utilized in the analysis of commutator machines, can be written as

$$n = \frac{E}{c_e \Phi_m} \qquad (7.11)$$

The speed of rotation of a commutator machine is proportional to the induced voltage, and inversely proportional to the main flux. The *no load speed* n_0 can be expressed by utilizing Eqs. (7.4) and (7.11) as

$$n_0 = \frac{V}{c_e \Phi_m} \qquad (7.12)$$

because at no load the armature current is equal to zero, there is no voltage drop across the armature resistance(s), and the applied voltage is equal to the induced voltage. The two modes of speed control of a commutator machine are based upon Eq. (7.12): the speed is varied either by changing the applied voltage, or the main flux. If the applied voltage is increased and the main flux remains unchanged, the rotor conductors have to rotate faster in order to cut the same number of lines of flux more frequently, i.e. to induce a higher voltage. If the applied voltage is constant and the main flux is decreased, the rotor conductors again have to rotate faster in order to cut the decreased number of lines of flux more times. In an extreme case when the main flux falls to zero, the rotor speed has to be infinite, so that the number of cut lines of flux in a given time period is equal to the applied voltage. Since high speed destroys the rotor mechanically, the commutator motor must never run without the field current.

The relationship between the main flux and the field current is determined by the parameters of the magnetic circuit of the machine. Therefore, the dependence of the induced voltage on the field current reflects a characteristic of the machine's magnetic circuit. This is illustrated in Fig. 7.4, in which the *open circuit characteristic,* i.e. the dependence of the induced voltage on the field current with speed of rotation n as a parameter, is given.

The *electromagnetic torque* of a commutator machine can be defined using the relationships introduced in the fourth chapter, where it was proved that the force on a conductor is proportional to the product of the current in the conductor and the air gap flux density at the current rotor position. The force on the conductors under an N-pole has the same direction as the force under an S-pole, because the currents in the rotor conductors under an N-pole have a direction opposite to the currents under an S-pole, and the direction of flux density under an N-pole is opposite to the direction of the flux density under an S-pole.

Only the constant component of the generated electromagnetic torque contributes to the permanent energy conversion in the machine. The pulsating components of torque, created by the spatial harmonics of flux density and the time harmonics of current, play no role in this analysis. Therefore, one may express the average force on a conductor F_c in terms of the average flux density B_{av} as

$$F_c = B_{av} I_c l \qquad (7.13)$$

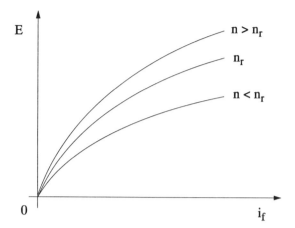

Fig. 7.4 *Open circuit characteristic of a D.C. commutator machine for various rotor speeds*

where the current per conductor I_c can be expressed in terms of the armature current I_a and the number of parallel circuits a as

$$I_c = \frac{I_a}{a} \tag{7.14}$$

The total average torque is equal to the product of the total average force and the rotor radius. The total average force F_{av} is z times greater than the average force on one conductor

$$F_{av} = \frac{z}{a} B_{av} I_a l \tag{7.15}$$

The average electromagnetic torque can be expressed as

$$T = \frac{z}{a} B_{av} I_a l \frac{D}{2} = \frac{pz}{a\pi} \Phi_m I_a = c_m \Phi_m I_a \tag{7.16}$$

The developed torque in a commutator machine is proportional to the product of the main flux and the armature current.

The relationship between the coefficients c_e and c_m, introduced in Eqs. (7.9) and (7.16), is

$$c_m = \frac{30}{\pi} c_e \tag{7.17}$$

If the armature current in Eq. (7.16) is expressed as a function of the applied and induced voltages and armature resistance, as shown in Eq. (7.2), the average torque of a commutator machine can be written as

$$T = c_m \Phi_m \frac{V - E}{R_a} = \frac{c_e c_m \Phi_m^2}{R_a} (n_0 - n) \tag{7.18}$$

The inverse function, i.e. the dependence of the rotor speed on the developed torque T, can be written as

$$n = n_0 - T \frac{R_a}{c_e c_m \Phi_m^2} \tag{7.19}$$

The torque–speed characteristic of a D.C. commutator machine with a constant main flux is given in Fig. 7.5.

This characteristic is linear for the complete range of operation of the machine. A D.C. commutator machine operates as a generator for a rotor speed greater than the no load speed, as a motor for the rotor speed between zero and the no load speed, and as a brake for a rotor speed less than zero.

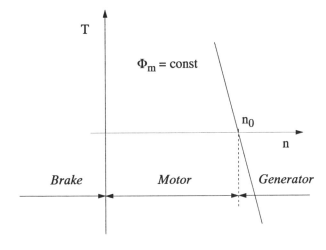

Fig. 7.5 *Torque–speed characteristic of a D.C. commutator machine with a constant main flux*

7.3 Armature Reaction

The air gap mmf and flux density distribution in a D.C. commutator machine was analyzed in the third chapter for various cases of interaction between the machine's windings. The resultant mmf in the air gap of a commutator machine is shown in Fig. 3.28. Such a spatial distribution of the mmf is basically caused by an electrical shift of 90 degrees between the axes of the armature and field winding. If the B–H curve of the machine's iron were linear, the air gap flux density distribution would be a copy of the mmf distribution. However, due to the saturation of the iron, the flux density curve cannot follow the air gap mmf at those portions of the main pole under which the mmf is very high. This can be seen in Fig. 3.28. An accurate form of the air gap flux density distribution for given ampereturns can be found only by the finite element method. The conclusions derived here are therefore qualitative, as shown in Fig. 7.6. Here the flux density distributions under one pole of a D.C. commutator machine with various B–H curves of the iron core and at various load are revealed.

The saturation of the iron cuts the peak value of the flux density under one pole and flattens it, as shown in Fig. 7.6. It should be mentioned here that the rated operating point of a commutator machine is usually on the knee of the B–H curve. If the machine had its rated operating point in the unsaturated portion of the B–H curve, or deep in the saturated section, the armature reaction could not cause any loss of the induced voltage because the air gap flux density distribution under the pole varies linearly as a function of the peripheral coordinate.

Only when the rated operating point is in the vicinity of the knee of the B–H curve is the induced voltage decreased due to the armature reaction.

Since the area in the (x, B) coordinate system can be interpreted as the air gap flux, one can see that the armature reaction causes the loss of the main flux in a D.C. commutator machine. The effect of decreasing the main flux due to armature reaction is reflected in the characteristics of a machine in a manner which depends on its operating mode. In a generator, less main flux means less induced voltage, as shown in Eq. (7.9). The loss of the induced voltage in a generator is proportional to the shaded area in Fig. 7.6. Therefore, the armature reaction leads to a loss of the generator's power. In a motor, the loss of the main flux increases the rotor speed, as shown in Eq. (7.11). The motor with a strong armature reaction can behave in an unstable manner when loaded – instead of decreasing, its speed can increase when the load on its shaft increases.

In addition to the loss of the induced voltage, the armature reaction causes some other negative effects on a D.C. commutator machine. The higher flux density in one pole half dissipates more iron core loss in the rotor teeth and yoke which belong to that pole half. Since the iron core loss is proportional approximately to the square of the amplitude of the flux density, an increase of the core loss under one half of the pole cannot be compensated for by a decrease of the core loss under the other half.

The armature reaction shifts the magnetic neutral in a commutator machine, as illustrated in the third chapter. The brushes are placed at magnetic neutral because here the flux density of an unloaded machine is zero. Therefore, a shift in the magnetic neutral introduces additional problems during commutation, as will be illustrated in the following section.

Last, but not least, the induced voltages in all the conductors which are placed under the main pole at a given time instant are equal when there is no armature reaction.

With an armature reaction, however, the induced voltage, being proportional to the air gap flux density at the current conductor position, varies from conductor to conductor. The difference between the induced voltages in conductors in adjacent slots is copied into the voltage difference between the commutator sectors, to which these conductors are connected. Under normal operating conditions this voltage difference does not play any role. If, however, the commutation is hard, the voltage difference between the commutator sectors can support sparking on the collector, causing severe damage to it.

The effects of the armature reaction are only negative. They only decrease the machine's performance; hence, the armature reaction should be minimized. There are two ways to minimize the negative effects of the armature reaction. One is to build compensating and/or commutating pole windings into the machine, and another is to place the rated operating point deep into the saturated region. The first measure increases the price of the machine; the second increases the size of the field winding enormously, at the expense of the rotor size, and therefore decreases the rated power of the machine.

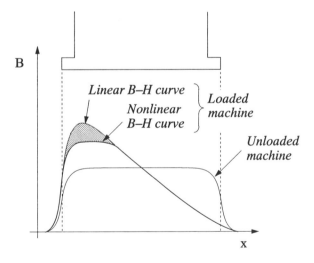

Fig. 7.6 *Air gap flux density distribution under one pole of a commutator machine for linear and nonlinear B–H curves of the iron core and at various loads*

7.4 Commutation

The commutator with brushes is the most important and the most sensitive part of a commutator machine. As shown earlier, the commutator with brushes is a mechanical converter – it converts A.C. current on the rotor winding side into D.C. current on the side of the brushes, and vice versa. Following Faraday's law, each change of current is accompanied by an induced voltage in the coil where the current is being commutated. The current carrying coils of the armature winding are periodically opened and closed during the commutation. When a current carrying coil opens, a voltage is induced in it which acts against the change of the current. This voltage is proportional to the linked flux which has to be changed during commutation. Therefore, the higher the flux linkage in the commutating coil, the larger the problems during commutation.

Commutation takes place in the coil, the ends of which are connected to the commutator sectors located under the brushes. The coil at this position links the following components of the machine's flux:

- Its own flux, created by the current in the conductors of the coil. To this component of flux, the flux created by the conductors placed in the same slot, but belonging to other coils, is added through the mutual inductance mechanism;
- Main flux, created by the field current in the main field winding;
- Armature flux, created by the armature current in all coils of the armature winding;
- Commutating pole flux, created by the armature current.

Any change in each of these fluxes induces voltage in the commutating coil. During commutation, some of these fluxes increase, some of them decrease. When designing a D.C. commutator machine, the goal is to obtain the minimum change of flux at the time instant when the commutating coil leaves the brushes, i.e. at the end of commutation time T_{comm}. In this way, the induced voltage which could cause sparking is minimized. This is illustrated in Fig. 7.7, where the current in a commutating coil as a function of time for various types of commutation is shown.

The curve denoted by a in Fig. 7.7 is characteristic for a commutator machine in which the commutating poles are reversely connected. Instead of acting against other fluxes linked by the commutating coil, the commutating poles delay the flux change at the beginning, causing a very high dI_c/dt at the end of commutation time T_{comm}. Such a variation of the current in the coil is called *strong undercommutation*. The current at the end of the commutation changes abruptly, causing sparks between the brushes and the commutator.

The curve denoted by b in Fig. 7.7 is typical for a machine in which no measures have been taken against the commutation problems. The current varies very slowly at the beginning of the commutation time T_{comm}, whereas, at the end, its time rate of change is high. Such a change of the coil current, known as *undercommutation*, again causes sparks at the trailing edges of the brushes.

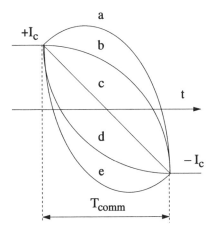

Fig. 7.7 *Curves of current change for various types of commutation*

The curve denoted by c in Fig. 7.7 is characteristic for *linear commutation*. This idealized curve is very often used to calculate the induced voltages in the coil during commutation. This curve is obtained by assuming that the inductance in the circuit of the commutating coil is much smaller than its resistance. The electrical time constant of the coil is therefore very short, and the current varies without the delay caused by the inductance.

The curve d in Fig. 7.7, which denotes *overcommutation*, satisfies commutation criteria optimally, because its slope at the end of the commutation time T_{comm} is equal to zero. The induced voltage in the coil at the end of the commutation is, therefore, also equal to zero in this case. The curve d is obtained by properly sizing the commutating poles.

If the commutating poles generate too many ampereturns in the magnetic neutral, the curve e in Fig. 7.7 is obtained. The commutating current falls to its negative value too fast, so that, at the end of the commutation time T_{comm}, it has to increase back to the negative value $-I_c$ of the coil current. The curve e is characteristic for *strong overcommutation*.

The assumption about linear commutation enables the calculation of the induced voltage in the commutating coil. This voltage must not be higher than 8V, which is an empirical limit, defined by the properties of spark in the atmosphere: an induced voltage higher than 8 V can cause and support sparking between the brush and commutator. If the current during commutation changes linearly, the induced voltage in this time interval is constant, as shown in Fig. 7.8.

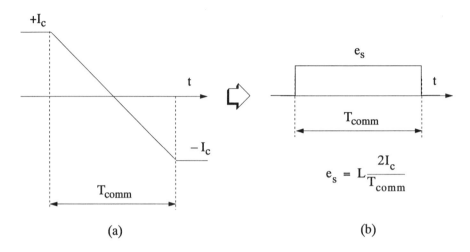

Fig. 7.8 *Current in a coil during linear commutation (a) and the corresponding induced voltage (b)*

7.5 D.C. Generators

In the past D.C. machines were often used as generators, feeding their own consumers. Depending on the field winding connection, four types of generators have been developed:
- separately excited;
- shunt excited;
- series excited, and
- compoundly excited.

The field winding in a **separately excited** generator is connected to its own D.C. source, independent of the armature circuit. This circuit connection is shown in Fig. 7.9 (a).

The open circuit characteristic of a D.C. generator represents the dependence of induced voltage on the field current. The curve denoted by E in Fig. 7.9 (b) represents the open circuit characteristic of a separately excited generator at no load.

At no load, the armature current is equal to zero, so that there is no decrease of the main flux due to armature reaction. At a given load current, the induced voltage is decreased for an equivalent voltage drop ΔE_{ar} due to armature reaction, and represented by the curve $E - \Delta E_{ar}$ in Fig. 7.9 (b). The voltage drop ΔE_{ar} increases as the field current increases because the machine is more saturated at a higher field current, and, therefore, the armature reaction effects are stronger.

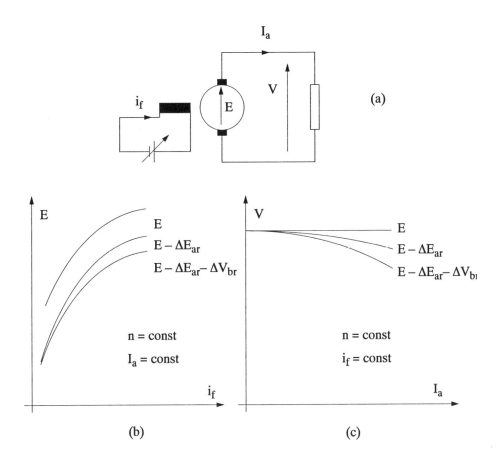

Fig. 7.9 *Separately excited D.C. generator: circuit connection (a), its open circuit characteristic (b), and external characteristic (c)*

The voltage across the machine's armature terminals is obtained when the voltage drop on the brushes ΔV_{br} is subtracted from the decreased induced voltage $E - \Delta E_{ar}$. In a wide range of armature current the voltage drop on the brushes is constant, and, for most types of brushes, equal to 1 V per brush. The terminal voltage as a function of the field current is represented in Fig. 7.9 (b) by the curve denoted by $E - \Delta E_{ar} - \Delta V_{br}$.

The external characteristic of a D.C. generator represents the dependence of the armature voltage on the load current at a constant speed of rotation. When the field current is constant, the induced voltage in a D.C. machine with a completely compensated armature reaction is independent of the load current. This characteristic is represented by a line, denoted by E in Fig. 7.9 (c). If the armature reaction effects are not compensated for by the armature current passing through the commutating pole and compensating windings, the induced voltage decreases as the load current increases, following the curve denoted by $E - \Delta E_{ar}$ in Fig. 7.9 (c). When the voltage drop on the brushes is subtracted from this curve, the curve $E - \Delta E_{ar} - \Delta V_{br}$ in Fig. 7.9 (c) is obtained, which represents the external characteristic of a separately excited D.C. generator. This characteristic is stiff – the terminal voltage of the generator changes slightly as the load increases. Therefore, the short-circuit current of a separately excited D.C. generator is highest among all types of D.C. generators.

The circuit connection and external characteristic of a **shunt** D.C. generator are shown in Fig. 7.10.

Unlike a separately excited generator, the shunt generator does not demand a separate source of D.C. voltage for excitation. Instead, the generator itself is a source of D.C. current for excitation – a component of the armature current flows through the field winding, building the main flux. The circuit connection of a shunt excited D.C. generator is shown in Fig. 7.10 (a).

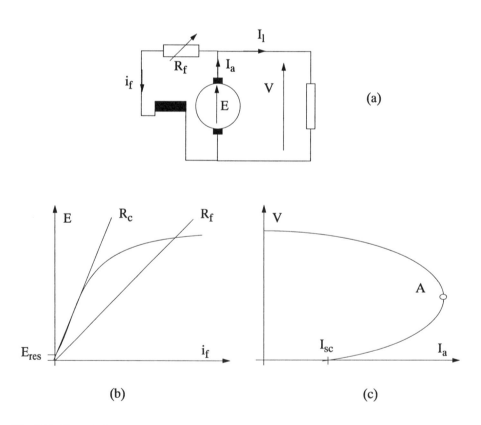

Fig. 7.10 *Shunt D.C. generator: circuit connection (a), its open circuit characteristic (b), and external characteristic (c)*

The field current is defined by the armature voltage V and the total resistance R_f in the field circuit. Since the resistance R_f is linear, it may be represented by a line, as shown in the open circuit diagram in Fig. 7.10 (b). The intersection of this line and the open circuit curve defines the operating point of the generator. By varying the resistance R_f, the terminal voltage of a shunt D.C. generator also varies. The maximum value of the field resistance at which the generator is still excited is denoted by R_c. Geometrically, the resistance R_c is equal to the slope of the tangent of the open circuit curve at the origin of the coordinate system (i_f, E). The triggering mechanism for the self excitation of a shunt D.C. generator is the residual magnetism in the machine's iron. Following the criterion for the minimization of the hysteresis loss, electric machines are built with the lamination of a narrow hysteresis loop. However, the hysteresis can never be eliminated from the iron core B–H curve. Therefore, the flux density in an unexcited machine is different from zero and equal to B_{res}. The only occasion when the flux density can be zero is after the demagnetization of the core with an A.C. current, the amplitude of which is gradually decreased to zero.

The residual flux density B_{res} induces voltages in the rotating conductors of the armature. Since the armature is connected in parallel to the field winding, the induced voltage drives current through the field winding. This current builds its own magnetic flux density which can support, or oppose, the residual flux density. When the magnetic flux created by the field current has the same direction in the main poles as in the residual flux, the resultant flux density becomes higher, inducing higher armature voltage, and driving more field current. This process of self excitation continues until the voltage drop across the total field resistance R_f becomes equal to the induced voltage. If, however, the magnetic flux created by the field current opposes the residual flux, the induced voltage cannot build up and the machine remains unexcited.

The external characteristic of a shunt D.C. generator is shown in Fig. 7.10 (c). The terminal voltage decreases as the load increases slightly faster than in a separately excited generator, because, in this case, the field current decreases along with the decrease of the terminal voltage. As the load increases, the terminal voltage decreases to point A on the external characteristic in Fig. 7.10 (c). At this voltage, the generator loses excitation, and the terminal voltage descends through the unstable part of the external characteristic down to zero. The current at a zero terminal voltage would have been zero without residual magnetism. With residual magnetism in the iron core, the residual voltage drives a certain short-circuit current I_{sc}. The advantage of the shunt generator over a separately excited D.C. generator is that it does not need an external source for the field current. The price for this is a poor voltage control of the shunt generator. The armature voltage can be decreased to approximately two thirds of its rated value at which operating point the resistance in the field circuit is close to the value of R_c. At a higher field resistance, the generator loses excitation.

If the magnetic circuit of the field coil contains an element which becomes saturated at a low flux, the voltage regulation of a shunt generator can be substantially improved. Such an element is obtained by narrowing the main pole, as shown in Fig. 7.11.

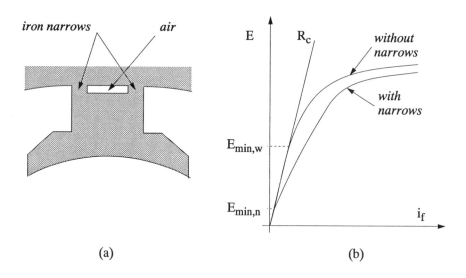

Fig. 7.11 *Main pole with narrows (a), and open circuit characteristics of a machine with and without narrows (b)*

The narrowed portion of the magnetic circuit shown in Fig. 7.11 (a) becomes saturated at much lower values of the main flux than the main poles and yokes. This is reflected in the machine's open circuit characteristic by a knee at low voltage ($E_{min,n}$), as shown in Fig. 7.11 (b). The minimum voltage $E_{min,n}$ which can be obtained by narrowing the main magnetic circuit is much lower than the minimum voltage $E_{min,w}$ in a regular machine.

The price for such a wider interval of voltage regulation is, however, a decrease in the rated power of the machine because more field current is needed to obtain the same induced voltage than in a regular machine.

Series excited D.C. generators are seldom manufactured. They can be found feeding a series D.C. motor in order to obtain suitable characteristics of a complete drive.

The circuit of a series generator is shown in Fig. 7.12 (a), and its external characteristics in Fig. 7.12 (b). The field current of a series generator is equal to the armature current. Therefore, the induced voltage is a direct function of the load current. At no load, the induced voltage of a series generator is equal to the residual voltage, as shown in Fig. 7.12 (b). The terminal voltage starts to increase along with the load current. This is a serious drawback of the series generator. Besides an unstable terminal voltage, the series generator cannot feed small loads because the resistance of small loads is greater than the critical resistance. These two disadvantages make the series D.C. generator useless in regular D.C. power systems.

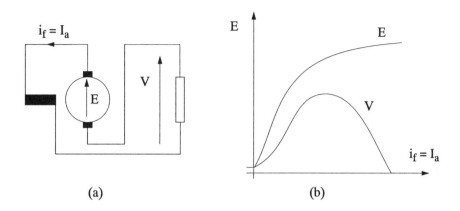

Fig. 7.12 *Circuit connection (a) and external characteristics (b) of a series D.C. commutator machine*

Compoundly excited D.C. generators have both series and shunt field winding, as shown in Fig. 7.13. A compoundly excited generator is derived from the shunt field generator, to which a series field winding is added in order to compensate for the voltage drop under the load.

Depending on the ampereturns of the series field winding, the external characteristics of a compoundly excited generator can vary over a wide range. The series field winding can either be connected to support the shunt field winding, when the generator is called a *cumulative compound*, or to oppose it, in a *differentially compound* generator. The series field winding in a cumulative compound generator can be designed to fully compensate at the rated point for the voltage drop in the armature circuit and the voltage loss due to the armature reaction. In this case, the machine is said to be a *level* compound one. When the series field winding ampereturns are too strong, the cumulative compounded generator is said to be *overcompounded*. The terminal voltage in an overcompounded generator increases with the load, which is very often not desirable. Too weak ampereturns of the series field winding, on the other hand, produce an *undercompounded* characteristic. The terminal voltage of an undercompounded generator decreases slightly with the load. The differentially compounded generator, similar to the series generator, is used very rarely, when feeding only one motor on its terminals. The external characteristics of compound generators with various connections of the series field winding are shown in Fig. 7.13 (c).

Besides the heteropolar generators mentioned in this section, D.C. generators are also built as **homopolar**. Homopolar D.C. generators give kiloamperes of current at a voltage, the order of magnitude of which is typically 10 V.

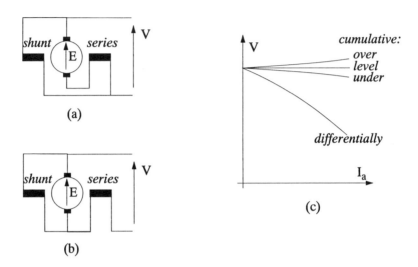

Fig. 7.13 *Compoundly excited D.C. generators: long shunt (a), and short shunt connection (b), and their external characteristics (c)*

The fact that the magnetic flux in a homopolar D.C. machine has the same direction throughout its whole active part allows one to build it without a commutator. A cross-sectional view of a homopolar D.C. generator is shown in Fig. 7.14.

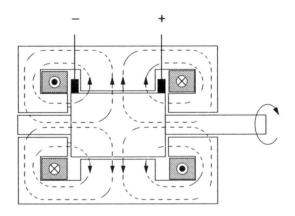

Fig. 7.14 *Cross-sectional view of a homopolar D.C. machine*

The field winding of a homopolar generator is manufactured in the form of two concentric coils on the stator: one on the left, and the other on the right side of the rotor. The rotor is made of iron with a copper cylinder on the outside, to which the brushes are pressed. The induced voltage in the copper rotor cylinder which plays the role of parallel connected conductors is perfectly D.C. – it does not contain any pulsating component. Homopolar D.C. generators are used in electrolysis plants, where high currents at low voltages are needed.

7.6 D.C. Motors

D.C. motors have always been used in applications in which good controllability is demanded. The possibility of controlling a large power in the armature circuit by varying a much smaller power in the field circuit made the D.C. motor the number one choice in earlier variable speed drives. Even special types of electromechanical power amplifiers, such as metadynes, amplidynes, etc., have been developed based upon the elementary principles of the operation of D.C. machines. However, with the advent of power semiconductor elements, electromechanical power amplifiers have lost their importance.

The external characteristics of D.C. machines, i.e. the dependence of the torque on the shaft on the rotor speed at constant armature voltage, are determined by the connection of the field winding(s). Similar to generators, D.C. motors can be grouped into:
- separately excited;
- shunt excited;
- series excited, and
- compoundly excited.

Separately and **shunt** excited D.C. motors have identical external characteristics, as long as the applied voltage is constant. The rotor speed is given in Eq. (7.11) as

$$n = \frac{E}{c_e \Phi_m} = \frac{V - I_a R_a - \Delta V_{br}}{c_e \Phi_m} \qquad (7.20)$$

because the induced voltage in a D.C. motor is lower than the applied voltage. The influence of the armature reaction is accounted for through the main flux Φ_m in the denominator of Eq. (7.20). If the motor is fully compensated, the main flux is constant. The induced voltage, being defined in Eq. (7.20) as the difference between the applied voltage and the voltage drops in the armature circuit, decreases as the load increases. The rotor speed in this case decreases as the load increases.

If the motor is not fully compensated, both the numerator and denominator of Eq. (7.20) decrease as the load (motor current) increases. The rotor speed can therefore either decrease, or increase as a function of the load, depending on whether the induced voltage, or the main flux, decreases faster. It is desirable that the rotor speed decreases as the load increases; the opposite behaviour is a characteristic of an unstable motor. The circuit connection of a separately excited D.C. motor is shown in Fig. 7.15 (a). Its torque–speed characteristic is described by Eq. (7.19), and shown in Fig. 7.5.

The circuit connection of a **series** excited D.C. motor is shown in Fig. 7.15 (b). The armature current of a series motor is at the same time its field current. This means that the speed of a series motor increases as the load decreases; at no load (zero armature current) the speed would, theoretically, become infinite. Since the series motor has inevitable friction and windage losses which are covered by a component of the armature current, the load can never go down to zero. However, the armature (field) current which covers the friction and windage losses is still small enough to cause the series motor to rotate at very high speed. As opposed to a robust squirrel cage rotor, the rotor of a D.C. machine is sensitive to centrifugal forces, and may not run at a speed typically 20% above its rated speed.

The reason for this is the commutator, which blows up at higher speeds. Therefore, all types of D.C. motors must be protected from too low a field current. Since series motors are mostly used in traction drives, where the load is always connected to the shaft, its overspeed protection is already built into the drive.

The torque equation of a D.C. motor is given in Eq. (7.16) as

$$T = c_m \Phi_m I_a \qquad (7.21)$$

When a series motor is unsaturated, its main flux is proportional to the field current, which is at the same time the armature current. In this case, one can express the torque equation (7.21) as

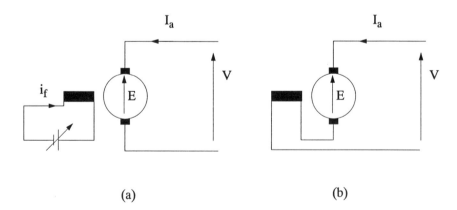

Fig. 7.15 *Circuit connection of a D.C. separately (a) and series (b) excited motor*

$$T = c_T I_a^2 \tag{7.22}$$

The rotor speed of an unsaturated D.C. series motor can be written as

$$n = \frac{E}{c_e \Phi_m} = \frac{V - I_a R_a}{c_n I_a} \tag{7.23}$$

Substituting the current from Eq. (7.22) into Eq. (7.23), one obtains

$$n = c \frac{V - R_a \sqrt{\dfrac{T}{c_T}}}{\sqrt{T}} \tag{7.24}$$

The torque–speed characteristic of a series D.C. motor is shown in Fig. 7.16.

Cumulative compound motors have both series and shunt field windings. Their external characteristic is controlled by dimensioning the series field winding, the ampereturns of which always support the shunt field winding ampereturns – a differentially compound motor is hard to find. Depending on the ratio between the ampereturns of the series and shunt field winding, the external characteristic of a compound motor is more like that of a series, or a shunt motor.

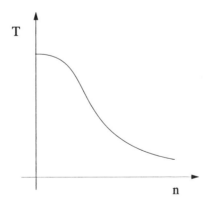

Fig. 7.16 *Torque–speed characteristic of a series D.C. motor*

7.7 A.C. Commutator Motors

Decades ago the very good controllability of D.C. commutator motors was the main reason to try to feed them from widely spread A.C. power systems. Wider application of A.C. series motors was limited by the commutation problems. Therefore, they have been used in two main areas: as traction motors, having a rated power up to 1000 kW, and as universal motors, for appliances. The use of A.C. commutator motors was so widely spread in railroad systems that in some countries separate power systems were built to feed them. The frequency of these systems is one third of the regular frequency, i.e. 16.66 Hz. Such low frequency is chosen in order to mitigate the commutation problems. Unlike D.C. motors, in commutating coils of A.C. commutator motors an additional component of voltage appears during commutation – the transformer voltage. The armature winding of an A.C. commutator motor is placed in an A.C. field created by the alternating current in the field winding. Due to the magnetic coupling between the field and armature winding, the field winding induces transformer voltage in the armature winding. The motional voltage, induced by the motion of the armature windings in the stator created magnetic field, is proportional to the speed, and maximum on the brushes. The speed voltage is in phase with the flux which induces it. The transformer voltage is *independent of the rotor speed*. It is maximum in the zone which is electrically perpendicular to the brushes, i.e. collinear with the axis of the field winding. The transformer voltage is shifted by $90°$ to the flux which induces it.

The commutating coil links the complete main flux because the commutation takes place in the magnetic neutral which is perpendicular to the axes of the main poles. Therefore, when a commutator machine is connected to an A.C. source, the transformer voltage is added to other induced voltages in the commutating coil, analyzed in Section 7.4. This voltage cannot be compensated for by commutating poles, as in a D.C. fed commutator machine. This is the most serious drawback of using a commutator machine connected to an A.C. power source. The inevitable uncompensated transformer voltage lowers the other components of voltage induced in the commutating coil, limiting the machine's power in that way. The solution found for traction D.C. motors, i.e. a frequency reduction to 16.66 Hz, limits the transformer voltage to one third of its value at 50 Hz.

Neglecting the problems related to commutation, a commutator machine develops less torque when connected to an A.C., than to a D.C. source. This is influenced by the fact that an A.C. current faces reactive voltage drop in the motor circuit in addition to the voltage drop on the resistance. Another power limiting factor for an A.C. connection is the relationship between the average and maximum value of flux. In both the A.C. and D.C. connections the saturation level must be the same. The saturation is determined by the maximum flux which in an A.C. fed machine is $\sqrt{2}$ times greater than the average flux. However, in a D.C. fed machine the maximum flux is equal to the average flux. The torque developed by an A.C. fed commutator machine is, as shown in Section 5.3, lower than the torque developed by an equal D.C. fed commutator machine by the factor $\cos\theta/\sqrt{2}$.

Another field of application of A.C. series motors are appliances. The motors for appliances have low rated power and are known as *universal motors*. Universal motors can be connected to both A.C. and D.C. sources. Their field winding has two connections: one for an A.C., and the other for a D.C. voltage. If a universal motor has to run at the same speed for a given torque when fed from an A.C. and from a D.C. source, the D.C. voltage has to be connected to the complete field winding, whereas for the A.C. voltage, connection with a lower number of turns of the field winding is foreseen.

7.8 Speed Control of Commutator Motors

The principles of the speed control of a commutator machine are based upon the relationship between the induced voltage, main flux, and speed, expressed in Eq. (7.11). According to this equation, the speed can be varied either by varying the applied voltage, which also causes a change of the induced voltage, or by varying the main flux. The speed control is usually performed so that the separate field winding of the motor at standstill is connected to its rated voltage. In this way the rated main flux is built into the machine. In the next step, a low voltage is applied to the armature. This voltage is typically only a few percent of the rated armature voltage, which is enough to start moving the rotor. The voltage across the armature is increased until the rated armature voltage is reached. At this operating point the rotor runs at the rated speed. In the whole speed interval from zero to the rated speed, the armature current is equal to its rated value. This demands external cooling of the machine, because at low speeds the quantity of cooling air delivered by the motor's own fan is insufficient to cool the machine.

If the speed has to be controlled above the rated value, the main flux is decreased. The armature voltage must not be increased above the rated value in order to obtain speeds above the rated because of commutation problems. The minimum value of the field current, or the main flux, is defined by the maximum allowed speed of the rotor.

The control principles of a commutator motor are illustrated in Fig. 7.17. The speed of a D.C. commutator motor can be controlled in the manner shown in Fig. 7.17, if variable D.C. voltage sources are available. The source of a variable D.C. voltage in the past was the so-called Ward–Leonard group. Nowadays, a variable D.C. voltage is almost exclusively provided by means of power electronics circuits. Depending on the primary energy source, the power electronic circuits are built as *choppers*, or as *rectifiers*. A chopper converts a constant D.C. voltage on input into a variable D.C. voltage on output. A rectifier converts a constant, or a variable A.C. voltage on input into a variable D.C. voltage on output. The variable voltage on output from a power electronic circuit contains, in both cases, higher harmonics.

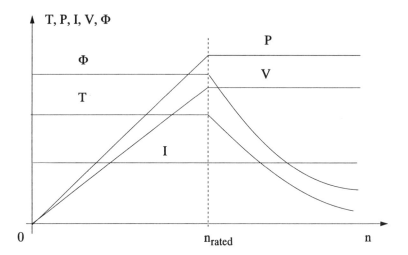

Fig. 7.17 *Speed control of a commutator motor*

When a D.C. separately excited motor is fed from a chopper, the source of energy is usually a constant voltage battery, as shown in Fig. 7.18 (a). The current waveform is shown in Fig. 7.18 (b). The chopper is symbolically represented as a switch which in the time interval t_{on} (position 1) connects the voltage source to the armature, and in t_{off} (position 2) short circuits the armature winding. At the switch position 1, one can write the following voltage equation for the armature circuit

$$V = iR_a + L_a \frac{di}{dt} + E \qquad (7.25)$$

where R_a is the total resistance, and L_a is the total inductance in the armature circuit (including the armature leakage inductance). When the switch is in position 2, one can write

$$0 = iR_a + L_a \frac{di}{dt} + E \qquad (7.26)$$

When the switch is in position 2, the motor is braking. The motor speed may be assumed constant in the complete period T, because the frequency at which the switch changes its position is too high to allow any observable changes of speed. Therefore, the induced voltage E may also be assumed constant. The transient state between energizing the motor and its reaching a constant speed is not important in this analysis; hence, only the steady state solution will be found. Denoting by τ the electrical time

Steady State Performance of Commutator Machines

constant of the armature circuit

$$\tau = \frac{L_a}{R_a} \quad (7.27)$$

one can write for the solution of Eq. (7.25) at steady state

$$i = i_{min} + \left(\frac{V-E}{R_a} - i_{min}\right)\left(1 - e^{-\frac{t}{\tau}}\right) \quad (7.28)$$

whereas the solution of Eq. (7.26) is

$$i = i_{max} e^{-\frac{t}{\tau}} - \frac{E}{R_a}\left(1 - e^{-\frac{t}{\tau}}\right) \quad (7.29)$$

Solution of Eqs. (7.28) and (7.29) at steady state is shown in Fig. 7.18 (b). Introducing the *duty cycle* ε

$$\varepsilon = \frac{t_{on}}{T} \quad (7.30)$$

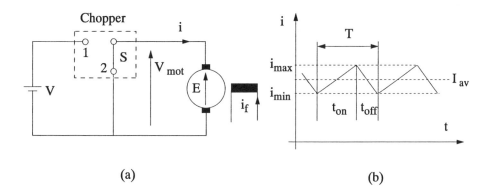

Fig. 7.18 *Principle of a chopper-fed separately excited D.C. motor (a), and the armature current waveform at steady state (b)*

one can define the current i_{min} in Eq. (7.28) as

$$i_{min} = \frac{V}{R_a} \frac{e^{\varepsilon \frac{T}{\tau}} - 1}{e^{\frac{T}{\tau}} - 1} - \frac{E}{R_a} \qquad (7.31)$$

because $i = i_{min}$ at $t = \varepsilon T$. Analogously, the current i_{max} in Eq. (7.29) is equal to

$$i_{max} = \frac{V}{R_a} \frac{1 - e^{-\varepsilon \frac{T}{\tau}}}{1 - e^{-\frac{T}{\tau}}} - \frac{E}{R_a} \qquad (7.32)$$

because $i = i_{max}$ at $t = T$. The average value of current I_{av} is obtained by integrating the current in interval $[0, T]$, which gives

$$I_{av} = \varepsilon \frac{V}{R_a} - \frac{E}{R_a} \qquad (7.33)$$

The average value of the armature current in a machine fed from a chopper, working with intermittence ε, is equal to the current in the same circuit in which the applied voltage is εV, instead of V. This means that the voltage across the motor terminals is varying continuously only by varying the on time t_{on} of the chopper. The ripple factor, defined in Eq. (7.1), is as small as the period T is short relative to the time constant τ of the armature circuit.

When a separately excited D.C. motor is connected to an m-phase bridge rectifier, the average value of the output voltage V_{av} being equal to

$$V_{av} = V_s \frac{m}{\pi} \sqrt{2} \sin \frac{m}{\pi} \cos \alpha = V_{av, 0} \cos \alpha \qquad (7.34)$$

where V_s is the R.M.S. value of the input voltage to the rectifier, and α is the firing angle, the voltage across the motor's armature terminals can be written as

$$V_m = V_{av, 0} \cos \alpha - \Delta V_{th} \qquad (7.35)$$

with ΔV_{th} denoting the voltage drop across thyristors. The motor speed is, therefore,

$$n = \frac{E}{c_e \Phi} = \frac{V_m - I_a R_a}{c_e \Phi} = \frac{V_{av,0} \cos\alpha - \Delta V_{th} - I_a R_a}{c_e \Phi} \qquad (7.36)$$

The torque–speed curve of a motor fed from a rectifier is similar to that of a motor fed from a constant voltage source, as long as the load is not too small. At small loads, the average value of the armature current is small; however, its instantaneous value at some time intervals can be equal to zero. At a small load, the current is supplied to the motor in the form of pulses. The number of current pulses in one period of the applied voltage is equal to the number of active phases on the A.C. side of the rectifier. The armature current between the pulses is equal to zero. Zero current means zero voltage drops in the circuit; hence, the induced voltage is equal to the output voltage of the rectifier. A higher induced voltage in zero current intervals means a higher no load speed. Therefore, the torque–speed characteristics of the motor in the discontinuous current range decline from lines in the manner shown in Fig. 7.19.

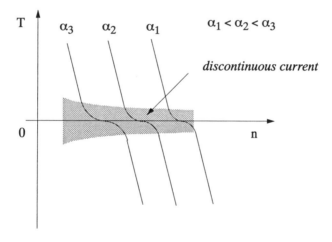

Fig. 7.19 *Torque–speed characteristics of a separately excited D.C. motor fed from a rectifier*

8 Steady State Performance of Synchronous Machines

Synchronous machines are the largest single unit machines manufactured. Their maximum rated power nowadays exceeds 1500 MVA. As opposed to D.C. commutator machines, where the maximum allowable induced voltage in the commutating coils limits the maximum rated power to several MW, the rated power of synchronous machines is determined only by the status of technology, expressed through the quality of electric insulation, mechanical strength of the material, cooling techniques, etc.

Besides their main application as sources of electric energy, synchronous machines have been utilized lately as permanent magnet and switched-reluctance motors in high efficiency electric drives. As opposed to induction machines, no copper losses are dissipated at rated load in rotors of permanent magnet and synchronous reluctance machines. Their efficiency is, therefore, better than the efficiency of induction machines of comparable size and rated power.

The fundamental component of the air gap flux density in a synchronous machine develops synchronous torque at only one speed. As long as the machine is in synchronism, its speed is independent of the load. Another advantage of the synchronous machine is that its rotor to stator angle can be controlled very accurately. This is important in compressor drives, in which the compressors are driven by two or more electric machines connected to the same source. In order to avoid mechanical resonance, the angle between the compressor shafts has to be kept within certain limits which is possible only if synchronous machines are used.

The torque of a synchronous machine is proportional to the applied voltage. A synchronous machine is less sensitive to power system disturbances than an induction machine because the maximum torque of an induction machine is proportional to the square of the applied voltage. Therefore, the same voltage decrease causes less torque loss in a synchronous than in an induction machine.

One advantage of a synchronous machine compared to an induction machine is its capability to deliver reactive energy to the power system. The reactive energy which transformers and induction machines consume from the power system is converted into the magnetic energy in them. An induction machine is always a consumer of reactive energy; a synchronous machine can consume, as well as deliver, the reactive energy. By delivering the reactive energy, a synchronous machine improves the power factor of the part of the power system where it is being used. Special types of synchronous machines, called *synchronous compensators*, are built only to improve the power factor of the power system to which they are connected. They operate as overexcited motors, consuming active energy from the power system to cover mechanical and electrical losses, and delivering reactive energy to it.

8.1 Constructional Features

Two basic types of synchronous generators were analyzed in the third chapter. Cylindrical rotor machines, or turbogenerators, are built as two, or four pole machines. They are named after the source of mechanical power which runs them. Two types of turbines are used as prime movers of turbogenerators: steam, and gas. Depending on the steam temperature and pressure, steam turbines are designed to run either at 3000 rpm for 50 Hz generators (3600 rpm for 60 Hz generators), or at 1500 rpm for 50 Hz generators (1800 rpm for 60 Hz generators). The rated speed of gas turbines is usually 3000 rpm at 50 Hz (3600 rpm at 60 Hz). Salient pole machines, or hydrogenerators, are built with numbers of poles which can exceed several dozen. The synchronous speed of a hydrogenerator is low, because its prime mover – a hydro turbine – usually has a very low rated speed.

The torque of a synchronous machine, as with any other type of machine, is defined by its rotor volume, and the air gap current sheet and flux density. The relationship between these quantities is given in Eq. (5.129)

$$T = VBA\cos\psi \tag{8.1}$$

where V is the rotor volume, B is the amplitude of the air gap flux density, A is the amplitude of the current sheet, and ψ is the angle between them. Keeping the synchronous speed constant, the rated power is proportional to the developed torque.

The maximum air gap flux density is limited by the magnetic properties of the stator and rotor iron cores. The flux density in the machine's iron in the largest machines is limited to 2.2–2.3 T at the highest magnetically loaded part of the machine. Considering that the tooth width at the air gap is approximately equal to the slot width, the average air gap flux density does not exceed 1.1–1.2 T. This limits the value of B in Eq. (8.1). If the maximum air gap flux density is held constant, the rated power of a machine can be increased by increasing its volume, and/or its specific current loading. In both cases, however, the temperature rise in the machine's windings and iron limits the power that the machine can permanently deliver. The machine's windings are electrically insulated to each other, and to the iron. The insulation material contains organic components which are temperature sensitive; hence, the windings and iron temperature has to be limited. The only way to limit the temperature in a machine is to cool it. A body in which the heat is dissipated can be cooled in three ways: by heat convection, conduction and/or radiation.

When a body is cooled by *heat convection*, a moving cooling medium is used. The low temperature cooling medium (air, hydrogen, water, or oil) is brought to the warm body, from which it takes out the heat. The warmed cooling medium is then transported to the coolers. In the coolers, the heat is taken from the cooling medium and delivered to the surrounding atmosphere. The amount of heat taken from a warm body is proportional to the area of the body from which the heat is being taken, and to the speed of the medium. Heat convection is applied in so-called *directly cooled machines*, where the rotor and/or stator conductors are cooled in this manner.

The heat is transported by the *heat conduction* mechanism when a body at a higher temperature has firm contact with a body at a lower temperature. The heat flows from the warmer to the cooler body as long as the temperature levels in both become equal. This kind of cooling takes place in those parts of electric machines to which the cooling medium has no direct access.

Radiation as a cooling mechanism of electric machines has no practical meaning because the temperatures which occur in machines are too low to produce observable radiation effects.

Following Eq. (8.1), the rated power of a machine increases as its volume increases, keeping the rated speed constant. The efficiency of large machines is very high, and increases very slowly as the rated power increases. Considering efficiency, as well as the specific magnetic (flux density B) and electric (current sheet A) loading constant, the losses increase proportionally to the rated power, or to the machine's volume. The surface from which the heat can be taken from the machine is, however, proportional to the square of the machine's linear dimensions. Therefore, the larger the electric machine, the more difficult it is to cool. The active part of small and medium size electric machines is cooled indirectly: the fan on the shaft forces the air to pass through the air gap, thus cooling the air gap facing surfaces of the stator and rotor teeth. The stator and rotor windings are cooled indirectly – the heat from them goes via a conduction mechanism to the adjacent teeth, and through them to the air gap. Here it is removed through the convection mechanism by air in the air gap.

As the rated power of the machine increases, the dissipated losses start to exceed the capabilities of indirect cooling. Therefore, direct cooling of the rotor and stator conductors must be applied. The cooling medium flows through the hollow conductors of the winding, removing the copper losses directly from the machine. This method of cooling, in addition, allows higher specific current loading, which is maximum when water is used as a cooling medium. The water in water cooled windings is demineralized, causing a drastic decrease in its electrical conductance. A very low value of electrical conductance of the water is necessary to minimize current flowing from the windings which are energized with the rated voltage, and the water tank which is on the earth potential.

8.2 Armature Reaction and the Synchronous Reactance Concept

The voltage applied to the terminals of a synchronous machine is equal to the sum of the induced voltage and the voltage drops on the machine's resistances and reactances. The voltage in the stator (armature) windings is induced by the time variation of the flux linkage. The flux linked by the stator windings is created by the rotor and stator mmfs. The rotor current is D.C.; however, when the rotor windings rotate at synchronous speed, the rotor mmf is recognized on the stator side as if it were produced by a stationary winding through which an A.C. current flows, the frequency of which is equal to the stator frequency. The stator mmf is created by the armature currents flowing through the stator windings.

The resultant mmf is a function of both the stator and rotor mmfs and the spatial shift between them. This is illustrated in the third chapter, where the resultant mmf for various spatial shifts between the stator and rotor mmfs in a turbogenerator (Fig. 3.35) and in a hydrogenerator (Fig. 3.38) is shown. One can see in these figures how the amplitude of the resultant mmf depends on the spatial angle between its stator and rotor components. Physically, such behaviour is a consequence of the stator armature reaction, i.e. of the action of the stator mmf which, along with the rotor mmf, build the resultant air gap mmf and air gap flux. Formally, the difference between the applied and induced voltage is represented as the sum of voltage drops on the stator resistance, leakage reactance and a fictitious reactance, called the *reactance due to armature reaction*. The sum of the leakage reactance and the reactance due to armature reaction is named *synchronous reactance*, and denoted as X_s.

The voltage drop across the stator leakage reactance and resistance is a few percent of the applied voltage except in very small machines where it can be higher. Therefore, the amplitude of the stator induced voltage is almost equal to the amplitude of the applied voltage. Neglecting saturation, the stator induced voltage is proportional to the resultant air gap mmf. When the applied voltage is constant, the stator induced voltage varies slightly in a wide range of change of the load. The amplitude and the phase shift of the resultant mmf are almost constant, although the amplitude and the phase shift of the stator current vary over a wide range. Since the resultant mmf is equal to the sum of the stator and rotor mmfs, the only way to keep it constant is to vary the rotor mmf, when the stator mmf varies. This is illustrated in Fig. 8.1, in which the mmfs in a synchronous machine at various stator current phase shifts φ are drawn. The vector of the applied voltage in this figure is denoted by V, and $E(F_{res})$ is the voltage induced by the resultant air gap flux. The stator current I_{st} creates the stator mmf F_{st}, and the rotor current creates the rotor mmf F_{rot}. The sum of the stator and rotor created mmfs is equal to the resultant mmf F_{res} which leads the induced voltage $E(F_{res})$ by 90°. The rotor mmf F_{rot} induces the voltage E , equal to

$$E = V + jI_{st}X_s + I_{st}R_{st} \qquad (8.2)$$

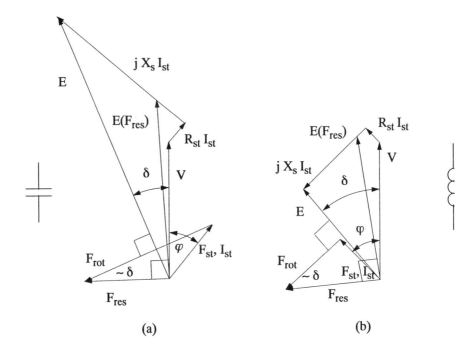

Fig. 8.1 *The mmfs and voltages in a synchronous generator at various stator current angles*

The induced voltage E is a nonlinear function of the rotor current. It depends on the magnetizing curve of the iron core and the dimensions of the magnetic circuit. The angle δ between the induced and applied voltage in Fig. 8.1 is called the *torque angle*.

When the stator current is equal to zero, the induced voltage E is equal to the applied voltage V. The field current which in this case induces the voltage E is called the *no load field current*.

When the stator current is lagging behind the applied voltage, as shown in Fig. 8.1 (a), the induced voltage E is greater than the applied voltage V. An induced voltage greater than the applied voltage is obtained by increasing the field current above its no load value, or by *overexciting* the machine. The reactive power Q, being defined as

$$Q = mVI_{st}\sin\varphi \qquad (8.3)$$

in this case is positive. An overexcited synchronous machine *delivers* reactive power to the power system. For the power system to which it is connected, an overexcited machine acts as a capacitor.

When the stator current is leading the applied voltage, as shown in Fig. 8.1 (b), the induced voltage is smaller than the applied voltage.

Steady State Performance of Synchronous Machines

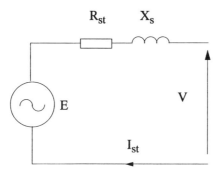

Fig. 8.2 *The equivalent circuit of a synchronous machine at steady state*

A field current smaller than the no load field current belongs to this value of the induced voltage. In this case, the machine is said to be *underexcited*. Following Eq. (8.3), the reactive power of an underexcited synchronous machine is negative. An underexcited synchronous machine is a *consumer* of reactive power, and acts as an inductance for the power system to which it is connected.

The reactive power generated in a synchronous machine is a function of the induced voltage, or the rotor current. The reactive and active powers in a synchronous machine are independent of each other. It is conventional to consider the active power delivered by a generator as positive. The active power of a motor is, therefore, negative. Both generator and motor can, however, operate in an overexcited or underexcited mode, i.e. deliver or consume the reactive power. The reactive power is controlled by varying the rotor current; the active power by varying the mechanical power on the machine's shaft.

The equivalent circuit of a synchronous machine at steady state is shown in Fig. 8.2. This equivalent circuit is described by the vector diagrams shown in Fig. 8.1 (a) and (b).

8.3 Performance of a Cylindrical Rotor Synchronous Machine Operating on an Infinite Bus

A synchronous generator is often connected to a power system which consists of a large number of generators and consumers. The total power which a power system can give exceeds by far the rated power of any of the generators connected to it. Therefore, the internal impedance of the power system is negligible compared to the generator's impedance. The power system acts for a generator as an ideal source, without internal impedance. Such a source can give to a generator, or take from it, any active or reactive power which the generator demands without changing the amplitude and frequency of the voltage on its terminals. It is said in this case that the generator is connected to an *infinite bus*. An infinite bus keeps the amplitude and frequency of the voltage V applied to the generator constant, determining in this way the generator's performance.

The distributions of the air gap mmfs in a synchronous generator shown in Fig. 3.35 are drawn by keeping the amplitudes of the stator and rotor mmf constant, and by varying the angle between them. The amplitude and phase of the resultant mmf are, therefore, in this case variable.

When a synchronous machine operates on an infinite bus, the applied voltage is constant and the resultant mmf varies insignificantly as a function of the stator current amplitude and angle. This is illustrated in Fig. 8.3 and Fig. 8.4, in which the vector diagrams of a synchronous generator operating on an infinite bus are shown.

The stator current I_{st} in Fig. 8.3 is lagging behind the applied voltage V. The armature mmf F_{st} is lagging behind the field winding mmf F_{rot}, the amplitude of which is greater than the amplitude of the resultant mmf F_{res}. The generator is overexcited, and supplies the reactive power to the infinite bus.

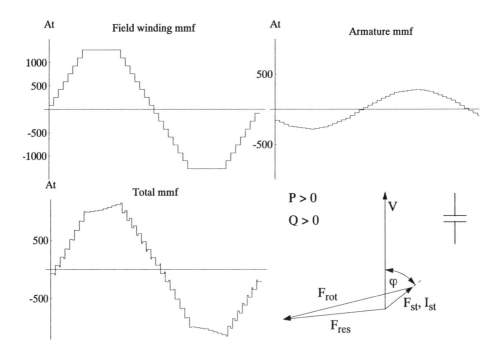

Fig. 8.3 *The mmfs in an overexcited synchronous generator*

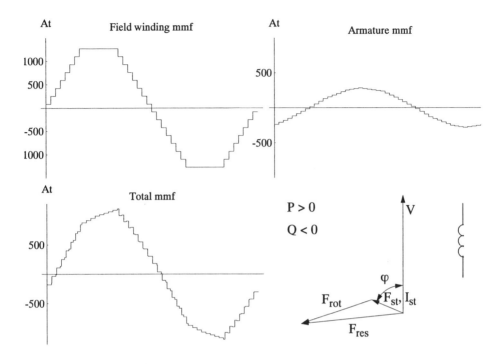

Fig. 8.4 *The mmfs in an underexcited synchronous generator*

The stator current in Fig. 8.4 is leading the applied voltage. The armature mmf is lagging behind the field winding mmf, the amplitude of which is smaller than the amplitude of the resultant mmf. The generator is underexcited and consumes reactive power.

The air gap mmfs in a synchronous motor are shown in Fig. 8.5 and Fig. 8.6. The component of the stator current which is in phase with the applied voltage V is negative in the motor mode of operation and positive in the generator mode.

By comparing the total mmf distributions in Fig. 8.3 – Fig. 8.6 one can see that the total mmf curve in an overexcited machine sags, and in an underexcited machine bulges out.

When a synchronous machine is loaded, the angle between the applied voltage V and induced voltage E increases. Neglecting the voltage drops on the stator resistance and leakage reactance, the voltage $E(F_{res})$ induced by the resultant air gap mmf F_{res} is identical to the applied voltage V.

The angle δ between the applied voltage V and induced voltage E is, under this assumption, identical to the angle between the induced voltages $E(F_{res})$ and E.

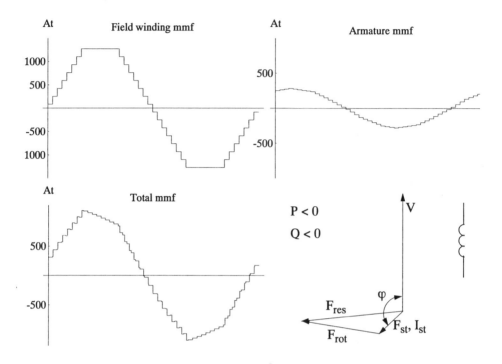

Fig. 8.5 *The mmfs in an underexcited synchronous motor*

Since the induced voltages are perpendicular to the mmfs which induce them, the angle between the resultant air gap F_{res} and rotor mmf F_{rot} is δ.

The amplitude and phase of the stator voltage V, and the induced voltage $E(F_{res})$ are firmly determined by the infinite bus. The maximum of the rotor mmf F_{rot}, on the other hand, coincides with the centre line of the rotor poles. Therefore, the torque angle δ is identical to the angle between the rotor centre line and the resultant mmf.

The dependence of the developed torque on the angle δ can be derived by utilizing the vector diagram in Fig. 8.7. When the stator resistance is not negligible, it is useful to define the *synchronous impedance* Z_s as

$$Z_s = \sqrt{X_s^2 + R_s^2} \qquad (8.4)$$

along with its angle ε

$$\varepsilon = \arctan\frac{R_{st}}{X_s} \qquad (8.5)$$

Steady State Performance of Synchronous Machines

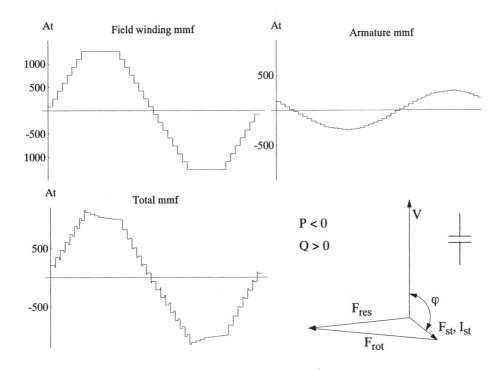

Fig. 8.6 *The mmfs in an overexcited synchronous motor*

The projection of the voltage vectors on the axis A–A in Fig. 8.7 can be expressed as

$$IZ_s \cos\varphi = E\sin(\varepsilon + \delta) - V\sin\varepsilon \tag{8.6}$$

The active power in an m-phase synchronous machine is equal to $mVI\cos\varphi$. By expressing the term $I\cos\varphi$ from Eq. (8.6), the active power can be written as

$$P = \frac{mV}{Z_s}[E\sin(\varepsilon + \delta) - V\sin\varepsilon] = P_{max}\left[\sin(\varepsilon + \delta) - \frac{V}{E}\sin\varepsilon\right] \tag{8.7}$$

where

$$P_{max} = \frac{mVE}{Z_s} \tag{8.8}$$

The electromagnetic torque as a function of the angle δ is obtained from Eq. (8.7) by substituting

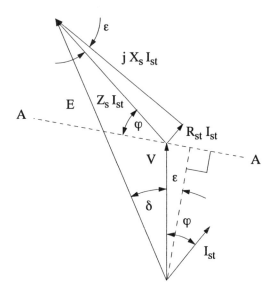

Fig. 8.7 *Illustrating derivation of the torque–angle characteristic*

$$T = \frac{P}{\omega_s} = T_{max}\left[\sin(\varepsilon + \delta) - \frac{V}{E}\sin\varepsilon\right] \qquad (8.9)$$

with

$$T_{max} = \frac{mVE}{\omega_s Z_s} \qquad (8.10)$$

The synchronous speed in Eq. (8.9) is denoted by ω_s.

The dependence of the developed torque on the torque angle in two machines is shown in Fig. 8.8 (a). The torque–angle curve of a machine the stator resistance of which is 30% of its synchronous reactance is gray in this figure. Following Eq. (8.9), the torque of such a machine depends on the synchronous impedance angle ε, as well as on the ratio between the applied and induced voltages.

The torque–angle curve in a machine in which the stator resistance is negligible compared to its synchronous reactance is solid in the same figure. This generator develops more torque at a given angle because in the former case a part of the generated power is dissipated on the stator resistance.

The torque–angle curve of a machine with negligible stator resistance is shown in

Steady State Performance of Synchronous Machines 385

Fig. 8.8 (b). In this figure two basic modes of operation of a synchronous machine are shown: stable and unstable. When a machine is operating in the stable mode, it reacts on the load changes only by changing the torque angle. A stable generator operates at a torque angle between 0 and 90°; the torque angle of a stable motor lies between −90° and 0°. Once a machine enters the unstable region, it can lose synchronism. The loss of synchronism is accompanied by very high stator currents and, therefore, must be avoided whenever possible.

A simple criterion which allows one to predict whether the machine will lose synchronism after a load change, called the *equal area criterion*, will be derived here. It is based on the behaviour of the torque angle δ after a load change: a torque angle which tends to a constant value after an infinite time indicates a stable process; a torque angle which keeps increasing or decreasing is a sign of unstable behaviour.

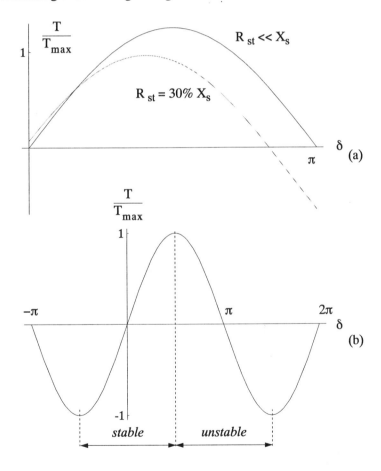

Fig. 8.8 *The dependence of developed torque in a synchronous machine on the torque angle δ (a) and stability limits (b)*

The mechanical power P_m on the shaft of a synchronous machine is equal to the sum of the power of the load P_l and the power which accelerates or decelerates the rotor

$$\frac{P_m - P_l}{\omega} = J\frac{d\omega}{dt} = J\frac{d^2\delta}{dt^2} \qquad (8.11)$$

where ω is the machine's angular speed, and J is its moment of inertia. After multiplying Eq. (8.11) by $2d\delta/dt$, and applying the rule for differentiation of the square of a derivative

$$\frac{d}{dt}\left(\frac{d\delta}{dt}\right)^2 = 2\frac{d\delta}{dt}\frac{d^2\delta}{dt^2} \qquad (8.12)$$

one can write Eq. (8.11) as

$$\left(\frac{d\delta}{dt}\right)^2 = \frac{2}{J\omega}\int_{\delta_1}^{\delta_2}(P_m - P_l)\,d\delta \qquad (8.13)$$

If the load angle δ is not a function of time, its second derivative with respect to time is zero. Therefore, the integral on the right hand side of Eq. (8.13) is equal to zero in a stable machine:

$$\int_{\delta_1}^{\delta_2}(P_m - P_l)\,d\delta = 0 \qquad (8.14)$$

The geometrical meaning of the integral in Eq (8.14) is the surface below the curve $(P_m - P_l)$ in the coordinate system (δ, P).

An application of the equal area criterion of stability of a synchronous machine is illustrated in Fig. 8.9 (a) and (b). In both cases the generator is running unloaded before being loaded with the power P_l. If the power of the load is not too large, as illustrated in Fig. 8.9 (a), the generator will operate at the power angle δ_l after the transient finishes.

The area A_1 between the load power line and sine function of the generator angle in Fig. 8.9 (a) is smaller than the area A_2. During the period of operation in interval $(0, \delta_l)$ the generator is decelerating.

Steady State Performance of Synchronous Machines

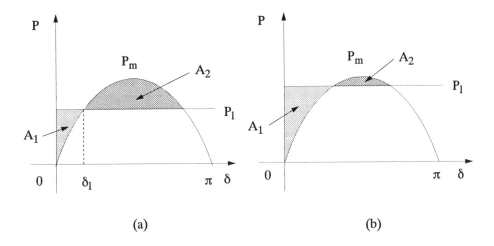

Fig. 8.9 *Application of the equal area criterion in a stable (a), and unstable (b) loading of a synchronous generator*

However, after exceeding the load angle δ_1, the mechanical power of the generator becomes greater than the power of the load, and the generator begins accelerating. It has enough time to reach the synchronous speed and overshoot it, so that it does not lose synchronism.

Loading with power P_1 in Fig. 8.9 (b) results in a loss of synchronism. Although the load demands less than the maximum power, the generator is not capable of compensating in the angle interval characterized by area A_2 the loss of speed realized while it was operating in the angle interval characterized by area A_1. Area A_2 is smaller than A_1, and the generator loses synchronism.

The vector diagram of a large synchronous machine in which the stator resistance is neglected is shown in Fig. 8.10.

The sine of angle φ can be expressed by applying the law of cosines on the voltage triangle in Fig. 8.10, which gives

$$\sin\varphi = \frac{E^2 - V^2 - (IX_s)^2}{2VIX_s} \qquad (8.15)$$

If the angle φ is kept constant, the relationship between the stator current I and the induced voltage E is obtained. The family of characteristics I(E) of a synchronous machine for various current shift angles φ is shown in Fig. 8.11. The applied voltage in this example was 220 V, and the machine's synchronous reactance was 10 Ω.

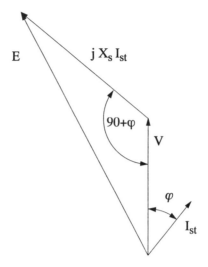

Fig. 8.10 *Simplified vector diagram of a cylindrical rotor synchronous machine*

The constant active power curves of an m-phase machine, or the V-curves, are obtained by solving modified Eq. (8.15), in which the sine of angle φ is replaced by the expression

$$\sin\varphi = \sqrt{1 - (\cos\varphi)^2} = \sqrt{1 - \left(\frac{P}{VI}\right)^2} \qquad (8.16)$$

After substituting Eq. (8.16) into Eq. (8.15) and solving it for the active power P, one obtains

$$P = mVI\sqrt{1 - \left[\frac{E^2 - V^2 - (IX_s)^2}{2VIX_s}\right]^2} \qquad (8.17)$$

The family of constant power characteristics for the same machine as in the previous example is shown in Fig. 8.12.

Equation (8.17) is biquadratic. When solved numerically, an error of computation has to be allowed for. This error of computation is clearly seen in Fig. 8.12 for small values of the active power, where the calculated solution oscillates around the line which represents the real solution. For this reason, the active power as a function of the induced voltage and stator current is shown again in Fig. 8.13, in the form of a 3D diagram.

Steady State Performance of Synchronous Machines

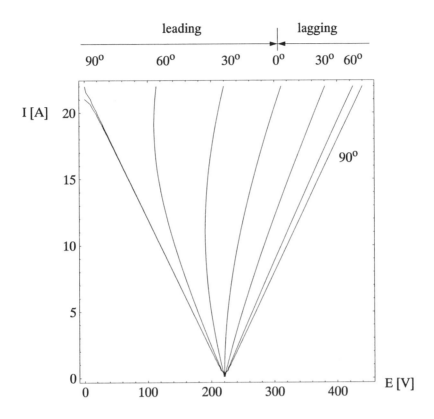

Fig. 8.11 *Constant power factor curves of a synchronous machine*

The V-curves shown in Fig. 8.12 are obtained from the three-dimensional power distribution in Fig. 8.13 by cutting it with planes parallel to the (E, I) plane.

The output performance of a synchronous generator is usually represented in the form of a *power chart*, shown in Fig. 8.14. The shaded area in this diagram defines allowed loci of the stator current vector.

Basic criteria which have to be satisfied when defining the power chart of a synchronous machine are the machine's heating and stability. Since the generator's iron core and mechanical losses are practically independent of the load, the only factors determining the heating at various loads are the stator and the rotor current, the latter being represented by the induced voltage vector E in the power chart. Another criterion, the machine's stability, is determined by the torque angle δ, as shown in Fig. 8.8. The maximum theoretical angle at which the machine can operate in a stable manner is $90°$; however, any disturbance at this angle pushes the machine into an unstable mode of operation.

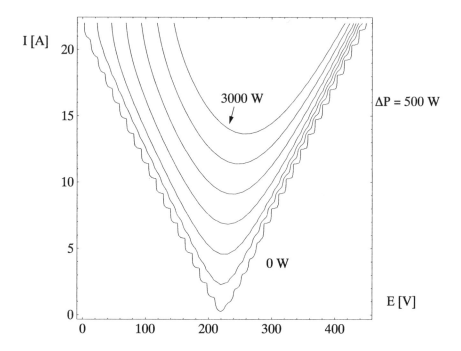

Fig. 8.12 *Constant active power curves of a synchronous machine*

Therefore, a practical stability limit is defined in such a manner that the maximum load power is not greater than (1 − s) times the maximum generator power at a given induced voltage E. The coefficient s is usually in the range between 5% and 10%.

In power systems it is customary to represent the physical quantities, such as power, voltage, current, impedance, etc., in terms of their percentile values. This so-called *per unit* (p.u.) representation has some practical value when a power system at different voltage levels has to be analyzed. The parameters of a synchronous generator are sometimes represented in per unit values. The relationships between the physical and per unit quantities are revealed in Table 8.1.

The per unit voltage of a synchronous generator corresponds to its rated voltage; the per unit current corresponds to its rated current. The rated impedance Z_r is then defined as

$$Z_r = \frac{V_r}{I_r} \tag{8.18}$$

Steady State Performance of Synchronous Machines

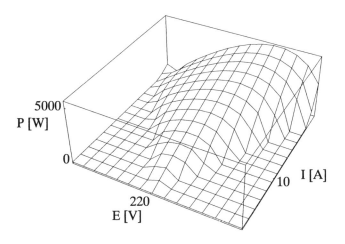

Fig. 8.13 *The active power in a synchronous machine as a function of the induced voltage E and stator current I*

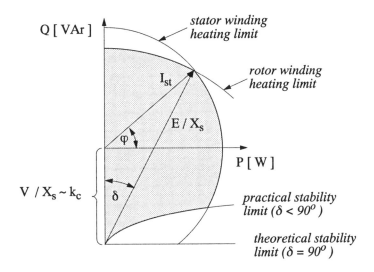

Fig. 8.14 *Power chart of a synchronous generator*

Table 8.1 Relationships between physical and p.u. quantities

	Physical dimension	p.u. notation
Rated apparent power	S_r [VA]	$S_{p.u.}$ [100%]
Rated active power	P_r [W]	$S_{p.u.}$ [100%]
Rated voltage	V_r [V]	$V_{p.u.}$ [100%]
Rated current	I_r [A]	$I_{p.u.} = S_{p.u.}/V_{p.u.}$
Rated impedance	Z_r [Ω]	$Z_{p.u.} = Z_r [Ω] \, S_{p.u.}/(V_{p.u.})^2$

If, for example, the synchronous reactance is two times greater than the rated impedance defined in Eq. (8.18), it is denoted by 200% in the per unit system.

The per unit value of synchronous reactance helps define the *short-circuit ratio* k_c of a synchronous generator, which is equal to

$$k_c = \frac{1}{X_{s,pu}} \tag{8.19}$$

The full name of the factor k_c is the no load to short-circuit ratio, because it is equal to the ratio between the field current which induced the rated voltage at no load and the field current which drives the rated stator current at a short circuit. Considering the applied voltage constant, the factor k_c is proportional to the length of the vector V/X_s in Fig. 8.14.

8.4 Salient Pole Machine

The air gap width of a salient pole machine is not constant, as in a cylindrical machine. It is small under the poles and extremely large between them. The air gap mmf distribution, calculated in the third chapter, is influenced by the variable gap geometry. The mmf created by the armature current drives an air gap flux, the amplitude of which is not only a function of the stator mmfs, but also of the relative position of the mmfs to the rotor. The armature reaction can be represented as a sum of the armature reaction along the poles, or in the d- axis, and the armature reaction in the inter-pole space, or in the q-axis. Since the armature reaction is represented in the machine's equivalent circuit by the voltage drop on a fictitious synchronous reactance, two synchronous reactances can be defined in a salient pole machine:

Steady State Performance of Synchronous Machines

- X_d, or direct (d -) axis synchronous reactance, and
- X_q, or quadrature (q -) axis synchronous reactance.

Furthermore, the armature current can be represented as a sum of its d-axis component I_d, and q-axis component I_q. The relationship between the two, shown in Fig. 8.15, can be expressed as

$$I_d = I_{st}\sin(\varphi + \delta); \qquad I_q = I_{st}\cos(\varphi + \delta) \qquad (8.20)$$

Relationships between the voltage drops across various components of synchronous reactances shown in the vector diagram in Fig. 8.15 are derived by utilizing relations

$$\cos\beta = \cos[90 - (\varphi + \delta)] = \sin(\varphi + \delta) \qquad (8.21)$$

$$\sin\beta = \sin[90 - (\varphi + \delta)] = \cos(\varphi + \delta) \qquad (8.22)$$

along with Eq. (8.20).

The torque–angle characteristic of a salient pole synchronous machine can be derived from the vector diagram in Fig. 8.15. One can write

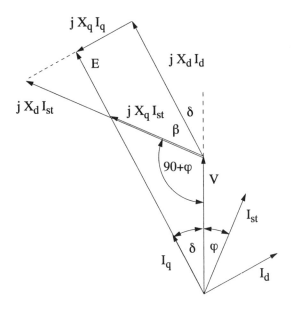

Fig. 8.15 *Vector diagram of a salient pole synchronous machine*

$$I_q X_q = V \sin\delta; \qquad\qquad I_d X_d = E - V\cos\delta \qquad (8.23)$$

Solving Eqs. (8.20) and (8.23) for $I_{st} \cos\varphi$, one obtains

$$I_{st}\cos\varphi = \frac{E}{X_d}\sin\delta + \frac{V}{2X_q}\sin 2\delta - \frac{V}{2X_d}\sin 2\delta \qquad (8.24)$$

and the power in a salient pole three phase machine

$$P = 3VI_{st}\cos\varphi = 3\left[\frac{VE}{X_d}\sin\delta + \frac{V^2}{2}\left(\frac{1}{X_q} - \frac{1}{X_d}\right)\sin 2\delta\right] \qquad (8.25)$$

The power-angle characteristic of a salient pole machine is shown in Fig. 8.16. The generated power has two components: the electromagnetic power, proportional to the sine of the angle δ, and the reluctance power, proportional to $\sin 2\delta$.

Unlike a cylindrical rotor machine, a salient pole synchronous machine develops power when the field current is equal to zero. The induced voltage in that case is equal to zero, but there still remains a component of power due to rotor saliency. This power is proportional to the product of the square of the applied voltage and the sine of the double torque angle δ. It is independent of the induced voltage. The maximum torque angle at which a salient pole machine operates in the stable mode is less than 90° electrically, however, the maximum developed power is greater than the electromagnetic power.

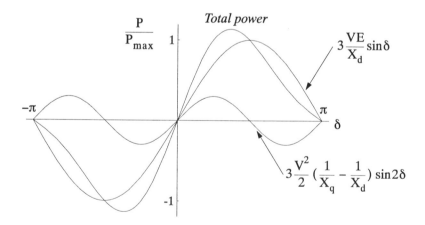

Fig. 8.16 *The components of electromagnetic power in a salient pole synchronous machine as functions of the torque angle δ in radians*

Steady State Performance of Synchronous Machines

Very often the q-axis synchronous reactance X_q in a cylindrical rotor machine is assumed to be equal to the d-axis synchronous reactance X_d. Yet there exists a small difference between them, caused by the rotor slots. These are placed in the quadrature axis of the machine, and, following the definition of the Carter factor in Chapter 3, increase the effective air gap width. Therefore, the q-axis synchronous reactance X_q in a turbogenerator is slightly smaller than the d-axis synchronous reactance X_d.

8.5 Permanent Magnet Synchronous Motor

The rotor winding of a synchronous motor can be replaced by permanent magnets. This results in higher efficiency of the motor, because the permanent magnet synchronous motor does not dissipate any rotor copper losses. Various rotor constructions of permanent magnet synchronous motors are shown in Fig. 8.17. Depending on the type of material used for the permanent magnets, the average value of the air gap flux density at no load varies from 0.3 to 0.7 T.

Rotors in which the permanent magnets are placed only in the d-axis, such as those shown in Fig. 8.17 (b) and (c), exhibit an interesting performance for small values of the torque angle δ. This performance is a consequence of the very low relative permeability of the permanent magnets around the rated operating point.

The relative permeability of permanent magnets is typically a single digit number. Therefore, the air gap width in the d-axis is fictitiously enlarged by the value h/μ_r, where h is the height of the magnet in radial direction.

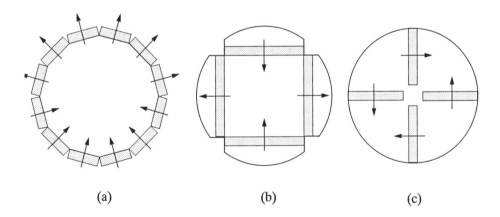

(a)   (b)   (c)

Fig. 8.17 *Rotor configurations of permanent magnet synchronous motors: a two pole motor with surface mounted magnets (a), a four pole motor with radial flux orientation (b), and a four pole motor with circumferential flux orientation (c)*

Depending on the rotor construction, the effective air gap in the d-axis can be larger than the effective air gap in the q-axis. A consequence of such geometrical relationships is that the synchronous reactance in the d-axis is smaller than the synchronous reactance in the q-axis, i.e. $X_d < X_q$! The developed torque in such a machine, calculated from Eq. (8.25) divided by the synchronous speed, is shown in Fig. 8.18. The characteristic local torque maximum at a small negative torque angle and the torque minimum at a small positive torque angle is caused by the relationship between the direct and quadrature components of the synchronous reactance. The machine cannot operate in a stable manner between these two points, making the no load torque angle significantly larger than zero.

Permanent magnet synchronous motors are usually made with a rotor cage. Such a cage can be built into the rotor poles of the machines shown in Fig. 8.17 (b) and (c). The cage is continuous, almost as in a squirrel cage induction machine, and has the same purpose – it enables line starting of the motor. Another role of the rotor cage is to protect the permanent magnets from demagnetization during starting. During starting, the cage currents create the magnetic field which is oriented against the main field, and the permanent magnets are not exposed to the high main field.

Besides the squirrel cage generated asynchronous torque during starting, a permanent magnet synchronous motor builds, at low speeds, a *braking* torque. The stator winding is short-circuited for permanent magnets in the rotor because the internal impedance of the stator connected source is zero. The voltage induced in the stator winding by rotation of the permanent magnets drives the currents which build a braking torque. Therefore, the starting performance of such a motor is deteriorated by this action of the permanent magnets.

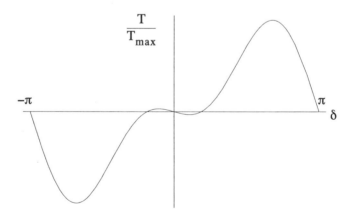

Fig. 8.18 *Torque–angle characteristic of a synchronous machine with permanent magnets in the rotor*

9 Fundamentals of Electric Machine Dynamics

In the analyses of steady states in various types of electric machines, carried out in previous chapters, it was assumed that the machine operates at a constant speed, without specifying how it reached that operating point. The steady state was characterized by constant values of the rotor speed, torque, voltages, currents, flux linkages, etc., in the machine. Considering that before energizing the machine's speed, torque, currents, fluxes, induced voltages, etc., are usually equal to zero, it is obvious that an operating point can be reached only after a transient state. Formally, any steady state operating point can be considered a special case of a transient state. At such an operating point the electromagnetic, mechanical and thermal quantities are constant. These quantities build up during the transient state and after a certain time they reach their steady state values. It is in the nature of electric machines that electromagnetic and mechanical transients usually take place simultaneously, whereas the thermal transients last much longer. From the point of view of thermal transients, the electromagnetic and mechanical variables reach their steady states immediately. From the point of view of electromagnetic and mechanical transients, the thermal state in a machine can be considered constant during the time in which the rotor speed, torque, voltages, currents, flux linkages, etc., in the machine reach their steady state values. Therefore, the analysis of transient states in electric machines is limited to electromagnetic and mechanical quantities, and interactions between them.

The basic reason why the quantities in an electric machine, as in any other physical system, change gradually is the energy stored in it. Neglecting the heat, an electric machine stores magnetic energy in its magnetic fields, and mechanical (kinetic) energy in the rotor. When any of these energy forms changes, the power has to be brought to the machine, or taken from it, satisfying the following equation:

$$dW = Pdt \qquad (9.1)$$

where dW is the differential of energy, P is the power, and dt is the differential of time.

An instantaneous change of quantity in an electric machine must be followed by an instantaneous change of accumulated energy in it, which is possible only if an infinite power is applied. Since a source of an infinite power is physically not possible, the only way to change the conditions in a machine is through a transient state.

The magnetic field in an electric machine is a storage of energy, as is the rotor inertia. During the transients, the energy accumulated in one of these storages is converted into the energy in the other. The way that this conversion takes place is dependent on the electrical, magnetic, and mechanical parameters of the system. Very often, the energy exchange is performed through oscillations.

A very important topic in the analysis of transient states of electric machines is the determination of losses. When a machine operates at its rated point, the amount of copper, iron and mechanical losses is only a fraction of the totally converted power. During transients, however, the electrical losses can substantially increase, thus limiting the thermal capability of the machine. This can be illustrated by the example of a squirrel cage induction machine which is by definition designed to start when connected directly to the rated voltage.

If the machine is unloaded, the developed electromagnetic moment during transients is used only to change the kinetic energy of the rotor. Newton's second law for rotational motion can be written in this case as

$$T_m = J\frac{d\omega}{dt} = -J\omega_s\frac{ds}{dt} \tag{9.2}$$

because the angular speed ω of an induction machine can be expressed as a function of the synchronous angular speed ω_s and the slip s as

$$\omega = \omega_s(1-s) \tag{9.3}$$

If the expression (9.2) is multiplied by $s\omega_s$, the result is

$$T_m\omega_s s\, dt = -J\omega_s^2 s\, ds \tag{9.4}$$

Since $T_m\omega_s$ is equal to the air gap power P_{ag}, and the product $P_{ag}s$ is equal to the power converted into the rotor copper losses P_{Cu2}, one can further write Eq. (9.4) as

$$P_{Cu2}dt = -J\omega_s^2 s\, ds \tag{9.5}$$

An integral of the power of losses P_{Cu2} in a given time interval from t_1 to t_2 is equal to the energy W_{Cu2} lost in the rotor copper

$$W_{Cu2} = \int_{t_1}^{t_2} P_{Cu2}dt = -J\omega_s^2\int_{s_1}^{s_2} s\, ds = J\frac{\omega_s^2}{2}(s_1^2 - s_2^2) \tag{9.6}$$

Fundamentals of Electric Machine Dynamics

During starting, the energy converted into heat in the rotor copper is equal to the change of the rotor kinetic energy! In this case the initial slip is equal to one, and the final slip is zero. The higher the speed which the rotor has to reach, and the higher the inertia on the shaft, the more energy is converted into rotor losses, and, consequently, the higher the temperature of the rotor cage. For this reason large squirrel cage machines with high inertia on the shaft may be started only a limited number of times per day.

When the machine is reversing, the energy dissipated into rotor losses is four times greater than during the starting transient because the slip changes from two to zero during reversal.

The additional rotor copper losses during transients can be reduced if the squirrel cage machine does not reach the final speed in one step, but gradually. This can be illustrated by an example of a two speed machine with a ratio of synchronous speeds 1:2. The motor is first accelerated to the low speed $\omega_s/2$, and then to the high speed ω_s. The rotor copper losses during starting to $\omega_s/2$ are only one quarter of the losses during starting to ω_s. When the motor is connected to the high speed winding, its initial slip is already 0.5, so that the dissipated energy is again one quarter of the dissipated energy during starting to the high speed from the starting slip 1. In total, 50% of energy lost when connecting the motor directly to high speed can be saved by starting it in two steps. The higher the number of speed steps, the lower the losses during transient. An infinite number of speed steps, realized by a gradual increase of the frequency of the applied voltage, causes zero dynamical losses in the rotor.

Rotor losses during transients increase if the load torque is different from zero. The developed torque T_m in this case covers the accelerating/decelerating torque, and the load torque T_l, according to the following equation:

$$T_m - T_l = J\frac{d\omega}{dt} = -J\omega_s \frac{ds}{dt} \qquad (9.7)$$

Repeating the derivation of the expression for the rotor losses from the beginning of the section, one obtains for a loaded machine

$$W_{Cu2} = J\omega_s^2 \int_{s_1}^{s_2} \frac{T_m}{T_m - T_l} s\, ds \qquad (9.8)$$

The ratio $T_m/(T_m - T_l)$ in Eq. (9.8) is always greater than one. Therefore, the rotor copper losses dissipated during transients in a loaded squirrel cage machine are always greater than the rotor copper losses in an unloaded machine.

Dissipation of additional losses during transient states is characteristic not only of induction machines, but of all types of machines. In a separately excited D.C. motor one can write from voltage equation (7.2)

$$i_a^2 R = Vi_a - Ei_a \tag{9.9}$$

where i_a is the armature current. Expressing the applied voltage V and the induced voltage E as functions of the rotor speed, see Eqs. (7.11) – (7.12), one can write Eq. (9.9) as

$$i_a^2 R = Vi_a - Ei_a = \frac{30}{\pi} c_e \Phi_m i_a (\omega_0 - \omega) = T_m (\omega_0 - \omega) \tag{9.10}$$

Multiplying Eq. (9.10) by the differential of time dt, expressed from Eq. (9.2) as

$$dt = J \frac{d\omega}{T_m} \tag{9.11}$$

one obtains

$$i_a^2 R \, dt = J(\omega_0 - \omega) \, d\omega \tag{9.12}$$

After integration from ω_1 to ω_2, one can express the dynamical losses of a D.C. shunt motor as

$$W_{Cu2} = J \left[\omega_0 (\omega_2 - \omega_1) - \frac{\omega_2^2 - \omega_1^2}{2} \right] \tag{9.13}$$

which is identical to the expression for a squirrel cage motor, in which the synchronous speed ω_s is replaced by the no load speed ω_0.

9.1 A Machine's Differential Equations

The relationships between the quantities in an electric machine at steady states are described by algebraic equations. These equations relate the currents and voltages in an equivalent circuit to the fluxes, mmfs, torque, and speed of the machine. They are usually grouped into electrical equations, and the equation of motion. The variables which appear in the electric equations are purely electric. An example of an electric equation is the current as a function of the applied voltage and parameters of the equivalent circuit of an induction machine. A typical equation of motion is the torque–speed, or torque-angle characteristic of a machine. In each of these equations both electrical (voltages, currents, fluxes, electrical power) and mechanical (angle, speed,

Fundamentals of Electric Machine Dynamics

torque, mechanical power) variables appear. As a rule, the algebraic equations of a machine are nonlinear. Only in special cases are the relationships between the variables in machines linear. The nonlinear character of algebraic equations is determined by various factors, such as:

– nonlinear magnetizing characteristic of the iron core B–H curve, which can be observed in all types of machines;

– nonlinear dependence of the developed torque on the product of the armature current and main flux in all types of machines;

– nonlinear dependence of the induced voltage on the product of the rotor speed and main flux in all types of machines;

– nonlinear dependence of the equivalent rotor resistance on the speed of an induction machine;

– nonlinear dependence of the developed torque on the speed of an induction machine, or on the rotor to stator angle in a synchronous machine;

Nonlinear relationships between physical variables also exist at transient states. Here, however, additional terms in electric equations and the equation of motion have to be introduced, in order to represent the conditions completely. One of these terms is a component of the induced voltage

$$e = \frac{d\Psi}{dt} \quad (9.14)$$

in electric equations, and the other term is the accelerating/decelerating torque

$$T_a = J\frac{d\omega}{dt} \quad (9.15)$$

in the equation of motion. Due to these additional terms, the relationships between the machine's variables in transient states have to be described in the form of *differential equations*. Steady state electrical equations, derived based upon relationships in an equivalent circuit, become voltage equations. Each circuit which takes place in electromechanical energy conversion and which is connected to a voltage source (even when its voltage is equal to zero!) is described by a voltage equation of the type

$$v = iR + \frac{d\Psi}{dt} \quad (9.16)$$

where v is the applied voltage, i is the current in the circuit, R is the circuit's resistance and Ψ is the flux linkage in the circuit. The equation of motion becomes

$$T_{em} = J\frac{d\omega}{dt} + T_l \qquad (9.17)$$

where T_{em} is the electromagnetic torque developed in the machine, J is the moment of inertia on the machine's shaft, and T_l is the load torque. The equation of motion has equal form in all types of electric machines, whereas the number and complexity of voltage differential equations varies as a function of the machine's type. Providing constant main flux, a separately excited D.C. machine is described by only one voltage differential equation, which allows one to solve the system of differential equations analytically. Voltage differential equations of A.C. machines are, however, more complex and they relate the phase flux linkages in a nonlinear manner.

The system of voltage differential equations is characteristic for each kind of machine. Its form does not depend on the type of transient. The solution of the system of voltage equations and equations of motion is a function of the initial conditions, which thus define the type of transient.

The variables, the derivatives of which appear in the machine's differential equations are called the *state variables*, because they completely define the state in the machine. Relationships between the state variables are nonlinear, as elaborated at the beginning of this section. These nonlinear relationships cause the system of the machine's differential equations to be nonlinear. The solution of a nonlinear system of differential equations cannot be found analytically. Therefore, the system of a machine's nonlinear differential equations has to be integrated numerically, step by step, in a given time interval.

9.2 Transient Phenomena in Commutator Machines

The connection of the field winding in a commutator machine defines its static as well as dynamic characteristics. Separately excited machines exhibit different dynamic performances than shunt connected ones, whereas the transients in series machines are firmly determined by the fact that the armature current flows through the field winding. An analysis of the dynamics of each type of commutator machine will be carried out separately in this section, and illustrated by an example of a starting transient. Any other transient (braking, loading, unloading, short-circuiting, etc.) can be calculated by utilizing the same differential equation(s) as for the starting, but with different initial conditions.

9.2.1 Separately Excited Motors

In the analysis of the dynamics of a separately excited motor it will be assumed that the main flux is constant. This means that the field winding was connected to its source much earlier than the armature winding, and that the field current is constant during the transient. The armature voltage equation of a separately excited commutator

Fundamentals of Electric Machine Dynamics

machine can be written as

$$V = i_a R + L_a \frac{di_a}{dt} + c_e \Phi_m n \tag{9.18}$$

and the circuit connection is shown in Fig. 7.15 (a). The resistance in the armature circuit is denoted by R in Eq. (9.18), and L_a is the total inductance in the armature. When there are no chokes connected in series with the armature, L_a is equal to the sum of the leakage slot inductance and the end winding leakage inductance of the armature winding.

The voltage differential equation (9.18) connects two state variables: the armature current i_a and the rotor speed n. This differential equation is linear because the state variables and their derivatives appear in linear form in it, and it has constant coefficients, because all its parameters are constant. After substituting the expression for electromagnetic torque from Eq. (7.16) in (9.17), one can write the equation of motion as

$$c_m \Phi_m i_a = J \frac{\pi}{30} \frac{dn}{dt} + T_l \tag{9.19}$$

As long as the main flux Φ_m is constant, the equation of motion is linear with constant coefficients. Therefore, the system of differential equations (9.18) – (9.19) can be solved analytically. Introducing the armature electrical time constant T_a

$$T_a = \frac{L_a}{R_a} \tag{9.20}$$

and the *electromechanical time constant* T_{em}

$$T_{em} = \frac{J R_a \pi}{30 c_e c_m} \tag{9.21}$$

one can write the two first-order differential equations (9.18) and (9.19) as one second-order differential equation with the armature current as the state variable

$$T_{em} T_{el} \frac{d^2 i_a}{dt^2} + T_{em} \frac{di_a}{dt} + i_a = I_l \tag{9.22}$$

or with the rotor speed as the state variable

$$T_{em} T_{el} \frac{d^2 n}{dt^2} + T_{em} \frac{dn}{dt} + n = n_l \tag{9.23}$$

The current I_l in Eq. (9.22) denotes the armature current at the steady state, and n_l is the steady state (load) speed.

The character of the solution of Eqs. (9.22) and (9.23) depends on the relationship between the electrical and electromechanical time constant. When $T_{em} < 4T_{el}$, the armature current and the rotor speed in transient states are damped harmonic functions; for other relationships between the time constants, the response is aperiodic.

State variables (armature current and rotor speed) during the starting transient of a separately excited D.C. motor are shown in Fig. 9.1 and Fig. 9.2.

The response of the machine's state variables in Fig. 9.1 is damped oscillatory. The energy in the motor oscillates during the starting transient between the motor's magnetic field and rotating masses. These oscillations are damped by the electrical resistance and mechanical friction.

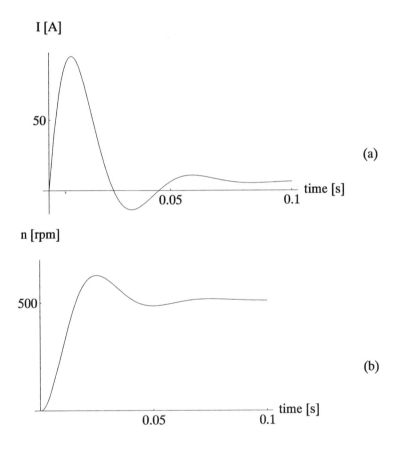

Fig. 9.1 *The armature current and the rotor speed of a separately excited D.C. motor during starting*

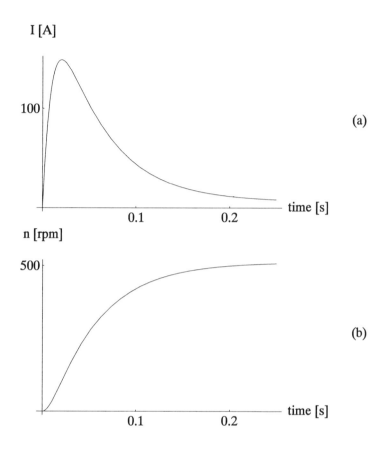

Fig. 9.2 *Same as Fig. 9.1, but with ten times greater inertia on the shaft*

Although the machine operates as a motor, without a source of mechanical energy on the shaft, the armature current during starting can be negative, as in a generator. As shown in Fig. 9.1, the armature current is negative exactly in the time interval in which the rotor speed is greater than the steady state speed defined by the load. In this time interval the machine operates as a generator, delivering the electrical power to the source at the expense of the kinetic energy of rotation. Following Eq. (9.19), zeros of the current during starting of an unloaded machine correspond to the extremes of the rotor speed.

The armature current during the transients can be much higher than its rated value. This can be dangerous for the commutator, which should not be exposed to currents greater than twice the rated current. Therefore, a separate and shunt excited D.C. motor are never directly connected to the source. They are either connected to the voltage via an additional armature resistance, which is an obsolete solution, or via a voltage regulator – a chopper, or a rectifier.

Additional inertia on the shaft increases the electromechanical time constant, without influencing the electrical time constant. At a certain value of additional inertia, the character of the machine's response changes from damped harmonic to aperiodic.

This is illustrated in Fig. 9.2 in which the current and speed in the machine is shown, having all parameters but the rotor inertia the same as the machine in Fig. 9.1, except that the rotor inertia in Fig. 9.2 is ten times greater than in Fig. 9.1. The starting transient in the latter case lasts longer than in the former, but the steady state values of the current and speed are equal in both cases. This result is based on the fact that the rotor inertia plays a role only during transient states, as shown in the equation of motion (9.17). Since a transient with a large inertia lasts longer than one with a small inertia, the rotor speed does not increase quickly, and the armature current has enough time to reach high values. In an extreme case with a very large inertia, the rotor current would, soon after the beginning of the transient, reach its short-circuit value which is defined only by the applied voltage and armature resistance, since the induced voltage is equal to zero (n = 0 !).

The saturation in the magnetic circuit of the main flux does not play any role in the analysis of the dynamics of a separately excited D.C. machine, as long as the main flux is held constant.

9.2.2 Shunt Motors

The field coil of a shunt motor is connected in parallel to the armature. Therefore, the transient in the armature circuit is simultaneous with the transient in the field circuit. In addition to differential equations (9.18) and (9.19), the voltage differential equation for the field circuit has to be solved

$$V = i_f R + L_f \frac{di_f}{dt} \qquad (9.24)$$

where index f denotes the field winding. The differential equation (9.24) is linear in an unsaturated machine. However, the system of voltage differential equations and the equation of motion is always nonlinear, because the electromagnetic torque in Eq. (9.19) is a nonlinear function (product!) of the main flux, i.e. field current, and armature current. The solution of the system of differential equations in this case cannot be found analytically, and has to be integrated step by step.

The response of an unsaturated shunt D.C. motor to the step of the applied voltage is shown in Fig. 9.3.

Since the field winding electrical time constant of a separately and shunt excited D.C. motor is typically one to two orders of magnitude longer than the armature electrical time constant, the rotor speed exhibits significant overshooting at the beginning of the transient.

Fundamentals of Electric Machine Dynamics

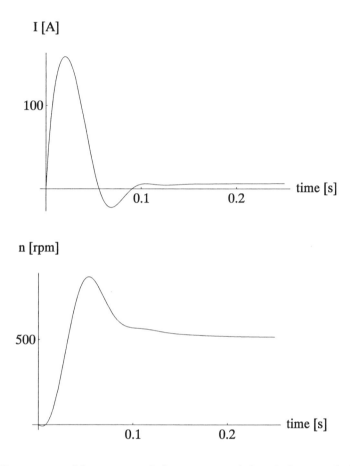

Fig. 9.3 *The current and the rotor speed of an unsaturated shunt D.C. motor during starting transient*

The field current, along with the main flux, slowly increases, and the corresponding no load speed is higher than for the rated field current.

Saturation of the main flux magnetic circuit is reflected in the transient performance of a shunt motor in such a manner that the transient states in the circuits containing saturated inductances are shorter. A saturated inductance is always smaller than an unsaturated inductance, and the electrical time constant is correspondingly shorter.

The open circuit characteristic of a saturated shunt motor, along with the dependence of the main field winding inductance on the field current, are shown in Fig. 9.4. Since the field winding inductance in a saturated machine is smaller than in an unsaturated machine, the starting transient of a shunt connected motor is shorter in a saturated than in an unsaturated machine.

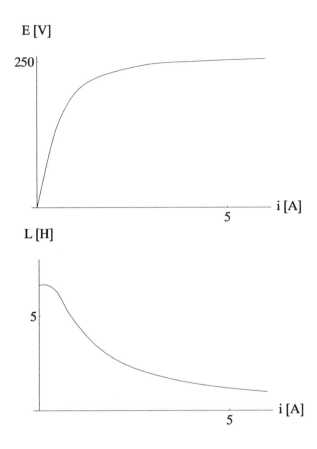

Fig. 9.4 *The open circuit characteristic and the dependence of the shunt field winding inductance on the field current in a saturated shunt connected motor*

The above is illustrated in Fig. 9.5, in which the armature current and the rotor speed during starting of a saturated shunt D.C. motor are shown. Due to the faster increase of the field current, the rotor speed in a saturated machine does not reach such high values as in an unsaturated machine.

The rotor steady state speed in a saturated shunt motor is higher than in an unsaturated one because the same field current creates less flux in a saturated than in an unsaturated machine. The rotor speed, being inversely proportional to the main flux, must be higher in a saturated motor in order to provide the same value of induced voltage as in an unsaturated motor.

Fundamentals of Electric Machine Dynamics

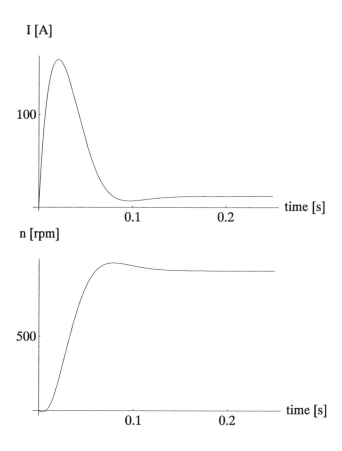

Fig. 9.5 *The current and the rotor speed of a saturated shunt D.C. motor during starting transient*

9.2.3 Series Motors

The system of differential equations of a series motor is nonlinear, no matter whether the B–H curve of the iron core is linear or nonlinear. The voltage equation of a series motor can be written as

$$V = i_a R_a + L_a \frac{di_a}{dt} + c_e \frac{L_f}{w_f} i_a n + i_a R_f + L_f \frac{di_a}{dt} \qquad (9.25)$$

with index f denoting the field winding quantities. The equation of motion is

$$T_{em} = c_m \frac{L_f}{w_f} i_a^2 = J\frac{d\omega}{dt} + T_l \qquad (9.26)$$

When the B–H curve of the iron core is linear, the coefficients of differential equations (9.25) – (9.26) are constant; otherwise, the inductance L_f is a function of the armature current i_a.

The speed and armature current during the starting transient in a series motor with nonlinear B–H curve are shown in Fig. 9.6. The armature current at the very beginning of the transient is high, in order to compensate for low induced voltage at low speed. As the motor speeds up, the current decreases to the value defined by the load.

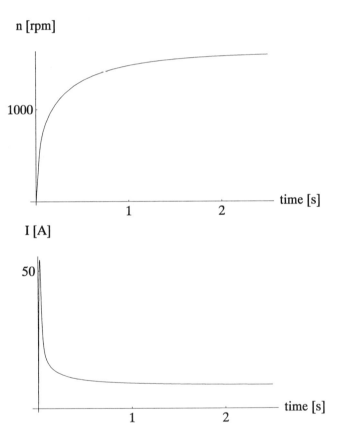

Fig. 9.6 *The speed and armature current during the starting of a series connected D.C. motor with nonlinear iron core B–H curve*

Fundamentals of Electric Machine Dynamics

9.3 Transient Phenomena in Induction Machines

A thorough analysis of transient states in induction machines which would include the influence of the nonlinearity of the B–H curve, the stator and rotor slotting, winding types and connections and many other important details would by far exceed the scope of this book, and has already been given in [6]. In this section a simplified, yet useful, dynamical model of an induction machine will be analyzed. The model is built based upon the assumptions that

- the B–H curve of the iron core is linear;
- the air gap is slotless and bilaterally cylindrical;
- the winding distribution in the air gap is continuous and sinusoidal.

Therefore, the self- and mutual inductances of the coils in this model are either constant or simple harmonic functions of the stator to rotor angle.

The induction machine in this model is represented by a set of magnetically coupled coils. The mutual inductances between the coils on the same side of the air gap are constant; the mutual inductances between the coils on opposite sides of the air gap are sinusoidal functions of the stator to rotor angle. Both stator and rotor are assumed to have three phases. A squirrel cage motor can be analyzed by utilizing this model if the electrical parameters of its squirrel cage are first reduced to a fictitious three phase winding on the rotor.

Denoting by indices a, b, and c the stator phase quantities, and by A, B, and C the rotor phase quantities, one can write the stator voltage differential equations

$$v_a = i_a R_a + \frac{d\Psi_a}{dt} \tag{9.27}$$

$$v_b = i_b R_b + \frac{d\Psi_b}{dt} \tag{9.28}$$

$$v_c = i_c R_c + \frac{d\Psi_c}{dt} \tag{9.29}$$

and the rotor voltage differential equations

$$v_A = i_A R_A + \frac{d\Psi_A}{dt} \tag{9.30}$$

$$v_B = i_B R_B + \frac{d\Psi_B}{dt} \tag{9.31}$$

$$v_C = i_C R_C + \frac{d\Psi_C}{dt} \qquad (9.32)$$

The stator phase flux linkages $\Psi_a - \Psi_c$ and the rotor phase flux linkages $\Psi_A - \Psi_C$ are functions of the machine's self- and mutual inductances, stator currents $i_a - i_c$ and rotor currents $i_A - i_C$

$$\begin{bmatrix} \Psi_a \\ \Psi_b \\ \Psi_c \\ \Psi_A \\ \Psi_B \\ \Psi_C \end{bmatrix} = \begin{bmatrix} L_a & L_{ab} & L_{ac} & L_{aA}(\gamma) & L_{aB}(\gamma) & L_{aC}(\gamma) \\ L_{ba} & L_b & L_{bc} & L_{bA}(\gamma) & L_{bB}(\gamma) & L_{bC}(\gamma) \\ L_{ca} & L_{cb} & L_c & L_{cA}(\gamma) & L_{cB}(\gamma) & L_{cC}(\gamma) \\ L_{Aa}(\gamma) & L_{Ab}(\gamma) & L_{Ac}(\gamma) & L_A & L_{AB} & L_{AC} \\ L_{Ba}(\gamma) & L_{Bb}(\gamma) & L_{Bc}(\gamma) & L_{BA} & L_B & L_{BC} \\ L_{Ca}(\gamma) & L_{Cb}(\gamma) & L_{Cc}(\gamma) & L_{CA} & L_{CB} & L_C \end{bmatrix} \begin{bmatrix} i_a \\ i_b \\ i_c \\ i_A \\ i_B \\ i_C \end{bmatrix} \qquad (9.33)$$

The self- and mutual inductances can be evaluated by utilizing the principles and procedures developed in the fourth chapter. When the windings are symmetrical, one can express the elements of the inductance matrix in Eq. (9.33) as

$$L_a = L_b = L_c = L_{\sigma, st} + L_{m, st} \qquad (9.34)$$

$$L_{ab} = L_{ba} = L_{ac} = L_{ca} = -\frac{L_{m, st}}{2} \qquad (9.35)$$

$$L_A = L_B = L_C = L_{\sigma, rot} + L_{m, rot} \qquad (9.36)$$

$$L_{AB} = L_{BA} = L_{AC} = L_{CA} = -\frac{L_{m, rot}}{2} \qquad (9.37)$$

where $L_{\sigma,st}$ is the stator phase leakage inductance, $L_{m,st}$ is the stator phase main inductance, $L_{\sigma,rot}$ is the rotor phase leakage inductance, and $L_{m,rot}$ is the rotor phase main inductance. Denoting by L_{sr} the maximum value of the stator to rotor mutual inductance per phase, one can further write

Fundamentals of Electric Machine Dynamics

$$L_{aA} = L_{sr}\cos\gamma \tag{9.38}$$

$$L_{aB} = L_{sr}\cos(\gamma + \frac{2\pi}{3}) \tag{9.39}$$

$$L_{aC} = L_{sr}\cos(\gamma - \frac{2\pi}{3}) \tag{9.40}$$

$$L_{bA} = L_{sr}\cos(\gamma - \frac{2\pi}{3}) \tag{9.41}$$

$$L_{bB} = L_{sr}\cos\gamma \tag{9.42}$$

$$L_{bC} = L_{sr}\cos(\gamma + \frac{2\pi}{3}) \tag{9.43}$$

$$L_{cA} = L_{sr}\cos(\gamma + \frac{2\pi}{3}) \tag{9.44}$$

$$L_{cB} = L_{sr}\cos(\gamma - \frac{2\pi}{3}) \tag{9.45}$$

$$L_{cC} = L_{sr}\cos\gamma \tag{9.46}$$

Denoting by $\underline{\Psi}$ the vector of the flux linkages on the left side of Eq. (9.33), by $\underline{L}$ the matrix of inductances on the right side of same equation, and by $\underline{i}$ the vector of the phase currents from the same equation, one can further write

$$\underline{\Psi} = \underline{L}\,\underline{i} \tag{9.47}$$

The voltage differential equations (9.27) – (9.32) can be written in matrix form as

$$\underline{v} = \underline{R}\,\underline{i} + \frac{d}{dt}\underline{\Psi} \tag{9.48}$$

where the diagonal matrix $\underline{R}$ of the stator and rotor resistances is defined as

$$\underline{R} = \begin{bmatrix} R_{st} & 0 & 0 & 0 & 0 & 0 \\ 0 & R_{st} & 0 & 0 & 0 & 0 \\ 0 & 0 & R_{st} & 0 & 0 & 0 \\ 0 & 0 & 0 & R_{rot} & 0 & 0 \\ 0 & 0 & 0 & 0 & R_{rot} & 0 \\ 0 & 0 & 0 & 0 & 0 & R_{rot} \end{bmatrix} \qquad (9.49)$$

The matrix $\underline{R}$ has no elements outside the main diagonal, because the windings do not interact with each other via their resistances. The matrix $\underline{L}$, on the other hand, has elements out of the main diagonal – the mutual inductances, because the mutual inductances provide the mechanism of interaction between the machine's windings.

Substituting Eq. (9.47) into Eq. (9.48) one obtains

$$\underline{v} = \underline{R}\,\underline{i} + \frac{d}{dt}(\underline{L}\,\underline{i}) = \underline{R}\,\underline{i} + \omega_{el}(\frac{d}{d\gamma}\underline{L})\underline{i} + \underline{L}\frac{d}{dt}\underline{i} \qquad (9.50)$$

where the rotor electrical angular speed ω_{el} is defined as

$$\omega_{el} = \frac{d\gamma}{dt} \qquad (9.51)$$

The system of voltage differential equations (9.50) contains six nonlinear differential equations with variable coefficients, and relates eight state variables: six phase currents, rotor to stator angle γ, and the rotor electrical angular speed ω_{el}. In addition to the system of voltage differential equations, two equations of motion have to be written in order to obtain the complete system of a machine's differential equations. The first equation of motion is given in (9.51); the second has to express the electromagnetic torque as a function of all state variables.

The torque expression can be derived from Eq. (5.8) which in matrix form yields

$$T_{em} = \frac{p}{2}\underline{i}^t(\frac{d}{d\gamma}\underline{L})\underline{i} \qquad (9.52)$$

where $\underline{i}^t$ is the transposed vector of phase currents, and p denotes the number of pole pairs.

Equations (9.50) – (9.52) describe the transient and steady states of the model of the induction machine introduced at the beginning of this section. The voltage differ-

Fundamentals of Electric Machine Dynamics

ential equations (9.50) are written from the standpoint of an observer who is fixed to the stator windings. If, however, the standpoint changes so that the observer rotates at the mechanical synchronous speed, the new frame of reference enables one to write much simpler voltage equations. The frame of reference which rotates at the mechanical synchronous speed is called the synchronous frame of reference. The currents, voltages and fluxes in the stator three phase system are transformed from the stationary frame of reference into the synchronous frame of reference by multiplying them by the transform matrix $\underline{C}_{st}$, defined as

$$\underline{C}_{st} = \begin{bmatrix} \cos \gamma_s & -\sin \gamma_s & 1 \\ \cos(\gamma_s - \frac{2\pi}{3}) & -\sin(\gamma_s - \frac{2\pi}{3}) & 1 \\ \cos(\gamma_s + \frac{2\pi}{3}) & -\sin(\gamma_s + \frac{2\pi}{3}) & 1 \end{bmatrix} \qquad (9.53)$$

where the angle γ_s can be written as

$$\gamma_s = p\omega_{s,mech} t + \gamma_{s,0} \qquad (9.54)$$

with $\omega_{s,mech}$ denoting the mechanical synchronous speed and $\gamma_{s,0}$ the initial value of the angle γ_s.

The currents, voltages and fluxes in the rotor three phase system are transformed from the rotor frame of reference which rotates at speed $(1-s)\omega_s$ to the synchronous frame of reference by multiplying them by the matrix $\underline{C}_{rot}$, defined as

$$\underline{C}_{rot} = \begin{bmatrix} \cos(\gamma_s - \gamma) & -\sin(\gamma_s - \gamma) & 1 \\ \cos(\gamma_s - \gamma - \frac{2\pi}{3}) & -\sin(\gamma_s - \gamma - \frac{2\pi}{3}) & 1 \\ \cos(\gamma_s - \gamma + \frac{2\pi}{3}) & -\sin(\gamma_s - \gamma + \frac{2\pi}{3}) & 1 \end{bmatrix} \qquad (9.55)$$

The currents in the stator phases a, b, and c are by this action transformed into new currents in the fictitious direct, quadrature, and zero (dq0) system. The angle between the d- and q-axis is 90° electrical. In a symmetrical machine, the zero component of the dq0 system does not exist. The stator phase currents transformed into direct, quadrature, and zero axes can now be written as

$$\underline{i}_{dq0} = \underline{C}_{st}^{-1} \underline{i}_{abc} \tag{9.56}$$

The rotor phase currents are analogously transformed into direct, quadrature, and zero axes by multiplication by the inverse matrix $\underline{C}_{rot}$

$$\underline{i}_{DQ0} = \underline{C}_{rot}^{-1} \underline{i}_{ABC} \tag{9.57}$$

The mutual inductances between the stator and rotor are functions of the rotor angle when observed from the stationary frame of reference. When observed from the synchronous frame of reference, the mutual inductances are constant because the stator dq axes travel at the same synchronous speed as the rotor DQ axes. This can be proven by writing Eq. (9.33) in the form of submatrices

$$\underline{\Psi}_{abc} = \underline{L}_s \underline{i}_{abc} + \underline{L}_{sr} \underline{i}_{ABC} \tag{9.58}$$

$$\underline{\Psi}_{ABC} = \underline{L}_{sr}^t \underline{i}_{abc} + \underline{L}_r \underline{i}_{ABC} \tag{9.59}$$

The structure of the submatrices and subvectors in Eqs. (9.58) – (9.59) follows from the original equation (9.33). If the fluxes and currents in the stationary frame of reference in Eqs. (9.58) – (9.59) are represented as functions of the direct, quadrature and zero axis components from the rotating frame of reference, one obtains

$$\underline{C}_{st} \underline{\Psi}_{dq0} = \underline{L}_s \underline{C}_{st} \underline{i}_{dq0} + \underline{L}_{sr} \underline{C}_{rot} \underline{i}_{DQ0} \tag{9.60}$$

for the stator, and

$$\underline{C}_{rot} \underline{\Psi}_{DQ0} = \underline{L}_{sr}^t \underline{C}_{st} \underline{i}_{dq0} + \underline{L}_r \underline{C}_{rot} \underline{i}_{DQ0} \tag{9.61}$$

or the rotor. After some matrix manipulations, the fluxes $\underline{\Psi}_{dq0}$ and $\underline{\Psi}_{DQ0}$ from Eqs. (9.60) and (9.61) can be written as

$$\underline{\Psi}_{dq0} = \underline{L}_{dq0} \underline{i}_{dq0} + \underline{L}_m \underline{i}_{DQ0} \tag{9.62}$$

$$\underline{\Psi}_{DQ0} = \underline{L}_m^t \underline{i}_{dq0} + \underline{L}_{DQ0} \underline{i}_{DQ0} \tag{9.63}$$

where

Fundamentals of Electric Machine Dynamics

$$\underline{L}_{dq0} = \begin{bmatrix} L_{s,\,tot} & 0 & 0 \\ 0 & L_{s,\,tot} & 0 \\ 0 & 0 & L_{\sigma,\,st} \end{bmatrix} \qquad \underline{L}_m = \begin{bmatrix} L_m & 0 & 0 \\ 0 & L_m & 0 \\ 0 & 0 & L_m \end{bmatrix}$$

$$L_{s,\,tot} = L_{\sigma,\,st} + \frac{3}{2} L_{m,\,st} \tag{9.64}$$

$$L_m = \frac{3}{2} L_{sr}$$

$$\underline{L}_{DQ0} = \begin{bmatrix} L_{r,\,tot} & 0 & 0 \\ 0 & L_{r,\,tot} & 0 \\ 0 & 0 & L_{\sigma,\,rot} \end{bmatrix}$$

$$L_{r,\,tot} = L_{\sigma,\,rot} + \frac{3}{2} L_{m,\,rot}$$

Since both the stator and rotor are cylindrical, and the B–H curve is unsaturated, the fluxes in perpendicular direct and quadrature axes are orthogonal, and the direct and quadrature magnetic circuits are decoupled. Therefore, the submatrices of the transformed inductances in Eq. (9.64) contain only diagonal elements.

The resistance matrix $\underline{R}$ introduced in Eq. (9.49) is invariant to the axis transform – it has the same form in the stationary as in the synchronous frame of reference.

If the system of voltage differential equations (9.50) is separated into stator and rotor parts, to which the transform matrices $\underline{C}_{st}$ and $\underline{C}_{rot}$ are then applied, one obtains

$$\underline{u}_{dq0} = \underline{C}_{st}^{-1} \underline{R}_s \underline{C}_{st}\, \underline{i}_{dq0} + \underline{C}_{st}^{-1} \frac{d}{dt} (\underline{C}_{st}\, \underline{\Psi}_{dq0}) \tag{9.65}$$

$$\underline{u}_{DQ0} = \underline{C}_{rot}^{-1} \underline{R}_r \underline{C}_{rot}\, \underline{i}_{DQ0} + \underline{C}_{rot}^{-1} \frac{d}{dt} (\underline{C}_{rot}\, \underline{\Psi}_{DQ0}) \tag{9.66}$$

After matrix manipulations, and assuming that the stator and rotor windings are symmetrical (no zero sequence components), and that the rotor windings are short circuited, one can write the system of voltage differential equations (9.65) – (9.66) in matrix form as

$$\begin{bmatrix} u_d \\ u_q \\ 0 \\ 0 \end{bmatrix} = \begin{bmatrix} \dfrac{R_s}{\sigma L_{s,tot}} & -\omega_s & -\dfrac{R_s L_m}{\sigma L_{r,tot} L_{s,tot}} & 0 \\ \omega_s & \dfrac{R_s}{\sigma L_{s,tot}} & 0 & -\dfrac{R_s L_m}{\sigma L_{r,tot} L_{s,tot}} \\ -\dfrac{R_r L_m}{\sigma L_{r,tot} L_{s,tot}} & 0 & \dfrac{R_r}{\sigma L_{r,tot}} & -s\omega_s \\ 0 & -\dfrac{R_r L_m}{\sigma L_{r,tot} L_{s,tot}} & s\omega_s & \dfrac{R_r}{\sigma L_{r,tot}} \end{bmatrix} \begin{bmatrix} \Psi_d \\ \Psi_q \\ \Psi_D \\ \Psi_Q \end{bmatrix} + \dfrac{d}{dt} \begin{bmatrix} \Psi_d \\ \Psi_q \\ \Psi_D \\ \Psi_Q \end{bmatrix}$$

(9.67)

The stator applied voltage u_d in the d-axis is equal to the sum of:
- the resistive voltage drop on the d-axis resistance due to the current i_d

$$\dfrac{R_s}{\sigma L_{s,tot}} \Psi_d \qquad (9.68)$$

- the motional induced voltage $-\omega_s \Psi_q$, due to the flux Ψ_q in the q-axis,
- the resistive voltage drop on the d-axis resistance due to the transformer induced current from the rotor D-axis

$$-\dfrac{R_s L_m}{\sigma L_{r,tot} L_{s,tot}} \Psi_D \qquad (9.69)$$

- and the transformer induced voltage $d\Psi_d/dt$, due to the flux Ψ_d in the d-axis,

Similar conclusions can be derived for the voltage in the stator q-axis, as well as for the voltages in the rotor D- and Q- axes.

The conversion from the original into direct and quadrature quantities is valid only if the power and torque remain invariant. Since the motor is connected to the source only on its stator terminals, the scalar of the applied electrical power p_{el} is equal to

Fundamentals of Electric Machine Dynamics

$$P_{el} = i^t_{abc} \, u_{abc} = i^t_{dq0} \, C^t_{st} \, C_{st} \, u_{dq0} \tag{9.70}$$

or

$$P_{el} = \frac{3}{2}(u_d i_d + u_q i_q) \tag{9.71}$$

if the symmetrical machine is fed from a symmetrical source. Analogously, one can express the scalar of the stator copper losses

$$P_{Cus} = \frac{3}{2} R_s (i_d^2 + i_q^2) \tag{9.72}$$

and the scalar of the rotor copper losses

$$P_{Cur} = \frac{3}{2} R_r (i_D^2 + i_Q^2) \tag{9.73}$$

The torque in Eq. (9.52) can be written as

$$T_{em} = \frac{P}{2} [i^t_{abc} (\frac{d}{d\gamma} L_{sr}) i_{ABC} + i^t_{ABC} (\frac{d}{d\gamma} L^t_{sr}) i_{ABC}] \tag{9.74}$$

or

$$T_{em} = \frac{P}{2} [i^t_{dq0} C^t_{st}(\frac{d}{d\gamma} L_{sr}) C_{rot} i_{DQ0} + i^t_{DQ0} C^t_{rot}(\frac{d}{d\gamma} L^t_{sr}) C_{st} i_{dq0}] \tag{9.75}$$

Since

$$C^t_{st}(\frac{d}{d\gamma} L_{sr}) C_{rot} = - C^t_{rot}(\frac{d}{d\gamma} L^t_{sr}) C_{st} = \begin{bmatrix} 0 & -\frac{3}{2}L_m & 0 \\ \frac{3}{2}L_m & 0 & 0 \\ 0 & 0 & 0 \end{bmatrix} \tag{9.76}$$

one can express the electromagnetic torque in terms of transformed currents as

$$T_{em} = \frac{3}{2} p L_m (i_q i_D - i_d i_Q) \tag{9.77}$$

or in terms of the transformed flux linkages as

$$T_{em} = \frac{3}{2} p \frac{L_m}{\sigma L_{r,tot} L_{s,tot}} (\Psi_q \Psi_D - \Psi_d \Psi_Q) \qquad (9.78)$$

Besides replacing the original currents with those in the synchronous frame of reference, the transformation of coordinates substitutes the state variables γ and dγ/dt for the mechanical equations with slip s and its derivative ds/dt. Therefore, the equation of motion in the synchronous frame of reference can be written as

$$J\omega_{s,mech} \frac{ds}{dt} = \frac{3}{2} p \frac{L_m}{\sigma L_{r,tot} L_{s,tot}} (\Psi_q \Psi_D - \Psi_d \Psi_Q) - T_l \qquad (9.79)$$

The differential equations (9.71) and (9.83) form the system of differential equations of an induction machine in transformed coordinates, which is valid for both transient and steady states. This system will be illustrated in the following example, see Fig. 9.7 – Fig. 9.10, in which the starting transient of an induction motor is calculated.

Before energizing, the fluxes and currents in the motor, as well as the rotor speed, are all equal to zero. The torque with which the motor starts must, therefore, be equal to zero.

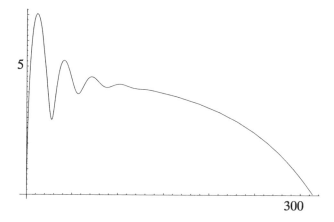

Fig. 9.7 *Torque [Nm] vs. angular speed [rad/sec] of an induction motor connected to a set of symmetrical sinusoidal voltages*

Fundamentals of Electric Machine Dynamics

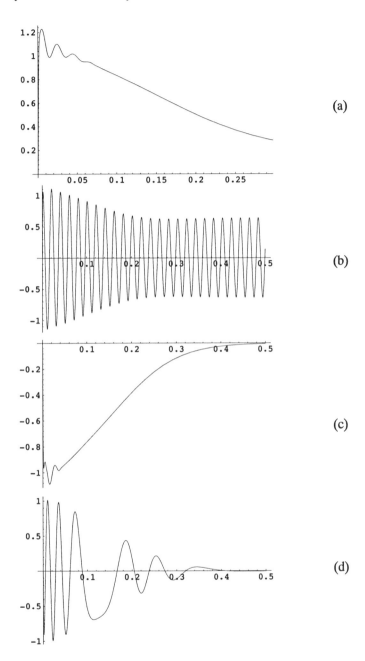

Fig. 9.8 *The motor currents [A] vs. time [sec] during starting: the transformed d-axis stator current (a), the current in the stator phase a (b), the transformed D-axis rotor current (c) and the current in the rotor phase A (d)*

Only when the motor currents build up does the torque become large enough to start the rotor. The average value of the transient torque at the beginning of the starting transient is equal to the steady state starting torque, calculated in the sixth chapter. The torque in Nm vs. angular speed in rad/s is shown in Fig. 9.7.

The currents in the motor during starting are shown in Fig. 9.8. The transformed stator current i_d, shown in Fig. 9.8 (a), is zero at the beginning, and after the transient reaches its steady state value, which is always different from zero. At steady state, the current i_d is constant, because the amplitudes of the real motor currents are constant. The stator current in phase a is shown in Fig. 9.8 (b). This current is sinusoidal, with the frequency equal to the frequency of the applied voltage. After the transient, its amplitude decreases and becomes constant at steady state.

The transformed rotor current i_D, shown in Fig. 9.8 (c) decreases to zero after the transient, because the load torque was assumed to be zero in this computation. The rotor current in phase A, shown in Fig. 9.8 (d), changes both the amplitude and frequency during starting. Its steady state value is equal to zero because the machine is unloaded. In a loaded machine, the steady state value of the rotor current is different from zero, and its frequency is a few hertz.

The starting torque of the same machine having a ten times smaller inertia is shown in Fig. 9.9. The currents in the motor which is analyzed in this example are shown in Fig. 9.10 The rotor in this case follows the changes in the fluxes and currents without delay, as a consequence of its small inertia.

In the vicinity of synchronism, the rotor has enough kinetic energy and mobility to briefly enter the generator mode of operation, and then it returns to below the synchronous speed. This mode of operation can be repeated several times, as represented by the curls around the synchronous speed in Fig. 9.9.

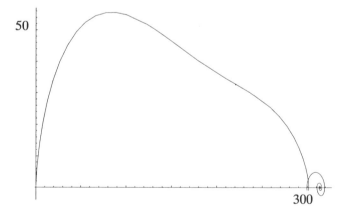

Fig. 9.9 *Torque [Nm] vs. angular speed [rad/sec] of the induction motor from Fig. 9.7, but with a ten times smaller inertia*

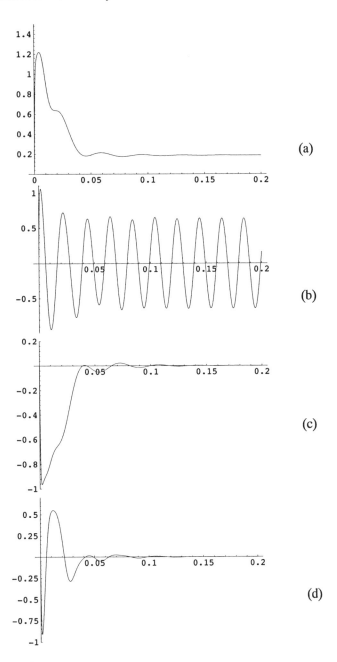

Fig. 9.10 *The currents [A] during starting of the motor in Fig. 9.9: the transformed d-axis stator current (a), the current in the stator phase a (b), the transformed D-axis rotor current (c) and the current in the rotor phase A (d)*

9.4 Transient Phenomena in Synchronous Machines

The very first analyses of transients in electric machines were prompted by problems in the operation of large synchronous generators on an infinite bus. The d–q representation of the machine's quantities, utilized in the previous section in the example of an induction machine, was originally developed to study the stability of synchronous generators. In such an analysis, it is necessary to calculate the performance of a generator connected to a power system in which a short circuit appears. The most important issue in stability analysis is whether the generator will lose synchronism, which can have severe consequences for the generator and other equipment.

9.4.1 Physics of Transients in a Synchronous Machine

Any change of the load current on the stator side, including a short circuit as the most drastic example, is followed by a change in the stator flux. This flux cannot instantaneously penetrate into the rotor because of the rotor reaction. In the very first periods of the load change, the penetration of the stator flux into the rotor is impeded by the damper winding. The damper winding (amortisseur) is built into the rotor poles of a salient pole machine as a squirrel cage. In cylindrical rotor machines, the complete rotor body, along with the slot wedges and the rotor end bell, play the role of the damper winding. In this so-called *subtransient* period, the additional stator flux is limited to the rotor damper winding, as illustrated in Fig. 9.11 (a).

After the stator created component of flux has linked the damper winding, it penetrates deeper into the rotor body, linking the turns of the field winding. This process can last several seconds and is called the *transient* period, as illustrated in Fig. 9.11 (b). At the end of the transient, steady state is reached and the flux distribution in the rotor assumes it stationary shape, as shown in Fig. 9.11 (c).

The physics of a sudden short circuit in a synchronous generator cannot be explained by its steady state equivalent circuit, introduced in the eighth chapter. This equivalent circuit, as well as any other steady state equivalent circuit, is valid only for stationary values of the electromagnetic quantities in the generator. Therefore, an extended equivalent circuit has to be developed, which will allow one to calculate the fluxes, voltages, currents, torque, etc., during transient states.

The equivalent circuit per phase of a synchronous machine which enables computation of the currents and the torque in transient states is shown in Fig. 9.12. Here the following elements are used:

- $X_{\sigma,st}$: stator leakage reactance per phase,
- X_h: stator fictitious reactance due to armature reaction,
- $X_{\sigma f}$: field winding leakage reactance,
- $X_{\sigma f}$: damper winding leakage reactance,
- R_{st}: stator phase resistance,
- R_f: field winding resistance,
- R_d: damper winding resistance.

Fundamentals of Electric Machine Dynamics

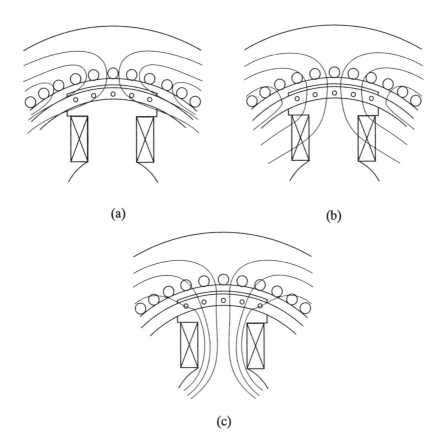

Fig. 9.11 *Stator flux penetration into the rotor of a four pole salient synchronous machine during a sudden short circuit: subtransient (a), transient (b), and steady state (c)*

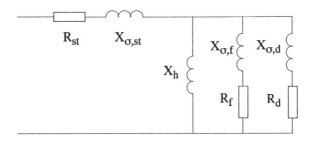

Fig. 9.12 *Equivalent circuit of a synchronous machine which allows for subtransient and transient effects*

Neglecting the influence of the resistances, one can define
(a) the subtransient reactance in the d-axis as

$$X_d'' = X_{\sigma, st} + \frac{X_h X_{tr}}{X_h + X_{tr}} \qquad (9.80)$$

(b) the subtransient reactance in the q-axis as

$$X_q'' = X_{\sigma, st} + \frac{X_h X_{\sigma, d}}{X_h + X_{\sigma, d}} \qquad (9.81)$$

(c) the transient reactance in the d-axis as

$$X_d' = X_{\sigma, st} + \frac{X_h X_{\sigma, f}}{X_h + X_{\sigma, f}} \qquad (9.82)$$

(d) the transient reactance in the q-axis as

$$X_q' = X_q \qquad (9.83)$$

where the reactance X_{tr} is equal to the parallel connection of reactances $X_{\sigma, f}$ and $X_{\sigma, d}$, i.e.

$$X_{tr} = \frac{X_{\sigma, f} X_{\sigma, d}}{X_{\sigma, f} + X_{\sigma, d}} \qquad (9.84)$$

These reactances are characteristic of symmetrical transients. If, however, the currents in the phases of a synchronous generator have different amplitudes, or phase shifts different from 120°, as in the case of a single phase short circuit, and a two phase short circuit with and without an earth connection, the influence of the inverse field in the air gap has to be allowed for.

As C.L. Fortescue proved at the beginning of the century, any asymmetric three phase system of voltages or currents can be represented as the sum of
– a symmetrical three phase system which rotates in the same direction as the original system (*positive sequence*),
– a symmetrical three phase system which rotates in a direction opposite to the original system (*negative sequence*), and
– the sum of three voltages or currents which are in-phase in all three phases (*zero

Fundamentals of Electric Machine Dynamics

sequence).

Introducing the complex operator α, defined as

$$\alpha = e^{j\frac{2\pi}{3}}; \qquad \alpha^2 = e^{j\frac{4\pi}{3}}; \qquad \alpha^3 = 1 \qquad (9.85)$$

one can define the amplitude of the positive sequence component of the currents as

$$I_p = \frac{I_a + \alpha I_b + \alpha^2 I_c}{3} \qquad (9.86)$$

the amplitude of the negative sequence component of the currents as

$$I_n = \frac{I_a + \alpha^2 I_b + \alpha I_c}{3} \qquad (9.87)$$

and the amplitude of the zero sequence component of the currents as

$$I_0 = \frac{I_a + I_b + I_c}{3} \qquad (9.88)$$

with I_a, I_b, and I_c denoting the currents in phases a, b, and c, respectively. Defining the transform matrix $\underline{T}$ as

$$\underline{T} = \begin{bmatrix} 1 & 1 & 1 \\ \alpha^2 & \alpha & 1 \\ \alpha & \alpha^2 & 1 \end{bmatrix}; \qquad \underline{T}^{-1} = \frac{1}{3}\begin{bmatrix} 1 & \alpha & \alpha^2 \\ 1 & \alpha^2 & \alpha \\ 1 & 1 & 1 \end{bmatrix}; \qquad (9.89)$$

one can express the relationship between the vector of the currents in phases $\underline{I}_{abc}$ and the vector of the symmetrical components of currents $\underline{I}_{pn0}$ as

$$\underline{I}_{abc} = \underline{T}\,\underline{I}_{pn0} \qquad (9.90)$$

Analogously, one can define the relationship between the vectors of asymmetric and symmetrical voltages as

$$\underline{V}_{abc} = \underline{T}\, \underline{V}_{pn0} \tag{9.91}$$

When a synchronous generator is loaded with an asymmetric load, Eq. (8.2) can be written in vector form as

$$\underline{E}_{abc} = \underline{V}_{abc} + \underline{Z}_u\, \underline{I}_{abc} \tag{9.92}$$

where $\underline{Z}_u$ is a matrix of impedances for an asymmetric load. All entries of the matrix $\underline{Z}_u$ can be different from zero, depending on the type of asymmetry; in a symmetrically loaded machine, the matrix $\underline{Z}_u$ is diagonal, with diagonal elements equal to the synchronous reactance of the machine per phase. Replacing the abc vectors in Eq. (9.96) by the pn0 vectors, one obtains

$$\underline{V}_{pn0} = \underline{T}^{-1}\, \underline{E}_{abc} - \underline{T}^{-1}\, \underline{Z}_u\, \underline{T}\, \underline{I}_{abc} \tag{9.93}$$

or, after matrix manipulations,

$$\begin{bmatrix} V_p \\ V_n \\ V_0 \end{bmatrix} = \begin{bmatrix} E \\ 0 \\ 0 \end{bmatrix} - \begin{bmatrix} Z_p & 0 & 0 \\ 0 & Z_n & 0 \\ 0 & 0 & Z_0 \end{bmatrix} \begin{bmatrix} I_p \\ I_n \\ I_0 \end{bmatrix} \tag{9.94}$$

It was assumed in deriving Eq. (9.98) that the induced voltages in all three stator phases are equal, and that their amplitude is equal to E. In the same equation, the positive sequence impedance of the generator is denoted by Z_p, the negative sequence impedance by Z_n, and the zero sequence impedance by Z_0.

Based upon the equivalent circuit in Fig. 9.12, the envelope of the sudden short-circuit current can be expressed as

$$i_{SC}(t) = i_{ss} + (i_t - i_{ss})\, e^{-\frac{t}{T_d'}} + (i_{st} - i_t)\, e^{-\frac{t}{T_d''}} + i_{st}\, e^{-\frac{t}{T_a}} \tag{9.95}$$

where the components of the sudden short-circuit current i_{st}, i_t, and i_{ss} depend on the type of short circuit, as shown in Table 9.1. The time constants in Eq. (9.95) are defined as:

- T_d'' : subtransient time constant,
- T_d' : transient time constant,
- T_a : D.C. current time constant

Table 9.1 Amplitudes of the components of a sudden short-circuit current in a three phase synchronous generator

Type	Subtransient i_{st}	Transient i_t	Steady state i_{ss}
Three pole	$\dfrac{V}{X_d''}$	$\dfrac{V}{X_d'}$	$\dfrac{V}{X_d}$
Two pole	$\sqrt{3}\dfrac{V}{X_d'' + X_n}$	$\sqrt{3}\dfrac{V}{X_d' + X_n}$	$\sqrt{3}\dfrac{V}{X_d + X_n}$
One pole	$3\dfrac{V}{X_d'' + X_n + X_0}$	$3\dfrac{V}{X_d' + X_n + X_0}$	$3\dfrac{V}{X_d + X_n + X_0}$

9.4.2 Park's Equations

Transients in a synchronous machine can be analyzed by applying the same technique as for an induction machine, introduced in the previous section. The stator quantities are transformed into a coordinate system which is fixed to the rotor d-axis. Such a transformation in a synchronous machine has a sound physical background, because the results in the transformed dq system can be directly interpreted physically.

Assume that on the stator side there is a symmetrical, three phase winding and that the inner stator surface is cylindrical. On the rotor there are salient poles, on which the field winding is placed. As shown in the fourth chapter, the self inductances of the stator windings in such a machine vary periodically with an integer multiple of double the stator to rotor angle γ. The same is true for the mutual inductances between the stator windings. Sometimes a term is added to the expression for mutual inductance between two stator phases, which represents the mutual flux between the stator phases in the end winding and in common slots. The mutual inductances between the stator and rotor windings are functions of integer multiples of the stator to rotor angle γ. Denoting the machine's self- and mutual inductances in the same manner as in an induction machine, one can express them as follows:

- the stator winding self inductances

$$L_a(2\gamma) = L_{\sigma, st} + L_{m, st} + L_2 \cos 2\gamma \qquad (9.96)$$

$$L_b(2\gamma) = L_{\sigma,st} + L_{m,st} + L_2 \cos(2\gamma + \frac{2\pi}{3}) \qquad (9.97)$$

$$L_c(2\gamma) = L_{\sigma,st} + L_{m,st} + L_2 \cos(2\gamma - \frac{2\pi}{3}) \qquad (9.98)$$

with L_2 denoting the second harmonic of the stator self inductance.
- the stator winding mutual inductances

$$L_{ab}(2\gamma) = -\frac{L_{m,st}}{2} + L_{2m} \cos(2\gamma - \frac{2\pi}{3}) \qquad (9.99)$$

$$L_{ac}(2\gamma) = -\frac{L_{m,st}}{2} + L_{2m} \cos(2\gamma + \frac{2\pi}{3}) \qquad (9.100)$$

$$L_{cb}(2\gamma) = -\frac{L_{m,st}}{2} + L_{2m} \cos 2\gamma \qquad (9.101)$$

with L_{2m} denoting the second harmonic of the mutual inductance between the stator phases.
- the rotor winding self inductance

$$L_f = L_{m,rot} \qquad (9.102)$$

- the rotor to stator mutual inductances

$$L_{af}(\gamma) = L_{sr} \cos\gamma \qquad (9.103)$$

$$L_{bf}(\gamma) = L_{sr} \cos(\gamma - \frac{2\pi}{3}) \qquad (9.104)$$

$$L_{cf}(\gamma) = L_{sr} \cos(\gamma + \frac{2\pi}{3}) \qquad (9.105)$$

The matrix relationship between the stator and rotor flux linkages and current can be written analogously to Eq. (9.33), or

$$\begin{bmatrix} \Psi_a \\ \Psi_b \\ \Psi_c \\ \Psi_f \end{bmatrix} = \begin{bmatrix} L_a(2\gamma) & L_{ab}(2\gamma) & L_{ac}(2\gamma) & L_{af}(\gamma) \\ L_{ba}(2\gamma) & L_b(2\gamma) & L_{bc}(2\gamma) & L_{bf}(\gamma) \\ L_{ca}(2\gamma) & L_{cb}(2\gamma) & L_c(2\gamma) & L_{cf}(\gamma) \\ L_{af}(\gamma) & L_{bf}(\gamma) & L_{cf}(\gamma) & L_f \end{bmatrix} \begin{bmatrix} i_a \\ i_b \\ i_c \\ i_f \end{bmatrix} \qquad (9.106)$$

Only the self inductance of the rotor field winding is constant in the machine. The voltage differential equation has the same form as that for a wound rotor induction machine, Eq. (9.50). In the case of a synchronous machine there are, however, only four instead of six differential equations from Eq. (9.50).

If the stator currents are transformed into the rotor coordinate system by multiplying them by the matrix $\underline{C}_{st}$ defined in Eq. (9.53), and if the same procedure for deriving the machine's differential equations is applied as for an induction machine in the previous section, one obtains

$$\begin{bmatrix} u_d \\ u_q \\ u_0 \\ u_f \end{bmatrix} = \begin{bmatrix} R_s + L_d \frac{d}{dt} & -\omega_s L_q & 0 & L_{sr} \\ \omega_s L_d & R_s + L_q \frac{d}{dt} & 0 & \omega_s L_{sr} \\ 0 & 0 & R_s + L_0 \frac{d}{dt} & 0 \\ \frac{3}{2} L_{sr} \frac{d}{dt} & 0 & 0 & R_f + L_f \frac{d}{dt} \end{bmatrix} \begin{bmatrix} i_d \\ i_q \\ i_0 \\ i_f \end{bmatrix} \qquad (9.107)$$

where

$$L_d = L_{\sigma,s} + \frac{3}{2}(L_{m,st} + L_2) \qquad (9.108)$$

$$L_q = L_{\sigma,s} + \frac{3}{2}(L_{m,st} - L_2) \qquad (9.109)$$

$$L_0 = L_{\sigma,s} + 2L_{s,ew} \qquad (9.110)$$

The mutual inductance between one rotor winding and three stator phases in the d-axis is not equal to the mutual inductance between three stator phases in the d-axis and one rotor winding. In the first case one has $3/2\, L_{sr}$, and in the second case only L_{sr}. The difference between the two is caused by the mode in which the stator mmf is reflected to the rotor and vice versa, as elaborated in the fourth chapter

The mutual inductance between two stator phases due to common flux between them out of the air gap (e.g. end winding, slots) is denoted by $L_{s,ew}$. Index f in the previous equations denotes the field quantities. It should be noted in Eq. (9.107) that the zero sequence component of the stator flux and currents is introduced, in order to allow for the transients in which it can be generated.

The flux linkages in the d–q system fixed to the rotor d-axis can now be written as

$$\Psi_d = L_d i_d + L_{sr} i_f \tag{9.111}$$

$$\Psi_q = L_q i_q \tag{9.112}$$

$$\Psi_0 = L_0 i_0 \tag{9.113}$$

$$\Psi_f = \frac{3}{2} L_{sr} i_d + L_f i_f \tag{9.114}$$

Substituting Eqs. (9.111) – (9.114) into Eq. (9.107) one obtains

$$u_d = R_s i_d + \frac{d\Psi_d}{dt} - \omega_s \Psi_q \tag{9.115}$$

$$u_q = R_s i_q + \frac{d\Psi_q}{dt} + \omega_s \Psi_d \tag{9.116}$$

$$u_0 = R_s i_0 + \frac{d\Psi_0}{dt} \tag{9.117}$$

$$u_f = R_f i_f + \frac{d\Psi_f}{dt} \tag{9.118}$$

Equations (9.115) and (9.116) are known as Park's equations. The complete system of a synchronous machine's differential equations consists of Eqs. (9.115) – (9.118) and the torque equation. The developed torque in a salient pole synchronous machine can be derived in the same manner as the torque in an induction machine. The machine

Fundamentals of Electric Machine Dynamics

is connected to electrical terminals on its stator and rotor terminals, so that Eq. (9.71) for electrical power can be extended to

$$P_{el} = \frac{3}{2}(u_d i_d + u_q i_q) + u_f i_f \qquad (9.119)$$

which finally gives the torque in the form

$$T_{em} = \frac{3}{2} p (\Psi_q i_d - \Psi_d i_q) \qquad (9.120)$$

Appendix

A 1.1

The curve in **Fig.1.2** is generated by executing the following sequence of *Mathematica* statements:

First, the function `Langevin` is defined

```
In[1]:= Langevin[M_,H_,Ms_,a_,alpha_]:=M/Ms-Coth[(H+alpha
M)/a]+a/(H+ alpha M)
```

Auxiliary plot arrays `Marray`, `MarrayPlotH` and `MarrayPlotM` are introduced:

```
In[2]:= Marray=Table[FindRoot[Langevin[M,H,Ms,a,alpha]==0,
{H,1}],{M,-14*10^5, 14*10^5,5*10^4}]
In[3]:=MarrayPlotH=Table[H /. Marray[[i]],{i,57}]
In[4]:=MarrayPlotM=Table[-14*10^5+(i-1)*5*10^4,{i,1,57}]
```

After executing the following statements, Langevin's function in Fig. 1.2 is drawn

```
In[5]:=Langevin=Table[0,{i,57},{j,2}];
In[6]:=Do[{Langevin[[i,1]]=MarrayPlotH[[i]],Langevin[[i,2]]
=Marray PlotM[[i]]},{i,57}]
In[7]:=ListPlot[Langevin,PlotJoined->True,PlotRange->{{-
6500, 6500}, Automatic},DefaultFont ->{"Times-Italic",9},
AxesLabel->{"H [A/m] "," M [A/m]"}, Ticks->{{-5000,5000},
{-10^6,10^6}}]
```

Appendix 435

When generating the hysteresis curve in **Fig.1.3**, the parameters of the magnetic material have to be defined first:

```
In[1]:= Ms=2 10^6;a=1100;kon=0.0004;alpha=1.6 10^(-3);
mu0=N[4 Pi 10^ ( -7)];c=0.2;Hmax=8000;omega=N[2 Pi];
```

Import the RungeKutta package to solve the B–H curve differential equation

```
In[2]:= <<RungeKutta.m
```

Define and solve the differential equation (1.11)

```
In[3]:= HystCurve=RungeKutta[{Hmax omega Cos[omega t],
((Ms(Coth[(H+ alpha M)/a]-a/(H+alpha M))-M)/((1+c)(Sign[Cos
[omega t]] kon/mu0 -alpha (Ms(Coth[(H+alpha M)/a]-a/(H+
alpha M))-M)))+c/(1+c)(Ms (a/(H+alpha M) ^2-Csch[(H+alpha
M)/a]^2/a)))Hmax omega Cos[omega t]},{t,H,M}, {0.,0.1,0.1},
1.3,10^(-2),InitialStepSize->0.001, MaximumStepSize-> 0.01,
ProgressTrace->True]
```

The solution of the previous equation is the curve M(H). The dependence of the flux density B on the field strength H is obtained from the curve M(H) by applying Eq.(1.9):

```
In[4]:= BHCurve=Table[0,{i,155},{j,2}];
In[5]:= Do[BHCurve[[i,2]]=mu0(HystCurve[[i,2]]+HystCurve[[
i,3]]),{i,155}]
In[6]:= Do[BHCurve[[i,1]]=HystCurve[[i,2]],{i,155}]
```

The curve in Fig.1.3 is now plotted by executing

```
In[7]:= ListPlot[ BHCurve, PlotJoined-> True, PlotRange->
{{-8000,8000},{-2.5,2.5}},DefaultFont ->{"Times-Italic",9},
AxesLabel->{"H [A/m]", "B [T]"},Ticks->{{-5000,5000}, Auto-
matic}]
```

A 1.2

The simultaneous plot of the flux density B and field strength H in **Fig.1.22** is obtained by executing the following statements

```
In[1]:= Ms=2 10^6;;a=1100;kon=0.0004;alpha=1.6 10^(-3);
mu0=N[4 Pi 10^ (-7)];c=0.2;Mmax=1.75 10^6;omega=N[2 Pi]

In[2]:= <<RungeKutta.m

In[3]:= HystCurve = RungeKutta[{Mmax omega Cos[omega t], 1/
((Ms(Coth[(H+alpha M)/a]-a/(H+alpha M))-M)/((1+c)(Sign[Ms
(Coth[(H+alpha M)/a]-a/(H+alpha M))-M] kon/mu0 -alpha
(Ms(Coth[(H+alpha M)/a]-a/(H+alpha M))-M)))+c/(1+c)(Ms (a/
(H+alpha M)^2-Csch[(H+alpha M)/a]^2/a)))Mmax omega Cos
[omega t]},{t,M,H},{0.,0.1,0.1},2,10^(-2),InitialStepSize->
0.001, MaximumStepSize->0.01,ProgressTrace->False]

In[4]:= Hoft=Table[0,{i,205},{j,2}];

In[5]:= Boft=Table[0,{i,205},{j,2}];

In[6]:= Do[(Hoft[[i,1]] =HystCurve[[i,1]];Hoft[[i,2]] =
HystCurve[[i,3]]) ,{i,205}]

In[7]:= Do[(Boft[[i,1]]=HystCurve[[i,1]]),{i,205}]

In[8]:= Do[Boft[[i,2]]=mu0(HystCurve[[i,3]]+HystCurve
[[i,2]]),{i,205}]
```

Plot first the curves H(t) and B(t) separately

```
In[9]:= HCurve=ListPlot[Hoft,PlotJoined->True,PlotRange->
{{0,2}, {-6500,6500}},DefaultFont ->{"Times-Italic",9},
AxesLabel->{"t/T"," "}, Ticks->{{1,2},None}]

In[10]:= BCurve=ListPlot[BoftRed,PlotJoined->True,
PlotRange->{{0,2},{-6500,6500}},DefaultFont -> {"Times-
Italic",9},AxesLabel->{"t/T"," "}, Ticks->{{1,2},None}]
```

Show the curves H(t) and B(t) in one coordinate system, as in Fig.(1.22)

```
In[11]:= Show[HCurve,BCurve]
```

The analysis of a nonlinear magnetic circuit sketched in **Fig.1.28** begins with the description of the iron core magnetization characteristic. Therefore, the table BHCurve should first be defined, the elements of which are discrete points of the B–H curve

Appendix

```
In[1]:=BHCurve={{0,0},{.1,12.5},{.2,25},{.3,37.5},{.4,50},
{.5,62.5},{.6,75},{.7,87.5},{.8,100},{.9,112.5},{1,125},
{1.1,160},{1.2,240},{1.3,420},{1.4,1100},{1.5,2650},{1.6,
5300},{1.7,9100},{1.8,14000},{1.9,21000},{1.95,27000},{2.,
40000},{3.,1170000},{4.,2380000},{5.,3550000},{6.,4720000},
{7.,5890000},{8.,7060000},{9.,8230000},{10.,9400000}};
```

Create the continuous function H(B), called HofB

```
In[2]:=HofB=Interpolation[BHCurve]
```

Define B(H), the inverse function to H(B), and name it HBCurve

```
In[3]:=HBCurve=Table[0,{i,30},{j,2}];
In[4]:=Do[{HBCurve[[i,1]]=BHCurve[[i,2]],HBCurve[[i,2]]=BHC
urve[[i,1]]},{i,30}]
```

Obtain the HBCurve in continuous form – the new function is called BofH

```
In[5]:=BofH=Interpolation[HBCurve]
```

Draw the B–H curve in **Fig. 1.30 (a)**

```
In[6]:=Plot[BofH[x],{x,0,40000}, DefaultFont-> {"Times-
Italic",9},Axes Label->{"H [At/m]","B [T]"},Ticks->
{{0,20000,40000}, Automatic}]
```

Define the reluctance RFe[phi], and the mmf drop Theta[phi] across an element, as functions of the flux through them

```
In[7]:=RFe[phi_]:=l HofB[Abs[phi]/S]/(Abs[phi]+10^(-9))
In[8]:=Theta[phi_]:=Sign[phi] l HofB[Abs[phi]/S]
```

Define the parameters l1, l2, l,S, Rsigma, Rgap, and w

```
In[9]:= Rsigma=10^6;Rgap=5*10^3;l1=.1;l2=.2;w=100;l=.1;
S=.001;
```

Evaluate the nonlinear characteristic of the magnetic circuit, Eq.(1.76), for the flux as an unknown:

```
In[10]:=Sol[phi_]:=i1 /. FindRoot[(i1 w - 11/1 Theta[phi])
(1/Rsigma+1/(Rgap+l2/1 RFe[phi-(i1 w - 11/1 Theta[phi])/
Rsigma])) ==phi,{i1,{-1.51 ,1.49}}]
```

Draw the characteristic in **Fig. 1.32 (b)**

```
In[11]:=Plot[Sol[0.0015 Sin[t]],{t,0,2 Pi},Ticks->{None,
Automatic},DefaultFont->{"Times-Italic",9},AxesLabel->
{None,"Current [A]"}]
```

Generate 161 points of the i_1–Φ characteristic of the magnetic circuit, and store them in the array Char

```
In[12]:=Char=Table[{phi,Sol[phi]},{phi,-0.002,0.002,
0.000025}]
```

Convert the previous set of data into a continuous curve CharInt

```
In[13]:=CharInt=Interpolation[Char]
```

Evaluate the magnetic circuit for the current as an unknown

```
In[14]:=Solinv[i1_]:=phi /. FindRoot[(i1 w - 11/1 Theta
[phi]) (1/Rsigma+1/(Rgap+l2/1 RFe[phi-(i1 w - 11/1 Theta
[phi])/Rsigma]))==phi, {phi,{-0.002,0.002}}]
```

Draw the characteristic in **Fig. 1.32 (a)** for the maximum value of the sinusoidal input current 3A

```
In[15]:=Plot[Solinv [ 3Sin[t]], {t,0,2Pi}, Ticks-> {None,
{-.001,0,.001}}, DefaultFont->{"Times-Italic",9},Axes
Label->{None,"Flux [Vs]"}]
```

Generate the $i_1(\Phi)$ characteristic CharInv from the given i_1–Φ curve Char

```
In[16]:=CharInv=Table[0,{i,161},{j,2}];
```

Appendix

In[17]:=Do[{CharInv[[i,1]]=Char[[i,2]],CharInv[[i,2]]=Char[[i,1]]},{i,161}]

Get the continuous interpolating curve $i_1(\Phi)$, called CharInvInt

In[18]:=CharInvInt=Interpolation[CharInv]

Draw the Φ–F characteristic of the magnetic circuit, shown in **Fig.1.30 (b)**

In[19]:=Plot[CharInvInt[x/w],{x,-5w,5w},DefaultFont->
{"Times-Italic" ,9},AxesLabel->{"Mmf[At]","Flux[Vs]"},
Ticks->{{-400,0, 400},{-0.0015,0 ,0.0015}}]

Plot the equivalent reluctance of the magnetic circuit as a function of the coil ampereturns, as shown in **Fig.1.31 (a)**

In[20]:=Plot[(x)/CharInvInt[x/w],{x,-8w,9w},DefaultFont->
{"Times Italic", 9},AxesLabel->{"Mmf[At]","Reluctance[At/Vs]"}, PlotRange->{0, 250000},Ticks->{{-500,250,0,250,500},
{0,250000}}]

Plot the inductance of the coil as a function of the coil ampereturns, as shown in **Fig. 1.31 (b)**

In[21]:=Plot[CharInvInt[x/w]w^2/x,{x,-7w,6w},PlotRange->
{0, 0.25}, DefaultFont ->{"Times-Italic",9},AxesLabel->
{"Mmf [At]", "Inductance [H]"},Ticks->{{-250,0,250},
{0,0.25}}]

Transient response of an RL circuit is evaluated in the following section. Define the parameters of a linear RL circuit – its inductance L, resistance R, time constant tau, and the value of the applied voltage U

In[22]:=L=w^2/(1/CharInvInt[.01]);R=0.2;U=3;tau=L/R;

Plot the current and flux responses of a linear RL circuit, as shown in **Fig. 1.33 (a)** and **(b)**

In[23]:=Plot[U/R*(1-Exp[-t/tau]),{t,0,5},DefaultFont ->

```
{"TimesItalic" ,9},AxesLabel->{"t [s]","Current [A]"}]
In[24]:=Plot[(L/w)*U/R*(1-Exp[-t/tau]),{t,0,5},DefaultFont-
> {"Times Italic",9},AxesLabel->{"t [s]","Flux[Vs]"}]
```

Define the differential equation of a nonlinear RL circuit for the applied constant voltage of U=3 V

```
In[25]:=NDSolve[{phi'[t]==(3-.2*CharInt[phi[t]])/w,
phi[0]==0},phi,{t, 0,0.1}]
```

Plot the current and flux from the solution of the previous differential equation, as shown in **Fig. 1.34**

```
In[26]:=Plot[Evaluate[phi[t]/.%],{t,0.,0.1},PlotRange->{0,
.002},DefaultFont->{"Times-Italic",9},AxesLabel->
{"t[s]","Flux [Vs]"},Ticks->{{0,0.025,0.05,0.075,0.1},
Automatic}]
In[27]:=Plot[Evaluate[CharInt[phi[t] /. %%]],{t,0.,0.1},
Default Font-> {"Times-Italic",9}, AxesLabel->{"t[s]",
"Current [A]"},Ticks ->{{0,0.025,0.05,0.075,0.1},Automatic
}]
```

Define the differential equation of a nonlinear RL circuit for the applied sinusoidal voltage with an amplitude of V=30 V

```
In[28]:=Transient=NDSolve[{phi'[t]==(30 Sin[314 t]-.2*
CharInt[phi[t]] )/w,phi[0]==0},phi,{t,0,0.2}]
```

Plot the current and the flux from the solution of the previous differential equation, as shown in **Fig. 1.35**

```
In[29]:=Flux=Plot[Evaluate[phi[t] /. Transient],{t,0.,0.2},
DefaultFont ->{"Times-Italic",9},AxesLabel->{"t[s]","Flux
[Vs]"}]
In[30]:=Current=Plot[Evaluate[CharInt[phi[t] /. Transient
]],{t,0., 0.2}, PlotRange->{-1,20},DefaultFont->{"Times-
Italic",9}, AxesLabel-> {"t [s]","Current [A]"}]
```

Define the differential equation of a nonlinear circuit with the current phase shift,

Appendix

shift, as a parameter

```
In[31]:=DiffEq[shift_]:=NDSolve[{phi'[t]==(30Sin[314 t-
shift]-.2* Char Int[phi[t]])/w,phi[0]==0},phi,{t,0,0.06}]
```

Define auxiliary arrays `Aux` and `LineData`

```
In[32]:=Aux=Table[0,{i,13},{j,60}];
In[33]:=LineData=Table[0,{i,13},{j,60}];
```

Repeat the calculation of transients in a nonlinear RL circuit for various values of the parameter shift, which correspond to the variation of the current phase shift between 0 and 90°

```
In[34]:=Do[Aux[[i]]=DiffEq[(i-1)Pi/24],{i,13}]
```

Prepare the obtained results for a 3D presentation – select the coil current first

```
In[35]:=Do[LineData[[i]]=Transpose[Table[CharInt[phi[t]/
.Aux[[i]]], {t,0,0.06,0.001}]][[1]],{i,13}]
```

Plot the current as s function of time in the first three periods and the phase shift, as shown in **Fig. 1.36**

```
In[36]:=ListPlot3D[LineData,AxesLabel->{"t[ms]","[rad]",
"I[A]" }, Shading->False,MeshRange->{{0,60},{0,Pi/2}},
PlotRange->{-1, 20} , DefaultFont ->{"Times-Italic",9},
Ticks->{Automatic,Automatic,{0,10,20}}]
```

Now prepare the flux data from the previously obtained solution of the nonlinear differential equation for a 3D representation

```
In[37]:=Do[LineData[[i]]=Transpose[Table[phi[t]/
.Aux[[i]],{t, 0,0.06, 0.001}]][[1]],{i,13}]
```

Plot the flux as a function of time in the first three periods and the phase shift, as shown in **Fig. 1.37**

```
In[38]:=ListPlot3D[LineData,AxesLabel->{"t[ms]","[rad]","
```

```
Flux [Vs]"},Shading->False,MeshRange->{{0,60},{0,Pi/2}},
DefaultFont -> {"Times-Italic",9},Ticks->{Automatic, Auto-
matic,{-0.001,0, 0.0015 }}]
```

A 1.3

The leakage permeance of a closed slot, the plot of which is shown in **Fig.1.41**, is defined and plotted by executing the following statements

```
In[1]:=Angle[x_]:=If[x<.6,((((-109.4 x+197.3)x-142.6)x+
54.36)x-12.03)x+1.593,If[x<4,((((-.005793x+.068)x-.3077)x
+.6857)x-.8018)x+.4888,.957 Exp[-x]+.01788]]
In[2]:=Plot3D[(100*ratio+.4)Tan[Angle[x]]+.73,{ratio,.001,1
},{x,.1,5},Shading->False,PlotLabel->"Specific slot bridge
permeance",AxesLabel->{"a/d","10 H [A/m]","lambda"},Default
Font->{"Times-Italic",9}]
```

The current distribution in the conductors in a slot, determined by skin effect, is evaluated in the following section. The current density distribution, introduced in Eq.(1.136), is obtained first by utilizing the auxiliary functions Convert and Sigma

```
In[1]:=Convert[t_,T_,n_]:=Floor[(t-Floor[t/T]*T)*n/T]+1
In[2]:=Sigma[beta_,ya_,h_,m_]:=2 I beta/((1+I)b)(Convert[
ya,h m,m] Cosh[beta(1+I) (ya-h(Convert[ya,h m,m]-1))]-
(Convert[ya,h m,m]-1)Cosh[beta(1+I)(Convert[ya,h m,m]h-ya)]
)/Sinh[beta(1+I)h]
```

with frequency f, copper conductivity kappa and conductor width b

```
In[3]:=f=50;kappa=56 10^6; mu=4 Pi 10^-7;beta=N[Sqrt[Pi f
kappa mu]]; b=.01
```

The plot of the real and imaginary part of the current density vector along the conductor height, shown in **Fig. 1.45**, is obtained by executing

```
In[4]:=ParametricPlot3D[{u Re[Sigma[beta,z,.05,1]]10^-6,u
Im[Sigma[beta, z,.05,1]]10^-6,z}, {z,0,.05}, {u,0,1},
BoxRatios->{1,1,.4}, Ticks-> None,DefaultFont->{"Times-
Italic",9},AxesLabel->{"Re{J}", "Im{J}","h"},BoxStyle->
```

Appendix

```
Dashing[{.02,.02}],PlotRange->{{-.001,.012},{-.0025,.012},
Automatic},AmbientLight->GrayLevel[.8]]
```

The field strength distribution along the conductor height r, as a function of the angular frequency omega of the unit current Islot through it, shown in **Fig. 1.46**, is obtained by executing

```
In[5]:=h=50;l=1;delta[omega_]:=Sqrt[2/(omega mu kappa)];Is-
lot=1;
In[6]:=Plot3D[ Islot/b Sqrt[(Cosh[2r/1000/delta[10^omega]]-
Cos[2r/1000/delta[10^omega]])/(Cosh[2h/1000/delta[10^omega
]]-Cos[2h/1000/delta[10^omega]])],{r,0,h},{omega,0.1,3},
Shading->False,AxesLabel-> {"r [mm]"," Log[omega]","H [A/m]
"},DefaultFont ->{"Times-Italic",9},PlotRange->{0,105},
Ticks->{{0,h},Automatic,{100}}]
In[7]:=Show[%,ViewPoint->{-2.744, -1.317, 0.731}]
```

The distribution of the amplitude of the current density, shown in **Fig. 1.47**, is the result of the following statements

```
In[8]:=Plot3D[ 10^(-6)/Sqrt[2] 10^omega mu kappa delta[
10^omega] Islot/b Sqrt[(Cosh[2r/1000/delta[10^omega]]+
Cos[2r/1000/delta[10^omega]])/(Cosh[2h/1000/delta[10
^omega]]-Cos[2h/1000/delta[10^omega]])],{r,0,h},{omega,
0.1,3},Shading->False,AxesLabel-> {"r [mm]"," Log[omega]",
"J [A/mm^2]"},DefaultFont ->{"Times-Italic" ,9},PlotRange->
{0,0.03},Ticks->{{0,h},Automatic,{0.002,0.03}}]
In[9]:=Show[%,ViewPoint->{-2.744, -1.317, 0.731}]
```

The variation of the instantaneous value of the current density, shown in **Fig. 1.48**, is obtained by executing

```
In[10]:=Plot3D[ 10^(-6)/Sqrt[2] 100 N[Pi] mu kappa .0095974
Islot Cos[100 N[Pi] t+ArcTan[Tanh[2r/1000/.0095974]Tan[2r/
1000/.0095974]]-ArcTan[Coth[2h/1000/.0095974]Tan[2h/1000/
.0095974]]]/b Sqrt[(Cosh[2r/1000/.0095974]+Cos[2r/1000/
.0095974])/(Cosh[2h/1000/.0095974]-Cos[2h/1000/.0095974])]
,{t,0,0.02},{r,0,h},Shading->False,AxesLabel->{"t [s]","r
[mm]","J [A/mm^2]"},DefaultFont ->{"Times-Italic",9},Plot
Range->{-0.015,0.015},Ticks->{{0,0.02},{0,h},{-0.015,0,
```

```
0.015}},PlotPoints->25]
In[11]:=Show[%,ViewPoint->{-2.850, 4.000, 0.990}]
```

The dependence of the ratio between the A.C. and D.C. slot leakage inductances, defined in Eq.(1.130) and shown in **Fig. 1.49**, is obtained by executing

```
In[12]:=Plot[3/(2x)(Sinh[2x]-Sin[2x])/(Cosh[2x]-Cos[2x]),
{x,0,5}, PlotRange->{0,1},DefaultFont ->{"Times-Italic",9}]
```

whereas the ratio between the A.C. and D.C. conductor resistances, defined in Eq.(1.133), is plotted by executing

```
In[13]:=Plot[x(Sinh[2x]+Sin[2x])/(Cosh[2x]-Cos[2x]),
{x,0,5},PlotRange ->{0,5},DefaultFont ->{"Times-Italic",9}]
```

The current densities in five conductors in a slot, shown in **Fig. 1.52**, are defined in Eq.(1.136). By varying the parameter p in this equation, one obtains the current density distribution along the hight of each conductor. Firstly, the real part of the current density is plotted:

```
In[14]:=Plot[Re[Sigma[105.14,y,.05,5]]10^-6,{y,0,.25},
DefaultFont-> {"Times-Italic",9},AxesLabel->{"r [m]","Re"},
PlotRange->{-0.06,0.06}, Ticks->{Automatic,{-0.05,0.05}}]
```

then the imaginary part

```
In[15]:=Plot[Im[Sigma[105.14,y,.05,5]]10^-6,{y,0,.25},
DefaultFont-> {"Times-Italic",9},AxesLabel->{"r [m]","Im"},
PlotRange->{-0.06,0.06}, Ticks->{Automatic,{-0.05,0.05}}]
```

and, finally, the absolute value of the current density

```
In[16]:=Plot[Abs[Sigma[105.14,y,.05,5]]10^-6,{y,0,.25},
DefaultFont-> {"Times-Italic",9},AxesLabel->{"r [m]","Abs"}
,PlotRange->{0,0.1}, Ticks->{Automatic,{0,0.1}}]
```

The simultaneous plot of the real and imaginary component of the current density, as a function of the conductor height, results in the trajectory of the current density vector. The trajectory of the current density vector, shown in **Fig. 1.53**, is obtained by executing the following statements

Appendix

```
In[17]:=RealPart=Table[{z,Re[Sigma[105.14,z,.05,5]]10^-6},
{z,.1,.15,.0025}]
In[18]:=RealPartInt=Interpolation[RealPart]
In[19]:=ImagPart=Table[{z,Im[Sigma[105.14,z,.05,5]]10^-6},
{z,.1,.15,.0025}]
In[20]:=ImagPartInt=Interpolation[ImagPart]
In[21]:=ParametricPlot3D[{u RealPartInt[z], u ImagPartInt
[z],z}, {z,0.1,.15},{u,0,1},BoxRatios->{1 ,1 ,.4},Default-
Font->{"Times-Italic",9},AxesLabel->{"Re","Im","r"},Ticks-
>{Automatic,Automatic,{.1, .15}},BoxStyle->Dashing[
{.02,.02}],PlotRange->{{-.025,.035},{-.025, .04},Automatic
},AmbientLight->GrayLevel[.8]]
```

A 2.3

The number of turns which minimize the content of the higher harmonics in the mmf, according to Eq.(2.37), is evaluated in this section. First, define the parameters of the winding to be optimized

```
In[1]:=tp=12 ts;r=11;w[1]=1;
```

Define the amplitude of a harmonic to be minimized, Theta[n_,i_]

```
In[2]:=Theta[n_,i_]:=4 w[i]/(Pi n) Sin[Pi n/2] Sin[Pi n ts
r*(1-2*(i-1)/r)/(2 tp)]
```

Apply Eq.(2.38) for weighing factors f1=f2=1

```
In[3]:=FindMinimum[Sum[(1-w[i])^2,{i,2,3}]+Sum[Sum[Theta
[n,i]^ 2, {n,3,17,2}],{i,1,3}],{w[2],1},{w[3],1}]
```

Apply Eq.(2.38) for weighing factors f1=0.3; f2=0.7

```
In[4]:=FindMinimum[0.3 Sum[(1-w[i])^2,{i,2,3}]+0.7 Sum[
Sum[Theta[n, i]^2,{n,3,17,2}],{i,1,3}],{w[2],1},{w[3],1}]
```

Apply Eq.(2.38) for weighing factors f1=0.7; f2=0.3

```
In[5]:=FindMinimum[0.7 Sum[(1-w[i])^2,{i,2,3}]+0.3 Sum[
Sum[Theta[n, i]^2,{n,3,17,2}],{i,1,3}],{w[2],1},{w[3],1}]
```

A 2.4

Define the electrical angle `alphae` between the slots

```
In[1]:=alphae=N[Pi/24];
```

Evaluate the distribution factor, `fd`

```
In[2]:=fd=Table[Sin[l n alphae/2.]/(l Sin[n alphae/2.]),
{n,1, 13,2}, {l,24}];
```

and plot it as a function of the number of in-series coils and the harmonic order, as shown in **Fig. 2.17**

```
In[3]:=ListPlot3D[fd,AxesLabel->{"coils"," harm. order",
"fd"}, Shading ->False,MeshRange->{{1,24},{1,13}},
PlotRange->{-0.5,1}, DefaultFont -> {"Times-Italic",9},
Ticks->{Automatic,{1,3,5,7,9, 11,13},{0,1}}]
```

Now evaluate the pitch factor, `fp`

```
In[4]:=fp=Table[Sin[k n Pi/2.],{k,0.6,1,0.05},{n,1,7,2}];
```

and plot it as a function of the coil pitch and the harmonic order, as shown in **Fig. 2.18**

```
In[5]:=ListPlot3D[fp,AxesLabel->{"harm. order","pitch","fp"
},Shading-> False,MeshRange->{{1,7},{0.6,1}}, PlotRange->
{-0.5,1 },DefaultFont-> {"Times-Italic",9},Ticks->{{1,3,
5,7},Automatic, {0,1}}]
```

A 2.5

Unprotect the symbol C first

Appendix 447

```
In[1]:=Unprotect[C];
```

Define the matrix C according to Eq.(2.66)

```
In[2]:=C[n_]:=Table[If[i<n,If[i==j,-1,If[i==j-1,1,0]],1],
{i,n} ,{j,n}]
```

as well as the matrix S in Eq.(2.71)

```
In[3]:=S[m_]:=Table[If[i<=m,If[i==j,1,0],If[i-m==j,-1,0]],
{i,2m}, {j,m}]
```

Generate matrix w defined in Eq.(2.72)

```
In[4]:=w2=Table[0,{i,12},{j,4}];
In[5]:=w2[[1]]={wa1,0,0,0}
In[6]:=w2[[2]]={wa2,wb2,0,0}
In[7]:=w2[[3]]={0,wb1,0,0}
In[8]:=w2[[4]]={0,wb1,0,0}
In[9]:=w2[[5]]={0,wb2,wa2,0}
In[10]:=w2[[6]]={0,0,wa1,0}
In[11]:=w2[[7]]={0,0,wa1,0}
In[12]:=w2[[8]]={0,0,wa2,wb2}
In[13]:=w2[[9]]={0,0,0,wb1}
In[14]:=w2[[10]]={0,0,0,wb1}
In[15]:=w2[[11]]={wa2,0,0,wb2}
In[16]:=w2[[12]]={0,0,0,0}
```

Define an auxiliary, time invariant matrix Auxmat2

```
In[17]:=Auxmat2=Inverse[C[12]].w2.S[2];
```

The vector of the stator created ampereturms Mmf2 is equal to the product of the matrix Auxmat2 and the vector of stator currents {Ia,Ib}

```
In[18]:=Mmf2[Ia_,Ib_]:=Auxmat2.{Ia,Ib}
```

Define the number of turns in each coil of the winding

```
In[19]:=wa1=wb1=30;wa2=wb2=25
```

and an auxiliary array Aux

```
In[20]:=Aux=Table[0,{j,60},{i,48}];
```

Calculate the mmf distribution created by the stator currents Ia=1 A and Ib=0

```
In[21]:=Ia=1;Ib=0;
In[22]:=Aux[[1]]=Mmf2[Ia,Ib];
```

Draw the mmf distribution shown in **Fig. 2.22 (a)**

```
In[23]:=ListPlot[Table[{Table[Floor[i/2]],Table[Aux[[1,
Floor[(i+1)/2 ]]]]},{i,1,24}], PlotJoined->True, Default-
Font-> {"Times-Italic",9},Axes Label->{"slot number","Mmf
[At]"},PlotRange->{-60,60}]
```

Calculate the mmf distribution created by the stator currents Ia=0.707 A and Ib=-0.707 A

```
In[24]:=Ia=0.707;Ib=-0.707;
In[25]:=Aux[[1]]=Mmf2[Ia,Ib];
```

Draw the mmf distribution shown in **Fig. 2.22 (b)**

```
In[26]:=ListPlot[Table[{Table[Floor[i/2]],Table[Aux[[1,
Floor[ (i+1)/2 ]]]]},{i,1,24}],PlotJoined->True, Default-
Font-> {"Times-Italic",9},Axes Label->{"slot number","Mmf
[At]"}]
```

The matrix w of the three phase double layer winding shown in **Fig. 2.23** is defined in Eq.(2.73). It is generated by executing the following statement

Appendix

```
In[27]:=BasicMatrix[Nt_,m_]:=Table[If[i<=Nt/(2m),
                                    If[j==1,wa,0],
                                    If[i<=2*Nt/(2m),
                                    If[j==3,-wc,0],
                                    If[i<=3*Nt/(2m),
                                    If[j==2,wb,0],
                                    If[i<=4*Nt/(2m),
                                    If[j==4,wa,0],
                                    If[i<=5*Nt/(2m),
                                    If[j==6,-wc,0],
                                    If[j==5,wb,0]]]]]]],
                                    {i,1,Nt},{j,1,2m}]
```

The winding has 3 phases, coil pitch equal to 5, and is placed into 12 slots

```
In[28]:=Nt=12;m=3;y=5;
```

The matrix wl which describes the distribution of conductors in the lower layer is defined as

```
In[29]:=wl=BasicMatrix[12,3];
```

and the matrix wu which describes the distribution of conductors in the upper layer is

```
In[30]:=wu=Table[0,{i,Nt},{j,2m}];
In[31]:=Do[If[i<=Nt-y,wu[[i+y]]=wl[[i]],wu[[i-(Nt-y)]]=
wl[[i ]]], {i,Nt}];
```

The complete winding matrix w is defined as

```
In[32]:=w=Table[0,{i,Nt},{j,2m}];
In[33]:=Do[w[[i]]=wl[[i]]-wu[[i]],{i,Nt}];
In[34]:=Do[w[[Nt,i]]=0,{i,2m}];
```

The time invariant matrix Auxmat is equal to

```
In[35]:=Auxmat=Inverse[C[12]].w.S[3];
```

The mmf distribution vector of the analyzed three phase winding Mmf [Ia,Ib,Ic] is

```
In[36]:=Mmf[Ia_,Ib_,Ic_]:=Auxmat.{Ia,Ib,Ic};
```

Define the number of turns per coil in each phase

```
In[37]:=wa=wb=wc=10.;
```

The mmf distribution is evaluated which is created by the winding carrying the three phase currents Ia, Ib, and Ic

```
In[38]:=Ia=N[Sqrt[3]/2];Ib=0;Ic=-N[Sqrt[3]/2];
In[39]:=Theta=Mmf[Ia,Ib,Ic]
```

These currents generate the mmf distribution shown in **Fig. 2.24 (a)**

```
In[40]:=ListPlot[Table[{Table[Floor[i/2]],Table[Theta[[
Floor[ (i+1)/2 ]]]]},{i,1,24}],PlotJoined->True, Default-
Font-> {"Times Italic",9},Axes Label->{"slot number","Mmf
[At]"}]
```

The mmf distribution in **Fig. 2.24 (b)** is created by the phase currents

```
In[41]:=Ia=1;Ib=-.5;Ic=-.5;
In[42]:=Theta=Mmf[Ia,Ib,Ic];
In[43]:=ListPlot[Table[{Table[Floor[i/2]],Table[Theta[[
Floor[ (i+1)/2] ]]]},{i,1,24}],PlotJoined->True, Default-
Font ->{"Times Italic",9},Axes Label->{"slot number","Mmf
[At]"}]
```

A 2.6

First generate the data for the 3D plot of the air gap mmf created by the symmetric two phase winding, shown in Fig. 2.21. The winding carries the currents which are in phase

Appendix

```
In[44]:=Do[{Ia=N[Cos[2 Pi/24 (k-1)]],Ib=N[Cos[2 Pi/24 (k-1)
]], Aux[[k]]=Mmf2[Ia,Ib]},{k,25}]
In[45]:=Output=Table[Table[Aux[[j,Floor[(i+3)/4]]],
{i,1,48}], {j,1,25}];
```

The mmf wave shown in **Fig. 2.26** is obtained by executing

```
In[46]:=ListPlot3D[Output,AxesLabel->{"tooth"," time","Mmf"
},Shading-> False,MeshRange->{{0,12},{0,1}},PlotRange->
Automatic, DefaultFont -> {"Times-Italic",9},Ticks->
{{1,2,3,4,5,6,7,8,9,10,11,12}, {0,0.25,0.5,0.75,1},Auto-
matic}]
```

A 2.7

The air gap mmf created by the same two phase winding as in the previous section, carrying the currents which are 90 degrees out of phase, is generated by executing

```
In[47]:=Do[{Ia=N[Cos[2 Pi/24 (k-1)]],Ib=-N[Cos[2 Pi/24 (k-
1)-Pi/2]], Aux[[k]]=Mmf2[Ia,Ib]},{k,25}]
In[48]:=Output=Table[Table[Aux[[j,Floor[(i+3)/4]]],
{i,1,48}], {j,1,25}];
```

The mmf waveform shown in **Fig. 2.28** is obtained by executing

```
In[49]:=ListPlot3D[Output,AxesLabel->{"tooth"," time","Mmf"
},Shading-> False,MeshRange->{{0,12},{0,1}},PlotRange->
Automatic, DefaultFont -> {"Times-Italic",9},Ticks->
{{1,2,3,4,5,6,7,8,9, 10,11,12} ,{0,0.25,0.5,0.75,1},Auto-
matic}]
```

The following function, Animcos[Nframe], generates a sequence of Nframe frames. Each frame represents the air gap mmf, created by the previously analyzed two phase winding connected to sinusoidal current sources. The time shift between the adjacent frames is $2\pi/\text{Nframe}$

```
In[50]:=Animcos[Nframe_]:=Do[{Ia=N[Cos[2 Pi/Nframe (k-1)]],
Ib= N[Cos[2 Pi/Nframe (k-1)]],Aux[[k]]=Mmf2[Ia,Ib],ListPlot
[Table [{Table[Floor [(i1/2)]],Table[Aux[[k,Floor[(i1+1)/
```

```
2]]]]},{i1, 1,24}],PlotJoined-> True, PlotRange -> {{0,12},
{-30,30}},Axes -> None]},{k,1,Nframe}]
```

A sequence of 24 frames is generated by executing

```
In[51]:=Animcos[24]
```

The 3D mmf distribution created by the three phase winding shown in Fig. 2.23, carrying symmetric currents, is obtained by executing

```
In[52]:=Do[{Ia=N[Cos[2 Pi/24 (k-1)]],Ib=N[Cos[2 Pi/24 (k-1)
-2Pi/3]], Ic=N[Cos[2 Pi/24 (k-1)-4Pi/3]],Aux[[k]]=Mmf[Ia,
Ib,Ic]} ,{k,25}]
In[53]:=Output=Table[Table[Aux[[j,Floor[(i+3)/4]]],
{i,1,48}], {j,1,25}];
```

The following statement results in the plot in **Fig. 2.29**

```
In[54]:=ListPlot3D[Output,AxesLabel->{"tooth"," time","Mmf"
}, Shading-> False, MeshRange->{{0,12},{0,1}}, PlotRange->
{-50,50} ,DefaultFont -> {"Times-Italic",9},Ticks->
{{1,2,3,4,5,6,7,8,9, 10,11,12},{0,0.25,0.5,0.75,1},{-50,
0,50}}]
```

Reverse the direction of rotation of the mmf and plot the 3D curve in **Fig. 2.30**

```
In[55]:=Do[{Ia=N[Cos[2 Pi/24 (k-1)]],Ib=N[Cos[2 Pi/24 (k-1)
-4Pi /3]], Ic=N[Cos[2 Pi/24 (k-1)-2Pi/3]],Aux[[k]]=Mmf[Ia,
Ib,Ic]}, {k,25}]
In[56]:=Output=Table[Table[Aux[[j,Floor[(i+3)/4]]],
{i,1,48}], {j,1,25}];
In[57]:=ListPlot3D[Output,AxesLabel->{"tooth"," time","Mmf"
}, Shading-> False, MeshRange->{{0,12},{0,1}}, PlotRange->
{-50,50}, DefaultFont -> {"Times-Italic",9},Ticks->
{{1,2,3,4,5,6,7,8,9, 10,11,12},{0,0.25,0.5,0.75,1},{-50,
0,50}}]
```

Change sign of the current in phase b and plot the 3D curve in **Fig. 2.31**

Appendix 453

```
In[58]:=Do[{Ia=N[Cos[2 Pi/24 (k-1)]],Ib=-N[Cos[2 Pi/24 (k-
1) -4 Pi/3]], Ic=N[Cos[2 Pi/24 (k-1)-2Pi/3]],Aux[[k]]=Mmf
[Ia,Ib,Ic]}, {k,25}]
In[59]:=Output=Table[Table[Aux[[j,Floor[(i+3)/4]]],
{i,1,48}], {j,1,25}];
In[60]:=ListPlot3D[Output,AxesLabel->{"tooth","time","Mmf"
}, Shading-> False, MeshRange->{{0,12},{0,1}}, PlotRange->
{-50,50}, DefaultFont -> {"Times-Italic",9},Ticks->
{{1,2,3,4,5,6,7,8,9, 10,11,12},{0,0.25,0.5,0.75,1},{-50,
0,50}}]
```

Draw the 3D mmf distribution created by in-phase currents in all three phases, shown in **Fig. 2.32**

```
In[61]:=Do[{Ia=N[Cos[2 Pi/24 (k-1)]],Ib=N[Cos[2 Pi/24 (k-1)
]], Ic=N[Cos[2 Pi/24 (k-1)]],Aux[[k]]=Mmf[Ia,Ib,Ic]},
{k,25}]
In[62]:=Output=Table[Table[Aux[[j,Floor[(i+3)/4]]],
{i,1,48}], {j,1,25}];
In[63]:=ListPlot3D[Output,AxesLabel->{"tooth"," time","Mmf"
}, Shading-> False, MeshRange->{{0,12},{0,1}}, PlotRange->
{-20,20}, DefaultFont -> {"Times-Italic",9},Ticks->
{{1,2,3,4,5,6,7,8,9, 10,11,12},{0,0.25,0.5,0.75,1},{-20,
0,20}}]
```

The air gap mmf distribution created by rectangular currents in the windings, shown in **Fig. 2.33**, is generated next. First, the function which generates a rectangular current Istep is defined

```
In[64]:=Istep[t_,T_,Imax_]:=If[t-Floor[t/T]*T<=T/6,0,
                    If[t-Floor[t/T]*T<=T/2,Imax,
                    If[t-Floor[t/T]*T<=2T/3,0,-Imax
]]]
```

and then drawn in **Fig. 2.33**

```
In[65]:=Plot[Istep[t+1/12,1,1],{t,0,2.5},DefaultFont ->
{"Times-Italic" ,9},AxesLabel->{"t/T","Current"}]
```

Next, the 3D mmf distribution created by rectangular currents in phases of a three phase machine is drawn, as shown in **Fig. 2.34**

```
In[66]:=Do[{Ia=Istep[k+.1,24,1],Ib=Istep[k+8.1,24,1],Ic=Ist
ep[k+16.1,24,1],Aux[[k]]=Mmf[Ia,Ib,Ic]},{k,25}]
In[67]:=Output=Table[Table[Aux[[j,Floor[(i+3)/4]]],
{i,1,48}], {j,1,25}];
In[68]:=ListPlot3D[Output,AxesLabel->{"tooth"," time","Mmf"
}, Shading-> False, MeshRange->{{0,12},{0,1}}, PlotRange->
{-50,50}, DefaultFont -> {"Times-Italic",9},Ticks->
{{1,2,3,4,5,6,7,8,9, 10,11,12},{0,0.25,0.5,0.75,1},{-50,
0,50}}]
```

The function Anim3unsymmcos generates a sequence of frames which represent the air gap mmf in an asymmetric winding connection

```
In[69]:=Anim3unsymmcos[Nframe_]:=Do[{Ia=N[Cos[2 Pi/Nframe
(k-1 )]], Ib=-N[Cos[2 Pi/Nframe (k-1)-4 Pi/3]],Ic=N[Cos[2
Pi/Nframe (k-1)-2Pi/3]],Aux[[k]]=Auxmat.{Ia,Ib,Ic},List
Plot[Table[( Table[Floor[( i1/2)]], Table[Aux[[k,Floor
[(i1+1)/2]]]]},{i1,1, 24}],PlotJoined->True,Plot Range ->
{{0,12},{-40,40}}, Axes -> None]},{k,1,Nframe }]
```

If 24 animated frames have to be created, one has to type

```
In[70]:=Anim3unsymmcos[24]
```

The function Anim3symmcos generates a sequence of frames which represent the air gap mmf in a symmetrical winding connection

```
In[71]:=Anim3symmcos[Nframe_]:=Do[{Ia=N[Cos[2 Pi/Nframe (k-
1) ]], Ib=N[Cos[2 Pi/Nframe (k-1)-4 Pi/3]],Ic=N[Cos[2 Pi/
Nframe (k-1)-2Pi/3]] ,Aux[[k]]=Auxmat.{Ia,Ib,Ic},ListPlot[
Table[{Table[ Floor[(i1/2)]], Table[Aux[[k,Floor[(i1+1)/
2]]]]},{i1,1,24}], PlotJoined->True, PlotRange -> {{0,12},
{-40, 40}},Axes ->None]} ,{k,1,Nframe}]
```

A sequence of 24 frames which can later be animated is obtained by typing

```
In[72]:=Anim3symmcos[24]
```

Appendix

The animation of the air gap mmf created by a three phase winding carrying rectangular currents is obtained by utilizing the function Anim3symmrect defined as

```
In[73]:=Anim3symmrect[Nframe_]:=Do[{Ia=Istep[k+.1,24,1],Ib=
Istep[k+8.1,24,1],Ic=Istep[k+16.1,24,1],Aux[[k]]=Inverse[C[
12]].w.S[3].{Ia,Ib,Ic},ListPlot[Table[{Table[Floor[(i1/
2)]],Table[ Aux[[k, Floor[(i1+1)/2]]]]},{i1,1,24}],Plot
Joined->True, PlotRange ->{{0,12},{-40,40}},Axes ->None]},
{k,1,Nframe}]
In[74]:=Anim3symmrect[24]
```

A 2.8

Convert the continuous quantity t with period T into the discrete quantity n

```
In[1]:=Convert[t_,T_,n_]:=Floor[(t-Floor[t/T]*T)*n/T]+1
```

Define the shift matrix Shift

```
In[2]:=Shift[t_,T_,n_]:=Table[If[i<=n-Convert[t,T,n]+1,
                If[i-(j-Convert[t,T,n]+1)==0,1,0],
                If[i-n+Convert[t,T,n]-1-j==0,1,0]],
                {i,n},{j,n}]
```

Rewrite the statements which help one find the mmf distribution of a three phase winding, introduced in the previous section

```
In[3]:=Unprotect[C];
In[4]:=C[n_]:=Table[If[i<n,If[i==j,-1,If[i==j-1,1,0]],1],
{i,n},{j,n}]
In[5]:=S[m_]:=Table[If[i<=m,If[i==j,1,0],If[i-m==j,1,0]],
{i,2m},{j,m}]
In[6]:=BasicMatrix[Nt_,m_]:=Table[If[i<=Nt/(2m),
                            If[j==1,wa,0],
                            If[i<=2*Nt/(2m),
                            If[j==3,-wc,0],
                            If[i<=3*Nt/(2m),
```

```
                        If[j==2,wb,0],
                        If[i<=4*Nt/(2m),
                        If[j==4,wa,0],
                        If[i<=5*Nt/(2m),
                        If[j==6,-wc,0],
                        If[j==5,wb,0]]]]]],
                        {i,1,Nt},{j,1,2m}]
In[7]:=wl=BasicMatrix[12,3];
In[8]:=Nt=12;m=3;y=5;
In[9]:=wu=Table[0,{i,Nt},{j,2m}];
In[10]:=w=Table[0,{i,Nt},{j,2m}];
In[11]:=Do[If[i<=Nt-y,wu[[i+y]]=wl[[i]],wu[[i-(Nt-y)]]=
wl[[i]]],{i,Nt}];
In[12]:=Do[w[[i]]=wl[[i]]-wu[[i]],{i,Nt}];
In[13]:=Do[w[[Nt,i]]=0,{i,2m}];
In[14]:=wa=wb=wc=10.;
```

Draw the mmf distribution shown in **Fig. 2.36**

```
In[15]:=Do[{Ia=N[Cos[2 Pi/24 (k-1)]],Ib=N[Cos[2 Pi/24 (k-1)
-2Pi /3]] ,Ic=N[Cos[2 Pi/24 (k-1)-4Pi/3]],Aux[[k]]=Inverse[
C[12]. Shift[k, 24,12]].w.S[3].{Ia,Ib,Ic}},{k,25}]
In[16]:=Output=Table[Table[Aux[[j,Floor[(i+3)/4]]],
{i,1,48}], {j,1,25}];
In[17]:=ListPlot3D[Output,AxesLabel->{"","time","Mmf"},
Shading-> False,MeshRange->{{0,12},{0,1}},PlotRange->{-50,
50},Default Font -> {"Times-Italic",9},Ticks->{{},
{0,0.25,0.5,0.75,1},{-50, 0,50}}]
```

Draw the mmf distribution shown in **Fig. 2.35**

```
In[18]:=Do[{Ia=N[Cos[2 Pi/24 (k-1)]],Ib=N[Cos[2 Pi/24 (k-1)
-4Pi /3]], Ic=N[Cos[2 Pi/24 (k-1)-2Pi/3]],Aux[[k]]=Inverse[
C[12]. Shift[k, 24,12]].w.S[3].{Ia,Ib,Ic}},{k,25}]
In[19]:=Output=Table[Table[Aux[[j,Floor[(i+3)/
4]]],{i,1,48}], {j,1,25}];
```

Appendix

```
In[20]:=ListPlot3D[Output,AxesLabel->{"","time","Mmf"},
Shading-> False,MeshRange->{{0,12},{0,1}},PlotRange->{-50,
50},Default Font -> {"Times-Italic",9},Ticks->{{},
{0,0.25,0.5,0.75,1},{-50, 0,50}}]
```

Repeat the same for the mmf distribution in **Fig. 2.37**

```
In[21]:=Istep[t_,T_,Imax_]:=If[t-Floor[t/T]*T<=T/6,0,
                        If[t-Floor[t/T]*T<=T/2,Imax,
                        If[t-Floor[t/T]*T<=2T/3,0,-Imax
]]]
In[22]:=Do[{Ia=Istep[k+.1,24,1],Ib=Istep[k+8.1,24,1],
Ic=Istep[k+16.1,24,1],Aux[[k]]=Inverse[C[12].Shift[k,24,12]
].w.S[3].{Ia,Ib,Ic}},{k,25}]
In[23]:=Output=Table[Table[Aux[[j,Floor[(i+3)/4]]],
{i,1,48}], {j,1,25}];
In[24]:=ListPlot3D[Output,AxesLabel->{"","time","Mmf"},
Shading-> False,MeshRange->{{0,12},{0,1}},PlotRange->{-50,
50},Default Font -> {"Times-Italic",9},Ticks->{{},
{0,0.25,0.5,0.75,1},{-50, 0,50}}]
```

The animation of the air gap mmf created by sinusoidal currents, as seen from the reference frame travelling at the speed of the fundamental of the mmf, is obtained by executing the following statements

```
In[25]:=Anim3symmcosref[Nframe_]:=Do[{Ia=N[Cos[2 Pi/Nframe
(k-1)]], Ib=-N[Cos[2 Pi/Nframe (k-1)-4 Pi/3]],Ic=N[Cos[2
Pi/Nframe (k-1)-2 Pi/3 ]],Aux[[k]]=Inverse[C[12].Shift[k,
24,12]].w.S[3]. {Ia,Ib,Ic},ListPlot [Table[{Table[Floor[
(i1/2)]],Table[Aux[[k, Floor[(i1+1)/2]]]]} ,{i1,1,24}],
PlotJoined->True, PlotRange -> {{0,12},{-40,40}},Axes ->
None]},{k,1,Nframe}]
In[26]:=Anim3symmcosref[24]
```

The animation of the air gap mmf created by rectangular currents, as seen from the reference frame travelling at the speed of the fundamental of the mmf is obtained by executing the following statements

```
In[27]:=Anim3symmrectref[Nframe_]:=Do[{Ia=Istep[k+.1,24,1],
```

```
Ib=Istep[k+8.1,24,1],Istep=I3[k+16.1,24,1],Aux[[k]]=Inverse
[C[12].Shift[-k,24, 12]].w.S[3].{Ia,Ib,Ic},ListPlot[Table[
{Table[Floor[ (i1/2)]],Table [Aux[[k,Floor[(i1+1)/2]]]]},
{i1,1,24}],Plot Joined->True, PlotRange -> {{0,12},{-45,
45}},Axes ->None]}, {k,1,Nframe}]
In[28]:=Anim3symmrectref[24]
```

A 2.9

First define the reduced identity matrix Uprime as

```
In[1]:=Uprime[n_]:=Table[If[i<n,If[i==j,1,0],0],{i,n},
{j,n}]
```

and the matrix S

```
In[2]:=S[n_]:=Table[If[i<=n/2,1,-1],{i,n}]
```

Define the vector of the mmf distribution of the commutator winding

```
In[3]:=Unprotect[C];
In[4]:=C[n_]:=Table[If[i<n,If[i==j,-1,If[i==j-1,1,0]],1],
{i,n},{j,n}]
In[5]:=Theta0[n_]:=Inverse[C[n]].Uprime[n].S[n]
```

Draw the mmf distribution in **Fig. 2.41**

```
In[6]:=ws1=10;Ia=1.;
In[7]:=Aux=Table[0,{j,60},{i,48}];
In[8]:=Do[Aux[[k]]=ws1 Ia Theta12,{k,25}]
In[9]:=Output=Table[Table[Aux[[j,Floor[(i+3)/4]]],
{i,1,48}], {j,1,25}];
In[10]:=ListPlot3D[Output,AxesLabel->{"tooth","time",
"Mmf"}, Shading-> False,MeshRange->{{0,12},{0,1}},
PlotRange->Automatic, DefaultFont -> {"Times-Italic",9},
Ticks->{{1,2,3,4,5,6,7,8,9, 10,11,12},{},Automatic }]
```

Appendix

The mmf distribution in **Fig. 2.42** is obtained by executing

```
In[11]:=Convert[t_,T_,n_]:=Floor[(t-Floor[t/T]*T)*n/T]+1
In[12]:=Shift[t_,T_,n_]:=Table[If[i<=n-Convert[t,T,n]+1,
                    If[i-(j-Convert[t,T,n]+1)==0,1,0],
            If[i-n+Convert[t,T,n]-1-j==0,1,0]],
{i,n},{j,n}]
In[13]:=Thetashift[t_,T_,n_]:=Inverse[C[n].Shift[-t,T,n]].
Uprime[n]. S[n]
In[14]:=Do[Aux[[k]]=wsl Ia Thetashift[k,24,12],{k,25}]
In[15]:=Output=Table[Table[Aux[[j,Floor[(i+3)/
4]]],{i,1,48}], {j,1,25}];
In[16]:=ListPlot3D[Output,AxesLabel->{"","time","Mmf"},
Shading-> False, MeshRange->{{0,12},{0,1}}, PlotRange->
Automatic,DefaultFont -> {"Times-Italic",9},Ticks->{{},
{0,0.25,0.5,0.75,1},Automatic}]
```

Draw the mmf distribution in **Fig. 2.43**

```
In[17]:=Do[{Ia=N[Cos[2 Pi/24(k-1)]],Aux[[k]]=wsl Ia
Theta12}, {k,25}]
In[18]:=Output=Table[Table[Aux[[j,Floor[(i+3)/4]]],
{i,1,48}], {j,1,25}];
In[19]:=ListPlot3D[Output,AxesLabel->{"tooth"," time","Mmf"
},Shading-> False,MeshRange->{{0,12},{0,1}},PlotRange->
Automatic,DefaultFont -> {"Times-Italic",9},Ticks->{{1,2,3,
4,5,6,7,8,9,10,11,12},{0,0.25,0.5,0.75,1},Automatic}]
```

Draw the mmf distribution in **Fig. 2.44**

```
In[20]:=Do[{Ia = Abs[N[Cos[2Pi/24(k-1)]]], Aux[[k]]= wsl Ia
Theta12} ,{k,25}]
In[21]:=Output=Table[Table[Aux[[j,Floor[(i+3)/4]]],
{i,1,48}], {j,1,25}];
In[22]:=ListPlot3D[Output,AxesLabel->{"tooth"," time","Mmf"
},Shading-> False,MeshRange->{{0,12},{0,1}},PlotRange->
Automatic, DefaultFont -> {"Times-Italic",9},Ticks->{{1,2,
3,4,5,6,7,8,9, 10,11,12}, {0,0.25,0.5,0.75,1},Automatic}]
```

Animation of the spatial distribution of the air gap in a reference frame which rotates with the period defined by the variable Period in a sequence of Nframe frames is obtained by executing the following statements

```
In[23]:=Animcommut[Nframe_,Period_]:=Do[{Ia=N[Cos[2 Pi/
Period (k-1)]] ,Aux[[k]]=wsl Ia Thetashift[k,Period,12],
ListPlot[ Table[{Table[Floor[ (i1/2)]],Table[Aux[[k,Floor[
(i1+1)/2]]]]}, {i1,1,24}],PlotJoined->True ,PlotRange->{
{0,12},{-40,40}},Axes -> None]},{k,1,Nframe}]
In[24]:=Animcommut[24,15]
```

A 2.10

Repeat the definitions of the matrices C and Uprime from the previous section

```
In[1]:=Unprotect[C];
In[2]:=C[n_]:=Table[If[i<n,If[i==j,-1,If[i==j-1,1,0]],1],
{i,n},{j,n}]
In[3]:=Uprime[n_]:=Table[If[i<n,If[i==j,1,0],0],{i,n},
{j,n}]
```

Define the vector of bar currents

```
In[4]:=Ib[n_]:=Table[N[Sin[2 Pi i/n]],{i,n}]
```

Apply the previous definition to a cage with 48 bars

```
In[5]:=Ibar=Ib[48]
Out[5]:={0.130526,0.258819,0.382683,0.5,0.608761,0.707107,
0.793353,0.866025,0.92388,0.965926,0.991445,1.,0.991445,
0.965926,0.92388,0.866025,0.793353,0.707107,0.608761,0.5,
0.382683,0.258819,0.130526,0,-0.130526,-0.258819,-0.382683
,-0.5,-0.608761,-0.707107,-0.793353,-0.866025,-0.92388,-
0.965926,-0.991445,-1.,-0.991445,-0.965926,-0.92388,-
0.866025,-0.793353,-0.707107,-0.608761,-0.5,-0.382683,-
0.258819,-0.130526,0}
```

Get the ring currents distribution from the known bar currents

Appendix   461

```
In[6]:=Iring=Inverse[C[48]].Uprime[48].Ib[48]
Out[6]:={-7.62853,-7.498,-7.23918,-6.8565,-6.3565,-5.74774
,-5.04063,-4.24728,-3.38125,-2.45737,-1.49144,-0.5,0.5,
1.49144,2.45737,3.38125,4.24728,5.04063,5.74774,6.3565,
6.8565,7.23918,7.498,7.62853,7.62853,7.498,7.23918,6.8565,
6.3565,5.74774,5.04063,4.24728,3.38125,2.45737,1.49144,0.5,
-0.5,-1.49144,-2.45737,-3.38125,-4.24728,-5.04063,-
5.74774,-6.3565,-6.8565,-7.23918,-7.498,-7.62853}
```

Plot the bar currents distribution, as shown in **Fig. 2.46 (a)**.

```
In[7]:=ListPlot[Ibar,DefaultFont ->{"Times-Italic",9},Axes
Label-> {"bar","Current"},PlotJoined-> True, PlotRange ->
{-8,8}]
```

Plot the bar currents distribution, as shown in **Fig. 2.46 (b)**.

```
In[8]:=ListPlot[Iring,DefaultFont ->{"Times-Italic",9},Axes
Label-> {"ring","Current"},PlotJoined-> True, PlotRange->
{-8,8}]
```

A 3.2

Repeat the statements to generate the stator mmf distribution

```
In[1]:=Unprotect[C];
In[2]:=Aux=Table[0,{i,24},{j,720}];
In[3]:=C[n_]:=Table[If[i<n,If[i==j,-1,If[i==j-1,1,0]],1],
{i,n},{j,n}]
In[4]:=S[m_]:=Table[If[i<=m,If[i==j,1,0],If[i-m==j,-1,0]],
{i,2m},{j,m}]
In[5]:=Convert[t_,T_,n_]:=Floor[(t-Floor[t/T]*T)*n/T]+1
In[6]:=Shift[t_,T_,n_]:=Table[ If[i<=n-Convert[t,T,n]+1,
                        If[i-(j-Convert[t,T,n]+1)==0,1,0],
                        If[i-n+Convert[t,T,n]-1-j==0,1,0]]
                        ,{i,n},{j,n}]
In[7]:=wa=wb=wc=100;
```

```
In[8]:=BasicMatrix[Nt_,m_]:=Table[If[i<=Nt/(2m),
                                  If[j==1,wa,0],
                                  If[i<=2*Nt/(2m),
                                  If[j==2,wb,0],
                                  If[i<=3*Nt/(2m),
                                  If[j==3,wc,0],
                                  If[i<=4*Nt/(2m),
                                  If[j==4,wa,0],
                                  If[i<=5*Nt/(2m),
                                  If[j==5,wb,0],
                                  If[j==6,wc,0]
                                  ]]]]],{i,1,Nt},{j,1,2m}]
In[9]:=Ns=12;ms=3;ys=5;
In[10]:=wus=Table[0,{i,Ns},{j,2ms}];
In[11]:=ws=Table[0,{i,Ns},{j,2ms}];
In[12]:=wls=BasicMatrix[Ns,ms];
In[13]:=Do[If[i<=Ns-ys,wus[[i+ys]]=wls[[i]],wus[[i-(Ns-ys)]]= wls[[i]]],{i,Ns}];
In[14]:=Do[ws[[i]]=wls[[i]]-wus[[i]],{i,Ns}];
In[15]:=Do[ws[[Ns,i]]=0,{i,2ms}];
In[16]:=Ias=1;
In[17]:=Ibs=0.5;
In[18]:=Ics=-0.5;
In[19]:=Thetas=Inverse[C[Ns]].ws.S[ms].{Ias,Ibs,Ics};
```

Do the same for the rotor mmf distribution

```
In[20]:=Nr=18;mr=3;yr=7;
In[21]:=wur=Table[0,{i,Nr},{j,2mr}];
In[22]:=wr=Table[0,{i,Nr},{j,2mr}];
In[23]:=wlr=BasicMatrix[Nr,mr];
In[24]:=Do[If[i<=Nr-yr,wur[[i+yr]]=wlr[[i]],wur[[i-(Nr-yr)]]= wlr[[i]]],{i,Nr}];
```

Appendix   463

```
In[25]:=Do[wr[[i]]=wlr[[i]]-wur[[i]],{i,Nr}];
In[26]:=Do[wr[[Nr,i]]=0,{i,2mr}];
In[27]:=Iar=N[Sqrt[3]/2];
In[28]:=Ibr=0;
In[29]:=Icr=-N[Sqrt[3]/2];
In[30]:=Thetar=Shift[0.2,1,Nr].Inverse[C[Nr]].wr.S[mr].{Iar
,Ibr,Icr};
```

In order to include the points on the rotor periphery which are placed within the slot openings, the complete interval of the double pole pitch is split into the pieces which divide both the slot opening and the tooth width an integer number of times. This is obtained by utilizing the function Pointss, defined as

```
In[31]:=Pointss[Nt_,os_]:=100/GCD[Round[os],100]*Nt
In[32]:=Npoints=Pointss[12,35];
In[33]:=Npointr=Pointss[18,15];
```

The minimum total number of points Totpoint along the peripheral coordinate x, at which the air gap flux density has to be calculated, is equal to the lowest common multiple of the number of points of the stator and rotor. In this example it is equal to

```
In[34]:=Totpoint=LCM[Npoints,Npointr];
In[35]:=m1=Totpoint/Ns;
In[36]:=m2=Totpoint/Nr;
```

Plot the stator, rotor and resultant mmfs, as shown in **Fig. 3.4**

```
In[37]:=ListPlot[Table[{Table[Floor[i/2]],Table[Thetas[[
Floor[(i+1)/2]]]]},{i,2 Ns}],PlotJoined->True,DefaultFont->
{"Times Italic",9},AxesLabel->{"tooth","At"}, PlotLabel->
"Stator mmf"]
In[38]:=ListPlot[Table[{Table[Floor[i/2]],Table[Thetar[[
Floor[(i+1)/2]]]]},{i,2Nr}],PlotJoined->True,DefaultFont->
{"Times Italic",9}, AxesLabel->{"tooth","At"}, PlotLabel->
"Rotor mmf", Ticks-> {{2,4,6,8,10,12,14,16,18},Automatic}]
In[39]:=Suma=Table[Thetas[[Floor[(i+m1-1)/m1]]]+Thetar[[
Floor[ (i+m2-1)/m2]]],{i,Totpoint}];
```

```
In[40]:=ListPlot[Table[{Table[Floor[i/2]],Table[Suma[[
Floor[ (i+1)/2]]]]},{i,2Totpoint}] , PlotJoined->True,
DefaultFont -> {"Times-Italic",9},AxesLabel->{"section",
"At"},PlotLabel->"Resultant mmf",Ticks-> {{0,500},Automatic
}]
```

Define the air gap functions from Eq. (3.6)

```
In[41]:=u[r_]:=r/2+N[Sqrt[1+(r/2)^2]]
In[42]:=beta[r_]:=(1-u[r])^2/(2(1+u[r]^2))
In[43]:=gamma[r_]:=4/N[Pi]((r/2)N[ArcTan[r/2]]-Log[N[-
Sqrt[1+(r/2)^2 ]]])
In[44]:=sigma[r_]:=gamma[r]/beta[r]/r
```

Draw the Carter factor, as shown in **Fig. 3.7**

```
In[45]:=Plot3D[If[gamma[r]/beta[r]<tausdelta,tausdelta/
(tausdelta-gamma[r]),1/(1-beta[r])],{tausdelta,5,100},
{r,0.01,20},Shading-> False,AxesLabel->{"taus/delta","o/
delta"," "},Plot Label-> "Carter factor",PlotRange->{1,2},
DefaultFont ->{"Times -Italic",9}]
```

Define and draw the effective air gap width in **Fig. 3.8**

```
In[46]:=deltaeff[o_,delta_,x_]:=delta/(1-beta[o/delta](1-N
[Cos [2Pi/(sigma[o/delta]*o)x]]))gapwidth1[x_,Ns_,os_,
taup_,delta_] :=(tauss= 2taup/Ns;arg1=x-Floor[x/tauss]*
tauss;xAs=os*(1+sigma [os/delta])/2;xBs=tauss-os(sigma[os/
delta]-1)/2;deltasl = del taeff[os,delta, arg1+os(sigma[os/
delta]-1)/2];deltasr=deltaeff[ os,delta,arg1-xBs];
If[arg1<=xAs,deltasl,If[arg1<=xBs,delta, deltasr]])
In[47]:=Plot[gapwidth1[x,12,.00525,.09,.001],{x,0,.18},Plot
Points-> 250,PlotRange->{0,0.003},DefaultFont ->{"Times-
Italic",9}, AxesLabel-> {"mm","m"}, PlotLabel->"Eff. gap
width",Ticks-> {{0,0.05,0.1,0.15},Automatic}]
```

Define the statements which draw the air gap flux density distribution in **Fig. 3.9**

```
In[48]:=m1=Npoints/Ns;
```

Appendix 465

```
In[49]:=Suma=Table[Thetas[[Floor[(i+m1-1)/m1]]],{i,Npoints
}];
In[50]:=mu0=4 N[Pi] 10^-7;
In[51]:=Bgap=Table[mu0 Suma[[i]]/gapwidth1[i/Npoints*.18,
12, 0.00525,0.09,.001],{i,Npoints}];
In[52]:=ListPlot[Table[{Table[Floor[i/2]],Table[Bgap[[
Floor[(i+1)/2] ]]]},{i,2Npoints}],PlotJoined->True,
PlotRange->{-.5, .5},DefaultFont->{"Times-Italic",9},Axes
Label->{"section","T" }, PlotLabel->"Air gap flux density"]
```

Define and draw the effective air gap width in **Fig. 3.10**

```
In[53]:=gapwidth[x_,shift_,Ns_,os_,Nr_,or_,taup_,delta_]:=(
tauss=2taup/Ns;tausr=2taup/Nr;arg1=x-Floor[x/tauss]*tauss;
arg2=x+ shift-Floor[(x+shift)/tausr]*tausr;xAs=os(1+sigma[
os/delta])/2 ;xBs=tauss-os(sigma[os/delta]-1)/2;xAr=or(1+
sigma[or/delta])/2 ;xBr=tausr-or(sigma [or/delta]-1)/
2;deltasl=deltaeff[os,delta, arg1+os(sigma[os/delta]-1)/
2];deltasr=deltaeff[os,delta, arg1 -xBs];deltarl=deltaeff[
or,delta,arg2+or(sigma[or/delta]-1)/2]; deltarr=deltaeff[
or,delta,arg2-xBr];If[arg1<=xAs,If[arg2<=xAr,deltasl
+deltarl-delta,If[arg2<=xBr,deltasl,deltasl+deltarr-delta
]],If[arg1<=xBs,If [arg2<=xAr,deltarl,If[arg2<=xBr,delta,
deltarr]],If[arg2<=xAr,deltasr+deltarl-delta, If[arg2<=xBr,
deltasr,deltasr+deltarr-delta]]]])
In[54]:=Plot[gapwidth[x,0.036,12,0.00525,18,0.0015,0.09,.00
1],{x,0,.09},PlotPoints->250,PlotRange->{0,0.003},Default-
Font->{"Times-Italic",9},AxesLabel->{"mm","m"},PlotLabel->
"Eff. gap width"]
```

Define and draw the air gap flux density distribution in **Fig. 3.11**

```
In[55]:=Thetas=Inverse[C[Ns]].ws.S[ms].{Ias,Ibs,Ics};
In[56]:=Thetar=Shift[0.2,1,Nr].Inverse[C[Nr]].wr.S[mr].{Iar
,Ibr,Icr};
In[57]:=m1=Totpoint/Ns;
In[58]:=Suma=Table[Thetas[[Floor[(i+m1-1)/m1]]]+Thetar[[
Floor[(i+m2-1)/m2]]],{i,Totpoint}];
In[59]:=Bgap=Table[mu0 Suma[[i]]/gapwidth[i/Totpoint*.18,
```

```
0.036 ,12,0.00525,18,.0015,0.09,.001],{i,Totpoint}];
In[60]:=ListPlot[Table[{Table[Floor[i/2]],Table[Bgap[[
Floor[ (i+1)/2] ]]]},{i,2Totpoint}],PlotJoined->True,
PlotRange->{-1,1} ,DefaultFont -> {"Times-Italic",9},Axes
Label->{"section","T"}, PlotLabel->"Air gap flux density",
Ticks->{{0,500},Automatic}]
```

The animation of the air gap flux density is obtained by executing

```
In[61]:=CinvNs=Inverse[C[Ns]];

In[62]:=CinvNr=Inverse[C[Nr]];

In[63]:=Anim3symmBgap[Nframe_]:=Do[{Ias=N[Cos[2 Pi/Nframe
(k-1)]], Ibs=-N[Cos[2 Pi/Nframe (k-1)-2 Pi/3]],Ics=N[Cos[2
Pi/Nframe (k-1)-4 Pi/3]],Iar=N[Sqrt[3]/2],Ibr=0,Icr=-
N[Sqrt[3]/2], Thetas=CinvNs.ws.S[ms ].{Ias,Ibs,Ics},-
Thetar=Shift[k/Nframe,1, Nr].CinvNr.wr.S[mr].{Iar,Ibr,Ic-
r},Aux[[k]]=Table[mu0 (Thetas[[ Floor[(i+m1-1)/m1]]]+Thetar
[[Floor[(i+m2-1)/m2]]])/gapwidth[i/Totpoint*.18,k/
Nframe*.18, 12,0.00525,18,0.0015,0.09,.001],{i, Tot-
point}];ListPlot[Table[{Table[ Floor[(i1/2)]],Table[Aux[[k,
Floor[(i1+1)/2]]]]},{i1,1,2Totpoint}], PlotJoined->True,
PlotRange->{-1.5,1.5},Axes ->None]},{k,1,Nframe}]

In[64]:=Anim3symmBgap[12]
```

A 3.3

Define and draw the main field unit mmf shown in **Fig. 3.13**

```
In[1]:=Thetafield[alpha_,Ns_]:=Table[If[i<=(1-alpha)*Ns/4,
0,If[i<=(1+alpha)*Ns/4,1,If[i<=(3-alpha)*Ns/4,0,If[i<=(3+
alpha )*Ns/4,-1 ,0]]]],{i,Ns}]

In[2]:=Thetatest1=Thetafield[12/18,36];

In[3]:=ListPlot[Table[{Table[Floor[i/2]],Table[Thetatest1[
[Floor[(i+1)/2]]]]},{i,1,72}], PlotJoined->True,Default
Font-> {"Times-Italic",9}, AxesLabel-> {"Ns","At"}, Plot
Label-> "Main field unit mmf"]
```

Define and draw the effective air gap width in **Fig. 3.14**

Appendix

```
In[4]:=u[r_]:=r/2+N[Sqrt[1+(r/2)^2]]
In[5]:=beta[r_]:=(1-u[r])^2/(2(1+u[r]^2))
In[6]:=gamma[r_]:=4/N[Pi]((r/2)N[ArcTan[r/2]]-Log[N[Sqrt[
1+(r/2)^2]]])
In[7]:=sigma[r_]:=gamma[r]/beta[r]/r
In[8]:=deltaeff[o_,delta_,x_]:=delta/(1-beta[o/delta](1-N[
Cos[ 2Pi/(sigma[o/delta]*o)x]]))
In[9]:=gapwidth1DCmp[xin_,alpha_,taup_,delta_]:=(o=(1-
alpha)taup;x=xin-Floor[xin/taup]*taup;xA=o*sigma[o/delta]/
2;xB=taup-xA;argl= x-xA;argr=x+xA-taup;If[x<xA,deltaeff[
o,delta,argl],If[ x<xB, delta,deltaeff[o,delta,argr]]])
In[10]:=Plot[gapwidth1DCmp[t,2/3,.18,.001],{t,0,.36},
PlotRange -> {0,.03},PlotPoints->500,DefaultFont ->{"Times-
Italic",9}, AxesLabel-> {"m","m"},PlotLabel->"Eff. gap
width", Ticks-> {{0,0.1,0.2,0.3}, {0.001,0.01,0.02,0.03}}]
```

Define and draw the air gap flux density created by the main field winding, shown in **Fig. 3.15**

```
In[11]:=m1=20;
In[12]:=Npoints=m1*36;
In[13]:=Suma=Table[Thetatest1[[Floor[(i+m1-1)/m1]]],{i,
Npoints}];
In[14]:=mu0=4 N[Pi] 10^-7;
In[15]:=wf=1000;
In[16]:=if=1;
In[17]:=Bgap=Table[mu0 if wf Suma[[i]]/gapwidth1DCmp[i/
Npoints *.36,2/3,.18,.001],{i,Npoints}];
In[18]:=ListPlot[Bgap,PlotJoined->True,DefaultFont ->
{"Times Italic" ,9},AxesLabel->{"section","T"},PlotLabel->
"Main field flux density" ,Ticks->{{360,720},{-1,0,1}}]
```

Define and draw the commutating pole unit mmf shown in **Fig. 3.16**

```
In[19]:=Thetacompole[beta_,Ns_]:=Table[If[i<=beta*Ns/4,1,
If[i<=(2- beta)*Ns/4,0,If[i<=(2+beta)*Ns/4,-1,If[i<=(4-bet-
a)*Ns/4,0,1]]]], {i,Ns}]
```

In[20]:=Thetatest2=Thetacompole[2/36,72];

In[21]:=ListPlot[Table[{Table[Floor[i/2]],Table[Thetatest2
[[Floor[(i+1)/2]]]]},{i,1,144}],PlotJoined->True,Default
Font ->{"Times-Italic",9}, AxesLabel ->{"Ns","At"}, Plot
Label-> "Comm. pole unit mmf",Ticks-> {{30,72},{-1,0,1}}]

Define and draw the effective air gap width in **Fig. 3.17**

In[22]:=gapwidth1DCmpcp[xin_,alpha_,beta_,taup_,delta_]:=
(o=(1-alpha-beta)taup/2;x=xin-Floor[xin/taup]*taup;opr=
o*sigma[o/delta]; xA= (o+taup*beta)/2;xB=taup-xA;x1=xA-opr/
2;x2=xA+opr/2; x3=xB-opr/2; x4=xB+opr/2;argl=x-x1;argr=x-
x3;If[x>x1,If[x<x2, deltaeff[o,delta,argl],If[x<x3,delta,
If[x<x4,deltaeff[o,delta,argr],delta]]],delta]);x=.

In[23]:=Plot[gapwidth1DCmpcp[x,2/3,1/24,.18,.001], {x,0,
.36},PlotRange->{0,.015},PlotPoints->500,DefaultFont->
{"TimesItalic ",9}, AxesLabel->{"m","m"},PlotLabel->"Eff.
gap width",Ticks-> {{0,0.1,0.2,0.3},
{0.001,0.005,0.01,0.015}}]

Define and draw the effective air gap width in **Fig. 3.18**

In[24]:=gapwidthcompslots[x_,tauss_,os_,taup_,delta_]:=
(argl=x-Floor [x/tauss]*tauss;xAs=os*(1+sigma[os/delta])/
2;xBs=tauss-os*(sigma[os/delta]-1)/2;deltasl=deltaeff[os,
delta,argl+os*(sigma[os/delta]-1)/2];deltasr=deltaeff[os,
delta,argl-xBs];If[argl<=xAs,deltasl,If [argl<=xBs,delta,
deltasr]])

In[25]:=gapwidth1DCmpcpcomp[xin_,alpha_,beta_,taup_,taus_,
os_,Nps_,delta_]:=(op=(1-alpha-beta)taup/2;x=xin-Floor[xin/
taup]* taup;oppr =op*sigma[op/delta];ospr=os*sigma[os/
delta];xA=(op+ taup*beta)/2;xB=taup-xA;x1=xA-oppr/2;
x2=xA+oppr/2;x3=xB-oppr/2;x4=xB+oppr/2;x7=taup*beta/2+op;
x8=(taup*alpha-(Nps-1)taus-os)/2+x7;x10=taup- x7;x9=taup-
x8;x5=x8-(ospr-os)/2;x6=taup-x5; argl=x-x1;argr=x- x3;If[
x<x1,delta, If[x<x5,deltaeff[op,delta, argl], If[x<x2,
deltaeff[op,delta,argl]+ gapwidthcompslots[x-x8, taus,os,
taup,delta]-delta, If[x<x3,gapwidthcompslots[x-x8,taus,
os,taup,delta], If[x<x6,deltaeff[op,delta,argr] + gap
widthcompslots[x-x8,taus,os,taup,delta]-delta, If[x<x4,del
taeff[op,delta, argr],delta]]]]]])

Appendix

```
In[26]:=Plot[gapwidth1DCmpcpcomp[x,2/3,1/
24,.18,.01,.004,12, .001],{x,0,.36},PlotRange-
>{0,.015},PlotPoints->500,Default Font-> {"Times-Itali-
c",9},AxesLabel->{"m","m"},PlotLabel-> "Eff. gap width",-
Ticks->{{0,0.1,0.2,0.3},{0.001,0.005,0.01, 0.015}}]
```

Define and draw the air gap flux density created by the main field winding in a machine with compensating winding slots on the main pole, shown in **Fig. 3.19**

```
In[27]:=Bgap=Table[mu0 if wf Suma[[i]]/gapwidth1DCmpcpcomp
[i/Npoints*.36,2/3,1/24,.18,.01,.004,12,.001],{i,Npoints}];
In[28]:=ListPlot[Bgap,PlotJoined->True,DefaultFont ->
{"Times-Italic",9},AxesLabel->{"section","T"},PlotLabel->
"Main field flux density",Ticks->{{360,720},{-1,0,1}}]
```

Define and draw the compensating winding unit mmf, shown in **Fig. 3.20**

```
In[29]:=Scompen[alpha_,Ns_]:=Table[If[i<=(1-alpha)*Ns/4,0,
If[i<=(1+ alpha)*Ns/4,1, If[i<=(3-alpha)*Ns/4,0, If[i<=(3+
alpha) *Ns/4,-1,0]]]], {i,Ns}]
In[30]:=Unprotect[C];
In[31]:=C[n_]:=Table[If[i<n,If[i==j,-1,If[i==j-1,1,0]],1],
{i,n},{j,n}]
In[32]:=Uprime[n_]:=Table[If[i<n,If[i==j,1,0],0],{i,n},
{j,n}]
In[33]:=Thetacompen[alpha_,Ns_]:=Inverse[C[Ns]].Uprime[Ns].
Scompen[ alpha,Ns]
In[34]:=Thetatest3=Thetacompen[12/18,36];
In[35]:=ListPlot[Table[{Table[Floor[i/2]],Table[Thetatest3
[[Floor[ (i+1)/2]]]]},{i,1,72}],PlotJoined->True,Default
Font->{"Times-Italic",9},AxesLabel->{"Ns","At"},PlotLabel-
> "Comp. winding unit mmf",Ticks-> {{18,36},{-5,0,5}}]
```

Obtain the air gap flux density created by the compensating winding mmf, shown in **Fig. 3.21**

```
In[36]:=Suma=Table[Thetatest3[[Floor[(i+m1-1)/m1]]],{i,
Npoints}];
```

```
In[37]:=wcompen=15;Iaq=10;

In[38]:=Bgap=Table[mu0 Iaq wcompen Suma[[i]]/gapwidth1DC
mpcpcomp[i/Npoints*.36,2/3,1/24,.18,.01,.004,
12,.001],{i,Npoints}];

In[40]:=ListPlot[Bgap,PlotJoined->True,DefaultFont ->
{"Times-Italic" ,9},AxesLabel->{"section","T"},PlotLabel->
"Comp. winding flux density" ,Ticks->{{360,720},{-1,0,1}}]
```

The resultant air gap mmf distribution created by all stator windings, shown in **Fig. 3.22**, is obtained by executing the following statements

```
In[41]:=Sumstat[if_,Iaq_,wf_,wcompen_,wcompole_]:=

In[42]:=Table[if wf Thetatest1[[Floor[(i+m1-1)/m1]]]+Iaq
wcompole Thetatest2[[Floor[(i+m1/2-1)/(m1/2)]]]+Iaq wcompen
Thetatest3[[Floor[ (i+m1-1)/m1]]],{i,Npoints}];

In[43]:=Thetastat=Sumstat[1,10,1000,15,-20];

In[44]:=ListPlot[Table[{Table[Floor[i/2]],Table[Thetastat
[[Floor[ (i+1)/2]]]]},{i,2Npoints}], PlotJoined->True,
DefaultFont->{"Times- Italic",9},AxesLabel->{"section","At"
},PlotLabel->"Stator air gap mmf",Ticks->{{360,720},{-1500,
0,1500}}]
```

whereas the resultant flux density, shown in **Fig. 3.23**, is obtained by executing the following statements

```
In[45]:=Bgap=Table[mu0 Thetastat[[i]]/gapwidth1DCmpcpcomp[
i/Npoints*.36,2/3,1/24,.18,.01,.004,12,.001],{i,Npoints}];

In[46]:=ListPlot[Bgap,PlotJoined->True,DefaultFont ->
{"Times-Italic", 9},AxesLabel->{"section","T"},PlotLabel->
"Stator gap flux density" ,Ticks->{{360,720},{-2,-1,0,
1,2}}]
```

The air gap mmf created by the rotor (armature) winding, shown in **Fig. 3.24**, is obtained by executing the following statements

```
In[47]:=S[n_]:=Table[If[i<=n/2,1,-1],{i,n}]

In[48]:=Thetaq[n_]:=Inverse[C[n]].Uprime[n].S[n]

In[49]:=Convert[t_,T_,n_]:=Floor[(t-Floor[t/T]*T)*n/T]+1
```

```
In[50]:=Shift[t_,T_,n_]:=Table[ If[i<=n-Convert[t,T,n]+1,
If[ i-(j-Convert [t,T,n]+1)==0,1,0],If[i-n+Convert[t,T,n]-
1-j==0,1,0] ],{i,n},{j,n}]
In[51]:=Thetashift[t_,T_,n_]:= Inverse[C[n].Shift[t,T,n]].
Uprime[n]. S[n]
In[52]:=Thetatest4=-Thetaq[48];
In[53]:=m2=Npoints/48;
In[54]:=Sumrot[Iaq_,wa_]:=Table[Iaq wa Thetatest4[[Floor[
(i+m2-1)/m2]]],{i,Npoints}];
In[55]:=Thetarot=Sumrot[10,11];
In[56]:=ListPlot[Table[{Table[Floor[i/2]],Table[Thetarot[[
Floor[ (i+1)/2]]]]},{i,2Npoints}],PlotJoined->True,Default
Font->{"Times-Italic",9},AxesLabel->{"section","At"},Plot
Label-> "Armature air gap mmf",Ticks->{{360,720},{-1000,0,
1000}}]
```

The total air gap mmf in the machine, shown in **Fig. 3.25**, is obtained by executing

```
In[57]:= Thetatot= Table[Thetastat[[i]]+Thetarot[[i]],{i,
Npoints}];
In[58]:=ListPlot[Table[{Table[Floor[i/2]],Table[Thetatot[[
Floor[ (i+1)/2]]]]},{i,2Npoints}],PlotJoined->True,Default
Font->{"Times-Italic",9},AxesLabel->{"section","At"},Plot
Label-> "Total air gap mmf",Ticks->{{360,720},{-1000,0,
1000}}]
```

The effective air gap width, shown in **Fig. 3.26**, is obtained by executing

```
In[58]:= gapwidthDC[xin_,shift_,alpha_,beta_,taup_,delta_,
Ns_,os_,tauss_,Nr_,or_]:= gapwidth1DCmpcpcomp[ xin,alpha,
beta,taup,tauss,os,Ns,delta/2]+gapwidth2DC[xin+shift,
Nr,or,taup,delta/2];x=.
In[59]:=Plot[gapwidthDC[x,.002,2/3,1/24,.18,.001,12,
.004,.01, 48,.003],{x,0,.36},PlotPoints->250,PlotRange->
{0,0.015}, DefaultFont-> {"Times-Italic",9},AxesLabel->
{"m","m"},Plot Label->"Eff. gap width",Ticks->{{0,0.1,0.2,
0.3},{0.001,0.005, 0.01,0.015}}]
```

and the corresponding resultant flux density, shown in **Fig. 3.27**, by executing

```
In[60]:=Bgap=Table[mu0 Thetatot[[i]]/gapwidthDC[i/
Npoints*.36,.0,2/3,1/24,.18,.001,12,.004,.01,48,.003],
{i,Npoints}];
In[61]:=ListPlot[Bgap,PlotJoined->True,DefaultFont ->
{"Times-Italic" ,9},AxesLabel->{"section","T"},PlotLabel->
"Resultant flux density",Ticks->{{360,720},{-1,0,1}}]
```

The air gap mmf in a commutator machine without compensating winding, shown in **Fig. 3.28**, is obtained by executing

```
In[62]:=Thetastatwoc=Sumstat[1,10,1000,0,-20];
In[63]:=Thetatotwoc=Table[Thetastatwoc[[i]]+Thetarot[[i]],
{i,Npoints}];
In[64]:= ListPlot[ Table[{Table[Floor[i/2]], Table[Theta
totwoc[[Floor[(i+1)/2]]]]},{i,2Npoints}],PlotJoined->
True,DefaultFont ->{"Times-Italic",9},AxesLabel->{"section
","At"},PlotLabel-> "Air gap mmf",Ticks->{{360,720},{-1000,
0,1000}}]
```

The air gap mmf created by a d-axis winding, shown in **Fig. 3.29**, is obtained by executing

```
In[65]:=Thetad[n_]:=Thetashift[1/4,1,n]
In[66]:=ListPlot[Table[{Table[Floor[i/2]],Table[Thetatest5
[[Floor[ (i+1)/2]]]]},{i,1,96}],PlotJoined->True,Default
Font->{"Times-Italic",9},AxesLabel->{"Ns","At"},PlotLabel-
> "d-axis unit mmf",Ticks-> {{24,48},{-10,0,10}}]
```

The air gap mmf in **Fig. 3.30** is created by executing the following statements

```
In[67]:=Thetatotdq=Table[Thetastat[[i]]+Thetarotdq[[i]],{i,
Npoints}];
In[68]:=ListPlot[Table[{Table[Floor[i/2]],Table[Thetatotdq
[[ Floor[(i+1)/2]]]]},{i,2Npoints}],PlotJoined-> True,
DefaultFont -> {"Times-Italic",9},AxesLabel->{"section","
At"},PlotLabel->"Total air gap mmf",Ticks->{{360,720},{-
1000,0,1000}}]
```

A 3.4

The field winding distribution shown in **Fig. 3.32** is generated by executing

```
In[1]:=Unprotect[C];
In[2]:=C[n_]:=Table[If[i<n,If[i==j,-1,If[i==j-1,1,0]],
1],{i,n},{j,n}]
In[3]:=S[m_]:=Table[If[i<=m,If[i==j,1,0],If[i-m==j,-1,0]],
{i,2m},{j,m}]
In[4]:=Uprime[n_]:=Table[If[i<n,If[i==j,1,0],0],{i,n},
{j,n}]
In[5]:=Sturbo[Nslot_,taup_,taus_]:=Table[If[i<=(1-Nslot*
taus/(2*taup) )*(2taup/taus)/4,0,If[i<=(1+Nslot*taus/
(2*taup))*(2taup/taus)/4, 1,If[i<=(3-Nslot*taus/
(2*taup))*(2taup/taus)/4,0, If[i<=(3+Nslot*taus/
(2*taup))*(2taup/taus)/4,-1,0]]]],{i,2taup/taus}]
In[6]:=Thetaturborotor[Nslot_,taup_,taus_]:=Inverse[C[2taup
/taus]].
In[7]:=Uprime[2taup/taus].Sturbo[Nslot,taup,taus]
In[8]:=Convert[t_,T_,n_]:=Floor[(t-Floor[t/T]*T)*n/T]+1
In[9]:=Shift[t_,T_,n_]:=Table[ If[i<=n-Convert[t,T,n]+1,
If[i- (j- Convert[t,T,n]+1)==0,1,0],If[i-n+Convert[t,T,n]-
1-j==0,1,0]],{i,n}, {j,n}]
In[10]:=Thetaturborotorshift[t_,T_,Nslot_,taup_,taus_]:=Inv
erse[C[2 taup/taus].Shift[t,T,2taup/taus]].Uprime[2taup/
taus].Sturbo[Nslot, taup,taus]
In[11]:=Nr=44;
In[12]:=Thetatest1=Thetaturborotorshift[-11/Nr,1,30,22,1];
In[13]:=Thetafield[if_,wf_]:=if wf Thetatest1
In[14]:=Thetarot=Thetafield[1,100];
In[15]:=ListPlot[Table[{Table[Floor[i/2]],Table[Thetarot[[
Floor[ (i+1)/2]]]]},{i,2Nr}],PlotJoined->True,Ticks->
{{22,44},{-500,500}}, DefaultFont ->{"Times-Italic",9},
AxesLabel->{"Nr", "At"},PlotLabel-> "Field winding mmf",
PlotRange->{-1000,1000}]
```

Draw the armature mmf, created by the given set of currents, as shown in **Fig. 3.33**

```
In[16]:=wa=wb=wc=10;
In[17]:=BasicMatrix[Nt_,m_]:=Table[If[i<=Nt/(2m),
                                  If[j==1,wa,0],
                                  If[i<=2*Nt/(2m),
                                  If[j==2,wb,0],
                                  If[i<=3*Nt/(2m),
                                  If[j==3,wc,0],
                                  If[i<=4*Nt/(2m),
                                  If[j==4,wa,0],
                                  If[i<=5*Nt/(2m),
                                  If[j==5,wb,0],
                                  If[j==6,wc,0]
                                  ]]]]],{i,1,Nt},{j,1,2m}]
In[18]:=Ns=48;m=3;y=19;
In[19]:=wu=Table[0,{i,Ns},{j,2m}];
In[20]:=wl=BasicMatrix[48,3];
In[21]:=w=Table[0,{i,Ns},{j,2m}];
In[22]:=Do[If[i<=Ns-y,wu[[i+y]]=wl[[i]],wu[[i-(Ns-y)]]=
wl[[i ]]],{i,Ns}];
In[23]:=Do[w[[i]]=wl[[i]]-wu[[i]],{i,Ns}];
In[24]:=Do[w[[Ns,i]]=0,{i,2m}];
In[25]:=Inv48=Inverse[C[48]];
In[26]:=Ia= 2 N[Sin[14.081/48*2*Pi]];Ib= -2 N[Sin[14.081/
48*2* Pi-2 Pi/3]];Ic= 2 N[Sin[14.081/48*2*Pi-4 Pi/3]];
In[27]:=Thetastat=Inv48.w.S[3].{Ia,Ib,Ic};
In[28]:=ListPlot[Table[{Table[Floor[i/2]],Table[Thetastat
[[Floor[(i+1)/2]]]]},{i,2Ns}],PlotJoined->True,Ticks->
{{24,48},{-500,500}},DefaultFont ->{"Times-Italic",9},
AxesLabel->{"Ns","At"},PlotRange->{-1000,1000},PlotLabel->
"Armature mmf"]
```

Generate the total mmf, created by the rotor and stator currents, as shown in **Fig. 3.34** and **Fig. 3.35 (a)**

```
In[29]:=Totpoint=LCM[Ns,Nr];
```

Appendix 475

```
In[30]:=m1=Totpoint/Nr;
In[31]:=m2=Totpoint/Ns;
In[32]:=Aux=Table[0,{i,24},{j,Totpoint}];
In[33]:= Suma= Table[ Thetarot[[Floor[(i+m1-1)/m1]]] +
Thetastat[[Floor[(i+m2-1)/m2]]],{i,Totpoint}];
In[34]:=ListPlot[Table[{Table[Floor[i/2]],Table[Suma[[
Floor[ (i+1)/2] ]]]},{i,2Totpoint}],PlotJoined->True,
PlotRange->{-1100,1100},Ticks-> {{264,528},{-500,500}},
DefaultFont ->{"Times-Italic",9},AxesLabel-> {"section","
At"},PlotLabel->"Total mmf"]
```

Repeat the previous procedure for various shifts between the stator and rotor mmf, as shown in **Fig. 3.35**

```
In[35]:=Ia= 2 N[Sin[14.081/48*2*Pi-Pi/2]];Ib= -2 N[Sin[
14.081/48*2*Pi-2 Pi/3-Pi/2]];Ic= 2 N[Sin[14.081/48*2*Pi-4
Pi/3-Pi/2]];
In[36]:=Thetastat=Inv48.w.S[3].{Ia,Ib,Ic};
In[37]:=ListPlot[Table[{Table[Floor[i/2]],Table[Thetastat
[[Flo or[ (i+1)/2]]]]},{i,2Ns}],PlotJoined->True,Ticks->
{{24,48},{-500,500}}, DefaultFont ->{"Times-Italic",9},
AxesLabel->{"Ns","At"},PlotRange->{-1000,1000},PlotLabel->
"Armature mmf"]
In[38]:= Suma= Table[ Thetarot[[ Floor[(i+m1-1)/m1]]]
+Thetastat[[ Floor[(i+m2-1)/m2]]],{i,Totpoint}];
In[39]:=ListPlot[Table[{Table[Floor[i/2]],Table[Suma[[
Floor[ (i+1)/2] ]]]},{i,2Totpoint}],PlotJoined->True,
PlotRange->{-1100,1100},Ticks-> {{264,528},{-500,500}},
DefaultFont ->{"Times-Italic",9},AxesLabel->{"section","
At"},PlotLabel->"Total mmf"]
In[40]:=Ia= 2 N[Sin[14.081/48*2*Pi-Pi]];Ib= -2 N[Sin[
14.081/48*2*Pi-2 Pi/3-Pi]];Ic=2 N[Sin[14.081/48*2*Pi-4 Pi/
3-Pi]];
In[41]:=Thetastat=Inv48.w.S[3].{Ia,Ib,Ic};
In[42]:=ListPlot[Table[{Table[Floor[i/2]],Table[Thetastat
[[Flo or[ (i+1)/2]]]]},{i,2Ns}],PlotJoined->True,Ticks->{{
24,48},{-500,500}}, DefaultFont ->{"Times-Italic",9},
AxesLabel->{"Ns","At"},PlotRange->{-1000,1000},PlotLabel->
"Armature mmf"]
```

```
In[43]:=Suma = Table[ Thetarot[[ Floor[(i+m1-1)/m1]]]
+Thetastat[[ Floor[(i+m2-1)/m2]]],{i,Totpoint}];

In[44]:=ListPlot[Table[{Table[Floor[i/2]],Table[Suma[[
Floor[ (i+1)/2] ]]]},{i,2Totpoint}],PlotJoined->True,
PlotRange->{-1100,1100},Ticks-> {{264,528},{-500,500}},
DefaultFont ->{"Times-Italic",9},AxesLabel-> {"section","
At"},PlotLabel->"Total mmf"]

In[45]:=Ia= 2 N[Sin[14.081/48*2*Pi-3 Pi/2]];Ib= -2 N[Sin[
14.081/48*2*Pi-2 Pi/3-3 Pi/2]];Ic= 2 N[Sin[14.081/48*2*Pi-4
Pi/3-3 Pi/2]];

In[46]:=Thetastat=Inv48.w.S[3].{Ia,Ib,Ic};

In[47]:=ListPlot[Table[{Table[Floor[i/2]],Table[Thetastat[
[Floor[ (i+1)/2]]]]},{i,2Ns}],PlotJoined->True,Ticks->{{ 24
,48},{-500,500}},DefaultFont->{"Times-Italic",9},AxesLabel
->{"Ns" ,"At"},PlotRange-> {-1000,1000},PlotLabel->"Arma-
ture mmf"]

In[48]:=Suma = Table[ Thetarot[[ Floor[(i+m1-1)/m1]]]
+Thetastat[[ Floor[(i+m2-1)/m2]]],{i,Totpoint}];

In[49]:=ListPlot[Table[{Table[Floor[i/2]],Table[Suma[[
Floor[(i+1)/2] ]]]},{i,2Totpoint}],PlotJoined->True,
PlotRange->{-1100,1100},Ticks-> {{264,528},{-
500,500}},DefaultFont ->{"Times-Italic",9},AxesLabel->
{"section","At"},PlotLabel->"Total mmf"]
```

The animation of the air gap mmf in the stator reference frame is performed by executing the following statements

```
In[50]:=Animturbommfstatpartial[Nstart_,Nfinal_,Nframe_]:=
Do[{Ia= 2 N[Sin[2 Pi/Nframe (k-1)+14.081/48*2*Pi-Pi/3]],Ib=
2 N[Sin[2 Pi/Nframe (k-1)+14.081/48*2*Pi-2 Pi/3-Pi/3]],Ic=
2 N[Sin[2 Pi/Nframe (k-1) +14.081/48*2*Pi-4 Pi/3-Pi/3]],
Thetastat = Inv48.w.S[3].{Ia,Ib,Ic}, Thetatest1 = Thetatur-
borotorshift[-11/Nr-(k-1)/Nframe,1,30,22,1], Thetarot=
Thetafield[1,100], Aux[k]=Table[Thetarot[[Floor[(i+m1-1)/
m1 ]]]+ Thetastat[[Floor[(i+m2-1)/m2]]],{i,Totpoint}],
ListPlot[Table[{ Table[Floor[i/2]],Table[Aux[[k,
Floor[(i+1)/2]]]]},{i,2Totpoint}], PlotJoined->True,
PlotRange->{-1200,1200},Axes ->None]}, {k,Nstart,Nfinal}]
```

The animation of the air gap mmf in the rotor reference frame is performed by executing the following statements

```
In[51]:=Thetatest1=Thetaturborotorshift[-11/Nr,1,30,22,1];
In[52]:=Thetarot=Thetafield[1,100];
In[53]:=Animturbommfrotpartial[Nstart_,Nfinal_,Nframe_]:=
Do[{Ia= 2 N[Sin[2 Pi/Nframe (k-1)+14.081/48*2*Pi-Pi/3]],Ib=
2 N[Sin[2 Pi/Nframe (k-1)+14.081/48*2*Pi-2 Pi/3-Pi/3]],Ic=
2 N[Sin[2 Pi/Nframe (k-1) +14.081/48*2*Pi-4 Pi/3-Pi/3]],
Thetastat=Inverse[C[Ns].Shift[(k-1)/Nframe,1,Ns]].w.S[3].
{Ia,Ib,Ic},Aux[[k]]= Table[Thetarot[[Floor[(i+m1-1)/m1]]]+
Thetastat[[Floor[(i+m2-1)/m2]]],{i,Totpoint}],ListPlot[
Table [{Table[Floor[i/2]],Table[ Aux[[k,Floor[(i+1)/2]]]]}
,{i,2Totpoint}], PlotJoined->True, PlotRange->{-1200,1200},
Axes ->None]},{k,Nstart, Nfinal}]
```

Animation of the air gap flux density in a turbogenerator is performed by executing the following statements

```
In[54]:=u[r_]:=r/2+N[Sqrt[1+(r/2)^2]]
In[55]:=beta[r_]:=(1-u[r])^2/(2(1+u[r]^2))
In[56]:=gamma[r_]:=4/N[Pi]((r/2)N[ArcTan[r/2]]-Log[N[Sqrt[
1+(r/2)^2 ]]])
In[57]:=sigma[r_]:=gamma[r]/beta[r]/r
In[58]:=mu0=4 N[Pi] 10^-7;
In[59]:=deltaeff[o_,delta_,x_]:=delta/(1-beta[o/delta](1-N
[Cos[2Pi/(o*sigma[o/delta])x]]))
In[60]:=gapwidth[x_,shift_,Ns_,os_,Nr_,or_,taup_,delta_]:=(
tauss=2 taup/Ns;tausr=2taup/Nr;arg1=x-Floor[x/tauss]*tauss;
arg2=x+ shift- Floor[(x+shift)/tausr]*tausr;xAs=os(1+sigma[
os/delta])/2; xBs=tauss- os(sigma[os/delta]-1)/2;xAr=or(1+
sigma[or/delta])/2;xBr=tausr-or ( sigma[or/delta]-1)/
2;deltasl=deltaeff[os,delta, arg1+os(sigma[os/delta]-1)/
2];deltasr=deltaeff[os,delta,arg1-xBs];deltarl=deltaeff[
or,delta,arg2+or(sigma[or/delta]-1)/2]; deltarr=deltaeff[
or,delta,arg2-xBr];If[arg1<=xAs,If[arg2<=xAr,deltasl
+deltarl-delta,If [arg2<=xBr,deltasl,deltasl+deltarr-delta
]],If[arg1<=xBs,If[arg2<=x Ar,deltarl,If[arg2<=xBr,delta,
deltarr]],If[arg2<=xAr,deltasr+deltarl- delta,If[arg2<=xBr,
deltasr,deltasr+deltarr-delta]]]])
In[61]:=AnimturboBgapstatpartial[Nstart_,Nfinal_,Nframe_]:=
Do[{Ia= 2 N[Sin[2 Pi/Nframe (k-1)+14.081/48*2*Pi-Pi/3]],Ib=
2 N[Sin[2 Pi/Nframe (k-1)+14.081/48*2*Pi-2 Pi/3-Pi/3]],Ic=
```

```
2 N[Sin[2 Pi/Nframe (k-1) +14.081/48*2*Pi-4 Pi/3-Pi/3]],
Thetastat =Inv48.w.S[3].{Ia,Ib,Ic}, Thetatest1=Thetatur-
borotorshift[-11/Nr-(k-1)/Nframe,1,30,22,1], Thetarot=-
Thetafield[1,100],Aux[[k]] =Table[mu0 (Thetarot[[Floor[
(i+m1-1)/m1]]]+Thetastat[[Floor[ (i+m2-1)/m2]]])/gapwidth
[i/Totpoint*.528,k/ Nframe*.528,48, 0.005,44,0.006,
0.264,.001],{i,Totpoint}],ListPlot[Table[{ Table[Floor[i/
2]],Table[ Aux[[k,Floor[(i+1)/2]]]]},{i,2 Totpoint}],Plot
Joined->True,PlotRange->{-1.2,1.2},Ticks->{{264,528},{-1,1
}},DefaultFont ->{"Times-Italic",9},AxesLabel->{"section",
"T"},PlotLabel->"Air gap flux density"]},{k,Nstart,Nfinal}]

In[62]:=AnimturboBgapstatpartial[1,22,22]
```

The distribution of the air gap quantities in a hydrogenerator is obtained as a result of the following statements

```
In[1]:=Unprotect[C];

In[2]:=C[n_]:=Table[If[i<n,If[i==j,-1,If[i==j-1,1,0]],1],
{i,n},{j,n}]

In[3]:=S[m_]:=Table[If[i<=m,If[i==j,1,0],If[i-m==j,-1,0]],
{i,2m},{j,m}]

In[4]:=Convert[t_,T_,n_]:=Floor[(t-Floor[t/T]*T)*n/T]+1

In[5]:=Shift[t_,T_,n_]:=Table[ If[i<=n-Convert[t,T,n]+1,If
[i-( j- Convert[t,T,n]+1)==0,1,0],If[i-n+Convert[t,T,n]-1-
j==0,1,0] ],{i,n}, {j,n}]

In[6]:=Hydrofield[alpha_,Ns_]:=Table[If[i<=(1-alpha)*Ns/
4,0, If[ i<=(1+alpha)*Ns/4,1,If[i<=(3-alpha)*Ns/4,0,If[i<=
(3+alpha)* Ns/4,-1,0] ]]],{i,Ns}]

In[7]:=Thetahydrorotorshift[t_,T_,Ns_,alpha_]:= Shift
[t,T,Ns] . Hydrofield[alpha,Ns]

In[8]:=Thetafield[if_,wf_,t_]:=if wf Thetahydrorotorshift[
t,1,Ns,2/3]

In[9]:=Ns=48;

In[10]:=Thetarot=Thetafield[3,250,0];

In[11]:=ListPlot[Table[{Table[Floor[i/2]],Table[Thetarot[[
Floor[ (i+1)/2]]]]},{i,2Ns}],PlotJoined->True,Ticks->{{22,
44},{-500,500}},DefaultFont->{"Times-Italic",9},AxesLabel-
> {"Nr","At"},PlotLabel-> "Field winding mmf",PlotRange->{-
1000,1000}]
```

Appendix 479

The total air gap mmf for various values of the stator to rotor mmf shift, shown in **Fig. 3.38**, is generated by executing

```
In[12]:=Thetarot=Thetafield[3,250,.25];
In[13]:=ListPlot[Table[{Table[Floor[i/2]],Table[Thetarot[[
Floor[ (i+1)/2]]]]},{i,2Ns}],PlotJoined->True,PlotRange->{-
1000,1000}]
In[14]:=BasicMatrix[Nt_,m_]:=Table[If[i<=Nt/(2m),
                                    If[j==1,wa,0],
                                    If[i<=2*Nt/(2m),
                                    If[j==2,wb,0],
                                    If[i<=3*Nt/(2m),
                                    If[j==3,wc,0],
                                    If[i<=4*Nt/(2m),
                                    If[j==4,wa,0],
                                    If[i<=5*Nt/(2m),
                                    If[j==5,wb,0],
                                    If[j==6,wc,0]
                                    ]]]]],{i,1,Nt},{j,1,2m}]
In[15]:=wa=wb=wc=10;Ns=48;m=3;y=19;
In[16]:=wu=Table[0,{i,Ns},{j,2m}];
In[17]:=wl=BasicMatrix[48,3];
In[18]:=w=Table[0,{i,Ns},{j,2m}];
In[19]:=Do[If[i<=Ns-y,wu[[i+y]]=wl[[i]],wu[[i-(Ns-
y)]]=wl[[i ]]], {i,Ns}];
In[20]:=Do[w[[i]]=wl[[i]]-wu[[i]],{i,Ns}];
In[21]:=Do[w[[Ns,i]]=0,{i,2m}];
In[22]:=Inv48=Inverse[C[48]];
In[23]:=Ia= 2 N[Sin[14.081/48*2*Pi]];Ib= -2 N[Sin[14.081/
48*2*Pi-2 Pi/3]];Ic= 2 N[Sin[14.081/48*2*Pi-4 Pi/3]];
In[24]:=Thetastat=Inv48.w.S[3].{Ia,Ib,Ic};
In[25]:=Suma=Table[Thetarot[[i]]+Thetastat[[i]],{i,Ns}];
In[26]:=ListPlot[Table[{Table[Floor[i/2]],Table[Suma[[
Floor[(i+1)/2] ]]]},{i,2Ns}],PlotJoined->True,PlotRange->
{-1100, 1100},Ticks-> {{24,48},{-1000,1000}},DefaultFont ->
```

{"Times-Italic",9},AxesLabel-> {"Ns","At"},PlotLabel->
"Total mmf"]

In[27]:=Ia= 2 N[Sin[14.081/48*2*Pi-Pi/2]];Ib= -2 N[Sin[
14.081/48*2*Pi-2 Pi/3-Pi/2]];Ic= 2 N[Sin[14.081/48*2*Pi-4
Pi/3-Pi/2]];

In[28]:=Thetastat=Inv48.w.S[3].{Ia,Ib,Ic};

In[29]:=Suma=Table[Thetarot[[i]]+Thetastat[[i]],{i,Ns}];

In[30]:=ListPlot[Table[{Table[Floor[i/2]],Table[Suma[[
Floor[(i+1)/2]]]]},{i,2Ns}],PlotJoined->True,PlotRange->
{-1100, 1100},Ticks-> {{24,48},{-1000,1000}},DefaultFont ->
{"Times-Italic",9},AxesLabel-> {"Ns","At"},PlotLabel->
"Total mmf"]

In[31]:=Ia= 2 N[Sin[14.081/48*2*Pi-Pi]];Ib= -2 N[Sin[
14.081/48*2*Pi-2 Pi/3-Pi]];Ic= 2 N[Sin[14.081/48*2*Pi-4 Pi/
3-Pi]];

In[32]:=Thetastat=Inv48.w.S[3].{Ia,Ib,Ic};

In[33]:=Suma=Table[Thetarot[[i]]+Thetastat[[i]],{i,Ns}];

In[34]:=ListPlot[Table[{Table[Floor[i/2]],Table[Suma[[
Floor[(i+1)/2]]]]},{i,2Ns}],PlotJoined->True,PlotRange->
{-1100, 1100},Ticks-> {{24,48},{-1000,1000}},DefaultFont ->
{"Times-Italic",9},AxesLabel-> {"Ns","At"},PlotLabel->
"Total mmf"]

In[35]:=Ia= 2 N[Sin[14.081/48*2*Pi-3 Pi/2]];Ib= -2 N[Sin[
14.081/48*2*Pi-2 Pi/3-3 Pi/2]];Ic= 2 N[Sin[14.081/48*2*Pi-4
Pi/3-3 Pi/2]];

In[36]:=Thetastat=Inv48.w.S[3].{Ia,Ib,Ic};

In[37]:=Suma=Table[Thetarot[[i]]+Thetastat[[i]],{i,Ns}];

In[38]:=ListPlot[Table[{Table[Floor[i/2]],Table[Suma[[
Floor[(i+1)/2]]]]},{i,2Ns}],PlotJoined->True,PlotRange->{
-1100,1100},Ticks-> {{24,48},{-1000,1000}},DefaultFont ->
{"Times-Italic",9},AxesLabel-> {"Ns","At"},PlotLabel->
"Total mmf"]

The effective air gap width in a hydrogenerator, shown in **Fig. 3.40**, is determined by the rotor saliency and stator slots. It is generated by executing

In[39]:=u[r_]:=r/2+N[Sqrt[1+(r/2)^2]]

In[40]:=beta[r_]:=(1-u[r])^2/(2(1+u[r]^2))

Appendix

```
In[41]:=gamma[r_]:=4/N[Pi]((r/2)N[ArcTan[r/2]]-Log[N[
Sqrt[1+(r/2)^2 ]]])
In[42]:=sigma[r_]:=gamma[r]/beta[r]/r
In[43]:=mu0=4 N[Pi] 10^-7;
In[44]:=deltaeff[o_,delta_,x_]:=delta/(1-beta[o/delta](1-N
[Cos[2Pi/(o*sigma[o/delta])x]]))
In[45]:=gapwidth1DCmp[xin_,alpha_,taup_,delta_]:=(o=(1-
alpha)taup; x=xin-Floor[xin/taup]*taup;xA=o*sigma[o/delta]/
2;xB=taup-xA;argl=x-xA ;argr=x+xA-taup;If[x<xA,deltaeff
[o,delta,argl],If[ x<xB,delta,deltaeff [o,delta,argr]]])
In[46]:=gapwidth2DC[xin_,Ns_,os_,taup_,delta_]:= (tauss=2
taup/Ns; x=xin-Floor[xin/taup]*taup;argl=x-Floor[x/tauss]*
tauss;xAs=os*(1 + sigma[os/delta])/2;xBs=tauss-os(sigma[os/
delta]-1)/2; deltasl=deltaeff [os,delta,argl+os*(sigma[os/
delta]-1)/2]; deltasr=deltaeff[os,delta, argl-xBs];If[
argl<=xAs,deltasl,If[ argl<=xBs,delta,deltasr]])
In[47]:=gapwidthhydro [xin_,shift_,alpha_,taup_,delta_,Ns_
,os_]: = gapwidth1DCmp[xin+shift,alpha,taup,delta/2]+gap-
width2DC[ xin,Ns,os, taup,delta/2]
In[48]:=Plot[gapwidthhydro[x,.002,2/3,.264,.001,48,.005],
{x, 0,.528}, PlotPoints->250,PlotRange->{0,0.04},Default
Font->{"Times-Italic",9},AxesLabel->{"m","mm"},PlotLabel->
"Eff. gap width",Ticks-> {{0,0.1,0.2,0.3,0.4,0.5},
{0.01,0.02,0.03,0.04}}]
```

The total air gap mmf for the given set of stator currents

```
In[49]:=Ia= 2 N[Sin[14.081/48*2*Pi- Pi/3]];Ib= -2 N[Sin[
14.081/48*2*Pi-2 Pi/3-Pi/3]];Ic= 2 N[Sin[14.081/48*2*Pi-4
Pi/3-Pi/3]];
```

is obtained by executing

```
In[50]:=Thetastat=Inv48.w.S[3].{Ia,Ib,Ic};
In[51]:=Totpoint=528
In[52]:=Suma=Table[Thetarot[[Floor[(i+10)/11]]]+Thetastat
[[Floor[ (i+10)/11]]],{i,Totpoint}];
In[53]:=ListPlot[Table[{Table[Floor[i/2]],Table[Suma[[
```

```
Floor[ (i+1)/2] ]]]},{i,2Totpoint}],PlotJoined->True,
PlotRange->{-1100 ,1100},Ticks-> {{264,528},{-1000,
1000}},DefaultFont ->{"Times-Italic",9},AxesLabel-> {"sec-
tion","At"},PlotLabel->"Total mmf"]
```

and shown in Fig. 3.41. The corresponding air gap flux density, shown in Fig. 3.42, is the result of the following statements

```
In[54]:=Bgap=Table[mu0 Suma[[i]]/gapwidthhydro[i*.001,.0,2/
3,.264, .001,48,.005],{i,Totpoint}];
In[55]:=ListPlot[Bgap,PlotJoined->True,DefaultFont ->
{"Times-Italic" ,9},AxesLabel->{"section","T"},PlotLabel->
"Resultant flux density", Ticks->{{264,528},{-1,0,1}}]
```

A 3.5

Repeat the definition of the air gap mmf created by a three phase winding, and of the reference frame matrices, in order to obtain the 3D mmf distribution shown in **Fig. 3.43**

```
In[1]:=Unprotect[C];
In[2]:=C[n_]:=Table[If[i<n,If[i==j,-1,If[i==j-1,1,0]],1],
{i,n},{j,n}]
In[3]:=S[m_]:=Table[If[i<=m,If[i==j,1,0],If[i-m==j,-1,0]],
{i,2m},{j,m}]
In[4]:=Convert[t_,T_,n_]:=Floor[(t-Floor[t/T]*T)*n/T]+1
In[5]:=Shift[t_,T_,n_]:= Table[If[i<=n- Convert[t,T,n]+1,
If[i-(j- Convert[t,T,n]+1)==0,1,0],If[i-n+Convert[t,T,n]-1-
j==0,1,0 ]],{i,n},{j,n}]
In[6]:=Aux=Table[0,{i,48},{j,50}];
In[7]:=wa=wb=wc=10;
In[8]:=BasicMatrix[Nt_,m_]:=Table[If[i<=Nt/(2m),
                             If[j==1,wa,0],
                             If[i<=2*Nt/(2m),
                             If[j==2,wb,0],
                             If[i<=3*Nt/(2m),
                             If[j==3,wc,0],
```

Appendix 483

```
                            If[i<=4*Nt/(2m),
                            If[j==4,wa,0],
                            If[i<=5*Nt/(2m),
                            If[j==5,wb,0],
                            If[j==6,wc,0]
                            ]]]]],{i,1,Nt},{j,1,2m}]
In[9]:=wls=BasicMatrix[12,3];
In[10]:=Ns=12;ms=3;ys=5;
In[11]:=wus=Table[0,{i,Ns},{j,2ms}];
In[12]:=ws=Table[0,{i,Ns},{j,2ms}];
In[13]:=Do[If[i<=Ns-ys,wus[[i+ys]]=wls[[i]],wus[[i-(Ns-ys)
]] =wls[[i]]],{i,Ns}];
In[14]:=Do[ws[[i]]=wls[[i]]-wus[[i]],{i,Ns}];
In[15]:=Do[ws[[Ns,i]]=0,{i,2ms}];
In[16]:=Do[{Ia=25 N[Cos[2 Pi/24 (k-1)]],Ib=-25 N[Cos[2 Pi/
24 (k-1) - 2 Pi/3]],Ic=25 N[Cos[2 Pi/24 (k-1) - 4 Pi/3]],
Aux[[k]]= Inverse[C[12]. Shift[-k,24,12]].ws.S[3].{Ia,Ib,
Ic}},{k,25}]
In[17]:=Output=Table[Table[Aux[[j,Floor[(i+3)/4]]],
{i,1,48}], {j,1,25}];
In[18]:=ListPlot3D[Output,Shading ->False, MeshRange->
{{0,12},{0,1}},AxesLabel->{"tooth","t/T","At"},PlotLabel->
"Air gap mmf at s=2",Shading->False,MeshRange->{{0,12},
{0,1}},DefaultFont -> {"Times- Italic",9},Ticks->{{0,2,4,6
,8,10,12},{0,0.25,0.5, 0.75,1},{-750,0,750 }}]
```

The air gap mmf distribution in **Fig. 3.44** is created by executing

```
In[19]:=Do[{Ia=25 N[Cos[2 Pi/24 (k-1)]],Ib=-25 N[Cos[2 Pi/
24 (k-1) - 2 Pi/3]],Ic=25 N[Cos[2 Pi/24 (k-1) - 4 Pi/
3]],Aux[[k]] =Inverse[C[12]. Shift[0,24,12]].ws.S[3].{Ia,
Ib,Ic}},{k,25}]
In[20]:=Output=Table[Table[Aux[[j,Floor[(i+3)/
4]]],{i,1,48}], {j,1,25}];ListPlot3D[Output,Shading ->
False, AxesLabel-> {"tooth","t/T","At"},PlotLabel->"Air gap
mmf at s=1",Shading-> False,MeshRange-> {{0,12},{0,1}},
DefaultFont->{"Times-Italic" ,9},Ticks-> {{0,2,4,6,8,
```

10,12},{0,0.25,0.5,0.75,1},{-750,0, 750}}]

and the mmf distribution in **Fig. 3.45**

```
In[21]:=Do[{Ia=25 N[Cos[2 Pi/24 (k-1)]],Ib=-25 N[Cos[2 Pi/
24 (k-1) - 2 Pi/3]],Ic=25 N[Cos[2 Pi/24 (k-1) - 4 Pi/3]],
Aux[[k]] =Inverse[C[12]. Shift[3k/4,24,12]].ws.S[3].{Ia,
Ib,Ic}},{k,25}]
In[22]:=Output=Table[Table[Aux[[j,Floor[(i+3)/4]]],{i,1,
48}],{j,1,25}];ListPlot3D[Output,Shading->False,AxesLabel-
> {"tooth"," t/T","At"},PlotLabel->"Air gap mmf at s=0.25",
Shading-> False,Mesh Range->{{0,12},{0,1}},DefaultFont ->
{"Times-Italic",9},Ticks-> {{0,2,4,6,8,10,12},
{0,0.25,0.5,0.75,1},{-750 ,0,750}}]
```

The mmf distribution in **Fig. 3.46** is the result of the following statements

```
In[23]:=Do[{Ia=25 N[Cos[2 Pi/24 (k-1)]],Ib=-25 N[Cos[2 Pi/
24 (k-1) - 2 Pi/3]],Ic=25 N[Cos[2 Pi/24 (k-1) - 4 Pi/3]],
Aux[[k]] =Inverse[C[12]. Shift[k,24,12]].ws.S[3].{Ia,
Ib,Ic}},{k,25}]

In[24]:=Output=Table[Table[Aux[[j,Floor[(i+3)/4]]],
{i,1,48}], {j,1,25}];

In[25]:=ListPlot3D[Output,Shading ->False,AxesLabel->{
"tooth"," t/T" ,"At"},PlotLabel->"Air gap mmf at s=0",Shad-
ing-> False,MeshRange-> {{0,12},{0,1}},DefaultFont->
{"Times-Italic" ,9},Ticks-> {{0,2,4,6,8,10,12},
{0,0.25,0.5,0.75,1},{-750,0, 750}}]
```

and the mmf distribution in **Fig. 3.47**

```
In[26]:=Do[{Ia=25 N[Cos[2 Pi/24 (k-1)]],Ib=-25 N[Cos[2 Pi/
24 (k-1) - 2 Pi/3]],Ic=25 N[Cos[2 Pi/24 (k-1) - 4 Pi/3]],
Aux[[k]] =Inverse[C[12]. Shift[5k/4,24,12]].ws.S[3].{Ia,
Ib,Ic}},{k,25}]

In[27]:=Output=Table[Table[Aux[[j,Floor[(i+3)/4]]],
{i,1,48}],{j,1,25}];

In[28]:=ListPlot3D[Output,Shading ->False, AxesLabel->{
"tooth"," t/T" ,"At"},PlotLabel->"Air gap mmf at s=-0.25",
Shading-> False,MeshRange-> {{0,12},{0,1}},DefaultFont ->
```

Appendix

```
{"Times-Italic",9},Ticks-> {{0,2,4,6,8,10,12},
{0,0.25,0.5,0.75,1},{-750 ,0,750}}]
```

The mmf distribution for the rotor slip equal to -1, shown in **Fig. 3.48**, is obtained by executing

```
In[29]:=Do[{Ia=25 N[Cos[2 Pi/24 (k-1)]],Ib=-25 N[Cos[2 Pi/
24 (k-1) - 2 Pi/3]],Ic=25 N[Cos[2 Pi/24 (k-1) - 4 Pi/3]],
Aux[[k]] =Inverse[C[12]. Shift[2k,24,12]].ws.S[3].{Ia,
Ib,Ic}},{k,25}]

In[30]:=Output=Table[Table[Aux[[j,Floor[(i+3)/4]]],
{i,1,48}], {j,1,25}];

In[31]:=ListPlot3D[Output,Shading ->False,AxesLabel->{
"tooth"," t/T" ,"At"},PlotLabel->"Air gap mmf at s=-1",
Shading-> False,MeshRange-> {{0,12},{0,1}},DefaultFont ->
{"Times-Italic",9},Ticks-> {{0,2,4,6,8,10,12},
{0,0.25,0.5,0.75,1},{-750 ,0,750}}]
```

Calculate the rotor ampereturns distribution represented in **Fig. 3.50**

```
In[32]:=wa=wb=wc=9;
In[33]:=wlr=BasicMatrix[14,2];
In[34]:=Nr=14;mr=2;yr=6;
In[35]:=wur=Table[0,{i,Nr},{j,2mr}];
In[36]:=wr=Table[0,{i,Nr},{j,2mr}];
In[37]:=Do[If[i<=Nr-yr,wur[[i+yr]]=wlr[[i]],wur[[i-(Nr-yr)
]]=wlr[[i]]],{i,Nr}];
In[38]:=Do[wr[[i]]=wlr[[i]]-wur[[i]],{i,Nr}];
In[39]:=Do[wr[[Nr,i]]=0,{i,2mr}];
In[40]:=Conmat14=Inverse[C[14]].wr.S[2];
In[41]:=Conmat12=Inverse[C[12]].ws.S[3];
In[42]:=Delta=Table[0,{i,128},{j,100}];
In[43]:=IaVector=Table[0,{j,100}];
In[44]:=IbVector=Table[0,{j,100}];
In[45]:=IcVector=Table[0,{j,100}];
In[46]:=Auxrot=Table[0,{i,128},{j,100}];
```

```
In[47]:=Do[{Iar=20 N[Cos[2 Pi/24 (k-1)]],Ibr=-20 N[Cos[2
Pi/24 (k-1)-Pi/2]],Auxrot[[k]]=Conmat14.{Iar,Ibr}},{k,25}]
In[48]:=Output=Table[Table[Auxrot[[j,Floor[(i+1)/2]]],
{i,2Nr}] ,{j,1,25}];
In[49]:=ListPlot3D[Output, Shading->False, AxesLabel->
{"tooth" ,"t/T" ,"At"},PlotLabel->"Rotor mmf",Shading->
False,MeshRange-> {{0,14}, {0,1}},DefaultFont ->{"Times-
Italic",9},Ticks->{{0,2, 4,6,8,10,12,14},
{0,0.25,0.5,0.75,1},{-750,0,750}}]
```

Calculate and plot the additional stator ampereturns, which compensate for the rotor ones, as shown in **Fig. 3.51**

```
In[50]:=Totpoint=LCM[Ns,Nr];

In[51]:=m1=Totpoint/Ns;

In[52]:=m2=Totpoint/Nr;

In[53]:=Reaction[Nstart_,Nfinal_,NofPoints_]:=Do[{Iar=20 N[
Cos[2 Pi/NofPoints (k-1)]],Ibr=-20 N[Cos[2 Pi/NofPoints (k-
1)-Pi/2]],Thetar=Conmat14.{Iar,Ibr},Thetarred=Table[ Thetar
[[Floor[(i+m2-1)/m2]]],{i,Totpoint}],Sol =FindMinimum [
Thetas = Conmat12.{Ia,Ib,Ic}; Thetasred=Table[Thetas[[
Floor[(i+m1-1)/m1]]] ,{i,Totpoint}];(Ia-Ib+Ic)^2+Sum[(-
Thetasred[[i]]+Thetarred[[i]])^2,{i,Totpoint/2}],
{Ia,{0,1}},{Ib,{0,1}},{Ic,{0,1}},MaxIterations->100, Accu-
racyGoal->10] ,temp=Chop[Drop[Sol,1]], Iasol=Ia /.
temp[[1]],Ibsol=Ib /. temp[[1]], Icsol=Ic /. temp[[1]],
IaVector[[k]]=Iasol, IbVector[[k]]= Ibsol,
IcVector[[k]]=Icsol,Thetasol=Conmat12.{Iasol,Ibsol,Icsol},
Aux[[k]] =Table[Thetasol[[Floor[(i+1)/2]]],{i,2 Ns}],
Thetasolred = Table[Thetasol [[Floor[(i+m1-1)/m1]]],{i,
Totpoint}],Delta[[k]]=Table[Thetasolred [[Floor[(i+1)/2]]]
+Thetarred[[Floor[(i+1)/2]]],{i,2 Totpoint}],}, {k,Nstart,
Nfinal}]

In[54]:=Reaction[1,25,24]

In[55]:=Output=Table[Table[Aux[[j,i]],{i,2 Ns}],{j,24}];

In[56]:=ListPlot3D[Output, Shading -> False,AxesLabel->{
"tooth"," t/T","At"},PlotLabel->"Stator mmf",Shading->
False, MeshRange-> {{0,12},{0,1}},DefaultFont ->{"Times-
Italic",9}, Ticks-> {{0,2,4,6,8,10,12},
{0,0.25,0.5,0.75,1},{-750,0,750}}]
```

Appendix

Obtain the current in phase *b*, shown in **Fig. 3.53**

```
In[57]:=Output=.
In[58]:=Ibcurr=Table[IbVector[[i]],{i,24}];
In[59]:=ListPlot[Ibcurr, PlotJoined->True,DefaultFont->
{"Times-Italic" ,9},AxesLabel->{"t/T","A"},PlotLabel->
"Phase b current"]
```

Obtain the difference between the stator and rotor mmfs, shown in **Fig. 3.52**

```
In[60]:=Output=Table[Table[Delta[[j,i]],{i,2 Totpoint}],
{j,24}];
In[61]:=ListPlot3D[Output,Shading->False,AxesLabel->{
"tooth","t/T","At"},PlotLabel->"Differential mmf",Shading->
False,MeshRange-> {{0,12},{0,1}},DefaultFont ->{"Times-
Italic" ,9},PlotRange->{-751, 751},Ticks->{{0,2,4,6,8,10,
12}, {0,0.25,0.5,0.75,1},{-750,0,750}}]
```

A 4.3

In order to obtain the open circuit characteristic of a machine with distributed winding, the following statements, introduced in the second chapter, have to be executed:

```
In[1]:=Unprotect[C];
In[2]:=C[n_]:=Table[If[i<n,If[i==j,-1,If[i==j-
1,1,0]],1],{i,n},{j,n}]
In[3]:=S[m_]:=Table[If[i<=m,If[i==j,1,0],If[i-m==j,-1,0]],
{i,2m}, {j,m}]
In[4]:=w2=Table[0,{i,12},{j,4}];
In[5]:=w2[[1]]={wa1,0,0,0};
In[6]:=w2[[2]]={wa2,wb2,0,0};
In[7]:=w2[[3]]={0,wb1,0,0};
In[8]:=w2[[4]]={0,wb1,0,0};
In[9]:=w2[[5]]={0,wb2,wa2,0};
In[10]:=w2[[6]]={0,0,wa1,0};
In[11]:=w2[[7]]={0,0,wa1,0};
```

```
In[12]:=w2[[8]]={0,0,wa2,wb2};
In[13]:=w2[[9]]={0,0,0,wb1};
In[14]:=w2[[10]]={0,0,0,wb1};
In[15]:=w2[[11]]={wa2,0,0,wb2};
In[16]:=w2[[12]]={0,0,0,0};
In[17]:=Auxmat2=Inverse[C[12]].w2.S[2];
In[18]:=Mmf2[Ia_,Ib_]:=Auxmat2.{Ia,Ib}
In[19]:=wa1=wb1=100;wa2=wb2=100;
In[20]:=ToothMmf=Mmf2[1,0]
```

The iron core magnetization curve is defined as

```
In[21]:=BHCurve={{0,0},{.1,12.5},{.2,25},{.3,37.5},{.4,50},
{.5,62.5},{.6,75},{.7,87.5},{.8,100},{.9,112.5},{1,125},{1.
1,160},{1.2,240},{1.3,420},{1.4,1100},{1.5,2650},{1.6,5300}
,{1.7,9100},{1.8,14000},{1.9,21000},{1.95,27000},{2.,40000}
,{3.,1170000},{4.,2380000},{5.,3550000},{6.,4720000},{7.,58
90000},{8.,7060000},{9.,8230000},{10.,9400000}}
In[22]:=HofB=Interpolation[BHCurve]
```

The reluctance of each part of the magnetic circuit is defined next. Rta is the tooth reluctance on the a-side of the air gap, Rya is the yoke reluctance on the a-side of the gap, Rtb is the tooth reluctance on the b-side of the air gap, and Ryb is the yoke reluctance on the b-side of the gap

```
In[23]:=Rta[phi_]:=lta HofB[Abs[phi]/Sta]/(Abs[phi]+10^(-
9))
In[24]:=Rtb[phi_]:=ltb HofB[Abs[phi]/Stb]/(Abs[phi]+10^(-
9))
In[25]:=Rya[phi_]:=lya HofB[Abs[phi]/Sya]/(Abs[phi]+10^(-
9))
In[26]:=Ryb[phi_]:=lyb HofB[Abs[phi]/Syb]/(Abs[phi]+10^(-
9))
```

The tooth and yoke length segments lta, ltb, lya and lyb=.015, along with the air gap reluctance Rgap are defined next

Appendix 489

```
In[27]:=lta=.015;ltb=.01;lya=.04;lyb=.015;Rgap=5.3 10^5;
In[28]:=Sta=.0005;Stb=.0004;Sya=.0014;Syb=.00125;
```

The linear values of tooth and yoke reluctances are obtained from the previously defined reluctance functions for small values of arguments

```
In[29]:=RtaLin=Rta[10^-6]
In[30]:=RyaLin=Rya[10^-6]
In[31]:=RybLin=Ryb[10^-6]
In[32]:=RtbLin=Rtb[10^-6]
```

The system of nonlinear algebraic equations (4.15), (4.19) and (4.20) is solved by executing the following statement, in which the unknowns x1, x2, and x3 are the branch fluxes $\Phi_{1,2}$, $\Phi_{2,3}$, and $\Phi_{3,4}$, respectively. These fluxes are defined in Fig. 4.6

```
In[33]:=Flux[F1_,F2_,F3_,F4_,AGoal_,MaxIt_]:=FindRoot[
{x1(Rgap+Rya[x1]+Rta[x1-x2]+Rtb[x1-x2]+Ryb[x1])-x2( Rta[
x1-x2]+Rgap+Rtb[x1-x2])-F1+F2==0,x2(2Rgap+Rta[x1-
x2]+Rya[x2]+Rta[x2-x3]+Rtb[x2-x3]+Ryb[x2]+Rtb[x1-x2])-
x1(Rta[x1-x2]+Rgap+Rtb[x1-x2])-x3(Rta[x2-x3]+Rgap+Rtb[x2-
x3])-F2+F3==0,x3(3Rgap+Rta[x2-x3]+Rya[x3]+2Rta[2x3]+2Rt-
b[2x3]+Ryb[x3]+Rtb[x2-x3])-x2(Rgap+Rta[x2-x3]+Rtb[x2-x3])-
F3+F4==0},{x1,{-10^-6,10^-6}},{x2,{-10^-6,10^-6}},{x3,{-
10^-6,10^-6}},MaxIterations->MaxIt,AccuracyGoal->AGoal]
```

The same system, when linearized, is solved by executing the following system of algebraic equations

```
In[34]:=FluxLin[F1_,F2_,F3_,F4_,AGoal_,MaxIt_]:=FindRoot[
{x1( Rgap+RyaLin+RtaLin+RtbLin+RybLin )-x2( RtaLin+
Rgap+RtbLin)-F1+F2==0,x2(2Rgap+RtaLin+RyaLin+RtaLin+RtbLin
+RybLin+RtbLin)-x1(RtaLin+Rgap+RtbLin)-x3(RtaLin+ Rgap+
RtbLin)-F2+F3==0,x3(3Rgap+RtaLin+RyaLin+2RtaLin+2RtbLin
+RybLin+RtbLin)-x2(Rgap+RtaLin+RtbLin)-F3+F4==0},{x1,{-10^-
6,10^-6}},{x2,{-10^-6,10^-6}},{x3,{-10^-6,10^-6}},MaxItera-
tions->MaxIt,AccuracyGoal->AGoal]
```

Define the matrices SolLin and Sol, into which the solutions of linear and nonlinear systems are stored

```
In[35]:=SolLin=Table[0,{i,10}];
In[36]:=Sol=Table[0,{i,100}];
```

Find the solution of the nonlinear magnetic circuit for given currents i

```
In[37]:=Do[{ToothMmf=Mmf2[i/20,0],Sol[[i]]=Flux[ ToothMmf
[[1]],ToothMmf[[2]],ToothMmf[[3]],ToothMmf[[4]],5,100]},{i,
100}]
```

List the solution matrix

```
In[38]:=Sol
```

Create auxiliary arrays which help obtain the curves in **Figs. 4.9** and **4.10**

```
In[39]:=Clear[Phi]
In[40]:=Phi=Table[0,{i,97},{j,3}];
In[41]:=Do[{Phi[[i,1]]=x1 /. Sol[[i,1]],Phi[[i,2]]=x2 /.
Sol[[i,2]],Phi[[i,3]]=x3 /. Sol[[i,3]]},{i,97}]
In[42]:=PhiGap=Table[0,{i,97},{j,4}];
In[43]:=Do[{PhiGap[[i,1]]=0,PhiGap[[i,2]]=Phi[[i,2]]-Phi[[-
i,1]],PhiGap[[i,3]]=Phi[[i,3]]-Phi[[i,2]],PhiGap[[i,4]]=-2
Phi[[i,3]]},{i,97}]
In[44]:=PhiTot=Table[0,{i,97},{j,6}];
In[45]:=Do[{PhiTot[[i,1]]=PhiGap[[i,1]],PhiTot[[i,2]]=Phi-
Gap[[i,2]],PhiTot[[i,3]]=PhiGap[[i,3]],PhiTot[[i,4]]=PhiGap
[[i,4]],PhiTot[[i,5]]=PhiGap[[i,3]],PhiTot[[i,6]]=PhiGap[[i
,2]]},{i,97}]
In[46]:=Psi=Table[0,{i,97}];
In[47]:=Do[Psi[[i]]=400(Sum[PhiTot[[i,j]],{j,2,5}]+Sum[PhiT
ot[[i,j]],{j,3,4}]),{i,97}]
In[48]:=PsiofI=Table[0,{i,97},{j,2}];
In[49]:=Do[{PsiofI[[i,2]]=Psi[[i]],PsiofI[[i,1]]=i/20},
{i,97}]
```

The dependence of the flux linkage on the current in a saturated machine is plotted by executing

Appendix

```
In[50]:=NonLin=ListPlot[PsiofI,PlotJoined->True, Default-
Font -> {"Times-Italic",9},AxesLabel->{"i [A]","Psi [Vs]"}]
```

The dependence of the inductance of the same coil on the current through it is obtained by typing

```
In[51]:=L=Table[0,{i,97}];
In[52]:=Do[L[[i]]=Psi[[i]]/(i/20),{i,97}]
In[53]:=ListPlot[L,PlotJoined->True,DefaultFont ->{"Times-
Italic",9}, AxesLabel->{"i [A]","L [H]"}]
```

To plot the curve in **Fig. 4.10**, the following statements have to be executed

```
In[54]:=LofI=Table[0,{i,97},{j,2}];
In[55]:=Do[{LofI[[i,2]]=L[[i]],LofI[[i,1]]=i/20},{i,97}]
In[56]:=ListPlot[LofI,PlotJoined->True,DefaultFont ->
{"Times-Italic",9},AxesLabel->{"i [A]","L [H]"},PlotRange->
{{0,5},{0,0.8}},Ticks-> {Automatic,{0,0.5,0.8}}]
```

The flux linkage of the winding as a function of the current through it, shown in **Fig. 4.9**, is obtained by typing

```
In[57]:=Do[{ToothMmf=Mmf2[i/2,0],SolLin[[i]]=FluxLin[
ToothMmf[[1]], ToothMmf[[2]],ToothMmf[[3]],ToothMmf[[4]],5,
100]},{i,10}]
In[58]:=PhiLin=Table[0,{i,10},{j,3}];
In[59]:=Do[{PhiLin[[i,1]]=x1 /. SolLin[[i,1]],
PhiLin[[i,2]]=x2 /. SolLin[[i,2]],PhiLin[[i,3]]=x3 /. Sol-
Lin[[i,3]]},{i,10}]
In[60]:=PhiGapLin=Table[0,{i,10},{j,4}];
In[61]:=Do[{PhiGapLin[[i,1]]=0,PhiGapLin[[i,2]]=
PhiLin[[i,2]]-PhiLin[[i,1]],PhiGapLin[[i,3]]=PhiLin[[i,3]]
- PhiLin[[i,2]], PhiGapLin[[i,4]]=-2 PhiLin[[i,3]]},{i,10}]
In[62]:=PhiTotLin=Table[0,{i,10},{j,6}];
In[63]:=Do[{PhiTotLin[[i,1]]=PhiGapLin[[i,1]], PhiTotLin[[
i,2]]=PhiGapLin[[i,2]],PhiTotLin[[i,3]]=PhiGapLin[[i,3]],
PhiTotLin[[i,4]]=PhiGapLin[[i,4]],PhiTotLin[[i,5]]=PhiGapLi
n[[i,3]],PhiTotLin[[i,6]]=PhiGapLin[[i,2]]},{i,10}]
```

```
In[64]:=PsiLin=Table[0,{i,10}];
In[65]:=Do[PsiLin[[i]]=400(Sum[PhiTotLin[[i,j]],{j,2,5}]+
Sum[PhiTotLin[[i,j]],{j,3,4}]),{i,10}]
In[66]:=ListPlot[PsiLin,PlotJoined->True,PlotRange->
{{0,10},{0,4}}]
In[67]:=PsiofILin=Table[0,{i,10},{j,2}];
In[68]:=Do[{PsiofILin[[i,2]]=PsiLin[[i]],PsiofILin[[i,1]]
=i/2}, {i,10}]
In[69]:=Lin=ListPlot[PsiofILin,PlotJoined->True,Default-
Font ->{"Times-Italic",9},AxesLabel->{"i [A]","Psi [Vs]"}]
In[70]:=Show[Lin,NonLin]
```

The 3D plot in **Fig. 4.8** is the result of the following statements

```
In[71]:=Sta=.0005
In[72]:=BTeetha=Table[0,{i,20},{j,7}];
In[73]:=Do[BTeetha[[20,j]]=PhiTot[[97,j]]/Sta,{j,6}]
In[74]:=ListPlot3D[BTeetha,AxesLabel->{"tooth"," i [A]","B
[T]"}, Shading->False,MeshRange->{{1,7},{0,5}},PlotRange->
{0,1.75},Default Font ->{"Times-Italic",9},Ticks->{{1,2,3,
4,5,6,7},{1,2,3,4,5},{0,1}}]
```

whereas the plot in **Fig. 4.7** is obtained by typing

```
In[75]:=BTeethaLin=Table[0,{i,10},{j,7}];
In[76]:=Do[BTeethaLin[[i,j]]=PhiTotLin[[i,j]]/Sta,
{i,1,10},{j,6}]
In[77]:=ListPlot3D[BTeethaLin,AxesLabel->{"tooth"," i
[A]","B [T]"}, Shading->False,MeshRange->{{1,7},{0,5}},
PlotRange->{0,4}, Default Font->{"Times-Italic",9},Ticks->
{{1,2,3,4,5,6,7},{1,2,3,4,5}, {0,2,4}}]
```

A 5.3

The results of trigonometric manipulations in Section 5.3 are obtained by executing the following statements. First, read in the trigonometry package

Appendix

```
In[1]:=<<Trigonometry.m
```

and then execute

```
In[2]:=Simplify[TrigReduce[Cos[w t -fsa](Sin[(1-s) w t]
Cos[s w t -fr] +Sin[(1-s) w t-2 Pi/3] Cos[s w t -fr+2 Pi/
3]+Sin[(1-s) w t-4 Pi/3] Cos[s w t -fr+4 Pi/3])+Cos[w t -
fsb](Sin[(1-s) w t+beta] Cos[s w t -fr] +Sin[(1-s) w t-2 Pi/
3+beta] Cos[s w t -fr+2 Pi/3]+Sin[(1-s) w t-4 Pi/3 +beta]
Cos[s w t -fr+4 Pi/3])]]
```

```
In[3]:=Simplify[TrigReduce[Cos[w t ](Sin[(1-s) w t] Cos[s w
t -fr] +Sin[(1-s) w t-2 Pi/3] Cos[s w t -fr+2 Pi/3]+Sin[(1-
s) w t-4 Pi/3] Cos[s w t -fr+4 Pi/3])+Cos[w t -Pi/2](Sin[(1-
s) w t+Pi/2] Cos[s w t -fr]+Sin[(1-s) w t-2 Pi/3+Pi/2] Cos[s
w t -fr+2 Pi/3] +Sin[(1-s) w t-4 Pi/3+Pi/2] Cos[s w t -fr+4
Pi/3])]]
```

```
In[4]:=Simplify[TrigReduce[Cos[w t ](Sin[(1-s) w t] Cos[s w
t -fr]+Sin[(1-s) w t-2 Pi/3] Cos[s w t -fr+2 Pi/3]+Sin[(1-
s) w t-4 Pi/3] Cos[s w t -fr+4 Pi/3])+Cos[w t +Pi/2](Sin[(1-
s) w t+Pi/2] Cos[s w t -fr]+Sin[(1-s) w t-2 Pi/3+Pi/2] Cos[s
w t -fr+2 Pi/3] +Sin[(1-s) w t-4 Pi/3+Pi/2] Cos[s w t -fr+4
Pi/3])]]
```

```
In[5]:=Simplify[TrigReduce[Cos[w t -fs](Sin[(1-s) w t]
Cos[s w t -fr]+Sin[(1-s) w t+2 Pi/3] Cos[s w t -fr+2 Pi/
3]+Sin[(1-s) w t+4 Pi/3] Cos[s w t -fr+4 Pi/3])+Cos[w t -fs-
2 Pi/3](Sin[(1-s) w t+2 Pi/3] Cos[s w t -fr]+Sin[(1-s) w t+4
Pi/3] Cos[s w t -fr+2 Pi/3]+Sin[(1-s) w t] Cos[s w t -fr+4
Pi/3])+Cos[w t -fs-4 Pi/3](Sin[(1-s) w t+4 Pi/3] Cos[s w t -
fr]+Sin[(1-s) w t] Cos[s w t -fr+2 Pi/3]+Sin[(1-s) w t+2 Pi/
3] Cos[s w t -fr+4 Pi/3])]]
```

```
In[6]:=Simplify[TrigReduce[Cos[w t -fs](Sin[(1-s) w t]
Cos[s w t -fr]+Sin[(1-s) w t-2 Pi/3] Cos[s w t -fr-2 Pi/
3]+Sin[(1-s) w t-4 Pi/3] Cos[s w t -fr-4 Pi/3])+Cos[w t -fs-
2 Pi/3](Sin[(1-s) w t-2 Pi/3] Cos[s w t -fr]+Sin[(1-s) w t-
4 Pi/3] Cos[s w t -fr-2 Pi/3]+Sin[(1-s) w t] Cos[s w t -fr-
4 Pi/3])+Cos[w t -fs-4 Pi/3](Sin[(1-s) w t-4 Pi/3] Cos[s w t
-fr]+Sin[(1-s) w t] Cos[s w t -fr-2 Pi/3]+Sin[(1-s) w t-2
Pi/3] Cos[s w t -fr-4 Pi/3])]]
```

```
In[7]:=Simplify[TrigReduce[Cos[w t -fs](Sin[(1-s) w t]
Cos[s w t -fr]+Sin[(1-s) w t-2 Pi/3] Cos[s w t -fr+2 Pi/
3]+Sin[(1-s) w t-4 Pi/3] Cos[s w t -fr+4 Pi/3])+Cos[w t -fs-
```

2 Pi/3](Sin[(1-s) w t-2 Pi/3] Cos[s w t -fr]+Sin[(1-s) w t-
4 Pi/3] Cos[s w t -fr+2 Pi/3]+Sin[(1-s) w t] Cos[s w t -fr+4
Pi/3])+Cos[w t -fs-4 Pi/3](Sin[(1-s) w t-4 Pi/3] Cos[s w t -
fr]+Sin[(1-s) w t] Cos[s w t -fr+2 Pi/3]+Sin[(1-s) w t-2 Pi/
3] Cos[s w t -fr+4 Pi/3])]]

In[8]:=Simplify[TrigReduce[Cos[w t -fs](Sin[(1-s) w t]
Cos[s w t -fr]+Sin[(1-s) w t-2 Pi/3] Cos[s w t -fr-2 Pi/
3]+Sin[(1-s) w t-4 Pi/3] Cos[s w t -fr-4 Pi/3])+Cos[w t -
fs+2 Pi/3](Sin[(1-s) w t-2 Pi/3] Cos[s w t -fr]+Sin[(1-s) w
t-4 Pi/3] Cos[s w t -fr-2 Pi/3]+Sin[(1-s) w t] Cos[s w t -
fr-4 Pi/3])+Cos[w t -fs+4 Pi/3](Sin[(1-s) w t-4 Pi/3] Cos[s
w t -fr]+Sin[(1-s) w t] Cos[s w t -fr-2 Pi/3]+Sin[(1-s) w t-
2 Pi/3] Cos[s w t -fr-4 Pi/3])]]

In[9]:=Simplify[TrigReduce[Cos[w t -fs](Sin[(1-s) w t]
Cos[s w t -fr]+Sin[(1-s) w t-2 Pi/3] Cos[s w t -fr+2 Pi/
3]+Sin[(1-s) w t-4 Pi/3] Cos[s w t -fr+4 Pi/3])+Cos[w t -
fs+2 Pi/3](Sin[(1-s) w t-2 Pi/3] Cos[s w t -fr]+Sin[(1-s) w
t-4 Pi/3] Cos[s w t -fr+2 Pi/3]+Sin[(1-s) w t] Cos[s w t -
fr+4 Pi/3])+Cos[w t -fs+4 Pi/3](Sin[(1-s) w t-4 Pi/3] Cos[s
w t -fr]+Sin[(1-s) w t] Cos[s w t -fr+2 Pi/3]+Sin[(1-s) w t-
2 Pi/3] Cos[s w t -fr+4 Pi/3])]]

In[10]:=Simplify[TrigReduce[Cos[w t -fs](Sin[(1-s) w t]
Cos[s w t -fr])+Cos[w t -fs-2 Pi/3](Sin[(1-s) w t-2 Pi/3]
Cos[s w t -fr])+Cos[w t -fs-4 Pi/3](Sin[(1-s) w t-4 Pi/3]
Cos[s w t -fr])]]

In[11]:=Simplify[TrigReduce[Cos[w t -fs](Sin[(1-s) w t]
Cos[s w t -fr]+Sin[(1-s) w t-2 Pi/3] Cos[s w t -fr+2 Pi/
3]+Sin[(1-s) w t-4 Pi/3] Cos[s w t -fr+4 Pi/3])]]

In[12]:=Simplify[TrigReduce[Cos[w t -f] Sin[w t] + Cos[w t
-f - 2 Pi/3] Sin[w t - 2 Pi/3] + Cos[w t - f -4 Pi/3] Sin[w
t - 4 Pi/3]]]

In[13]:=Simplify[TrigReduce[Cos[w t -f] Sin[w t] + Cos[w t
-f - 2 Pi/3] Sin[w t + 2 Pi/3] + Cos[w t - f -4 Pi/3] Sin[w
t + 4 Pi/3]]]

In[14]:=Simplify[TrigReduce[(Cos[w t -f1])^2 Sin[2 w t] +
(Cos[w t -f1 - 2 Pi/3])^2 Sin[2 w t - 2 Pi/3] + (Cos[w t -
f1 -4 Pi/3])^2 Sin[2 w t - 4 Pi/3]]]

In[15]:=Simplify[TrigReduce[(Cos[w t -f1])^2 Sin[2 w t] +
(Cos[w t -f1 - 2 Pi/3])^2 Sin[2 w t + 2 Pi/3] + (Cos[w t -
f1 -4 Pi/3])^2 Sin[2 w t + 4 Pi/3]]]

Appendix 495

The stator current in **Fig. 5.12 (a)** and the torque function in **Fig. 5.12 (b)** are plotted by executing

```
In[16]:=w=314.15926;s=0.05;fr=Pi/12;
In[17]:=Tooth[x_,P_]:=If[x-Floor[x/P]*P<=P/6,0,If[x-Floor[
x/P]*P<=P/2 ,1,If[x-Floor[x/P]*P<=2P/3,0,-1]]]
In[18]:=Plot[Tooth[x,0.02],{x,0,0.025},DefaultFont ->
{"Times-Italic" ,9},AxesLabel->{"t [s] ",""}, PlotRange->
{-1.2,1.2},PlotLabel->"Stator phase current"]
In[19]:=Plot[Tooth[t,0.02](Sin[(1-s) w t] Cos[s w t -
fr]+Sin[ (1-s) w t-2 Pi/3] Cos[s w t -fr+2 Pi/3]+Sin[(1-s) w
t-4 Pi/3] Cos[s w t -fr+4 Pi/3])+Tooth[t-.02/
3,0.02](Sin[(1-s) w t-2 Pi/3] Cos[s w t -fr]+ Sin[(1-s) w t-
4 Pi/3] Cos[s w t -fr+2 Pi/3]+Sin[(1-s) w t] Cos[s w t -fr+4
Pi/3])+Tooth[t-0.04/3,0.02](Sin[(1-s) w t-4 Pi/3] Cos[s w t
-fr]+Sin[(1-s) w t] Cos[s w t -fr+2 Pi/3]+Sin[(1-s) w t-2
Pi/3] Cos[s w t -fr+4 Pi/3]),{t,0,0.02},DefaultFont ->
{"Times-Italic",9},AxesLabel-> {"t [s] ",""},PlotRange->
{0,3},PlotLabel->"Torque function"]
```

The torque function in **Fig. 5.15 (a)** is obtained by executing

```
In[20]:=Plot[Tooth[t+0.004,0.02] Sin[ w t] + Tooth[t+0.004-
.02/3,0.02] Sin[w t - 2 Pi/3] + Tooth[t+0.004-0.04/3,0.02]
Sin[w t - 4 Pi/3], {t,0,0.02},DefaultFont->{"Times-Italic"
,9},AxesLabel-> {"t [s] ",""},PlotRange->{0,2},PlotLabel->
"Pure electromagnetic"]
```

and the torque function in **Fig. 5.15 (b)**

```
In[21]:=Plot[(Tooth[t+0.004,0.02])^2 Sin[2 w t] + (Tooth[t+
0.004-.02/3,0.02])^2 Sin[2 w t + 2 Pi/3] + (Tooth[t+0.004-
0.04/3,0.02])^2 Sin[2 w t + 4 Pi/3],{t,0,0.02},DefaultFont
-> {"Times-Italic",9},AxesLabel-> {"t [s] ",""},PlotRange->
{0,2},PlotLabel->"Reluctance"]
```

The torque function in **Fig. 5.17 (b)** is plotted by typing

```
In[22]:=Rect[x_,P_]:=If[x-Floor[x/P]*P<=P/2,1,-1]
In[23]:=n=9;
```

```
In[24]:=Plot[ Sum[Rect[x-(i-1)2Pi/n,2Pi] Sin[ x -(i-1)2Pi/
n],{i,1,n}], {x,0,2Pi},PlotLabel->"Torque function",
PlotRange->{0,7.5},DefaultFont ->{"Times-Italic",9}]
```

and the torque function in **Fig. 5.17 (a)**

```
In[25]:=Plot[4/Pi Sum[Sin[x-(i-1)2Pi/n] Sin[ x -(i-1)2Pi/
n],{i,1,n}], {x,0,2Pi},PlotLabel-> "Torque function",
PlotRange->{0,7.5},DefaultFont ->{"Times-Italic",9}]
```

A 6.2.2

First import the ReIm package, in order to explore the complex character of the induction machine's impedances

```
In[1]:=<<ReIm.m
```

Specify the type of the constants which appear in computation

```
In[2]:=Im[R1]= 0;Im[R2c]= 0;Im[X1]= 0;Im[X2c]= 0;Im[RFe]=
0;Im[X0]= 0; Im[U]= 0
```

Define the impedances from the equivalent circuit

```
In[3]:=Z1[R1_,X1_,f_]:=R1+I f X1
In[4]:=Z0[RFe_,X0_,f_]:=I RFe f X0/(RFe+I f X0)
In[5]:=Z2c[R2c_,s_,X2c_,f_]:=-R2c/s+ I f X2c
```

The parameters of the equivalent circuit are

```
In[6]:=X1=1.02;X2c=1.32;R1=.91;R2c=.73;X0=53.3;RFe=984;f=1.
```

The curve in **Fig. 6.3 (a)** is now obtained by executing

```
In[7]:=Plot[Abs[1/(1+Z1[R1,X1,f]/Z0[RFe,X0,f]+Z1[R1,X1,f]/
Z2c[R2c, s,X2c,f])],{s,-1,1},PlotRange->{Automatic,
{0,1.2}},DefaultFont -> {"Times-Italic",9},AxesLabel->
```

{"slip","| Es/V |"},Ticks->{None, {0.5,1}}]

The phase of the ratio between the induced and applied stator voltages, shown in **Fig. 6.3 (b)**, is obtained by executing

In[8]:=ParametricPlot[{Re[1/(1+Z1[R1,X1,f]/Z0[RFe,X0,f]+
Z1[R1,X1,f]/Z2c[R2c,s,X2c,f])],Im[1/(1+Z1[R1,X1,f]/
Z0[RFe,X0,f]+ Z1[R1,X1,f]/Z2c[R2c,s,X2c,f])]},{s,-1,1},
DefaultFont ->{"Times-Italic",9}, PlotRange->{{0,1.2},{-
0.2,0.5}},AxesLabel->{"Re{Es/V}","Im{Es/V}" },Ticks->
{{0.5,1},{0,0.5}}]

The reduced rotor current in **Fig.6.4** is the result of the following statements

In[9]:=U=220;
In[10]:=Sigma[R1_,X1_,RFe_,X0_,f_]:=1+Z1[R1,X1,f]/
Z0[RFe,X0,f]
In[11]:=I2[U_,R1_,X1_,RFe_,X0_,R2c_,s_,X2c_,f_]:=U/((R1-
Sigma[R1,X1,RFe,X0,f]R2c/s)+I(X1+Sigma[R1,X1,R-
Fe,X0,f]X2c))
In[12]:=Plot[Abs[I2[U,R1,X1,RFe,X0,R2c,s,X2c,f]],{s,-
2,1},PlotRange-> {{-2,1},Automatic},DefaultFont ->{"Times-
Italic",9},AxesLabel-> {"slip","Ir' [A]"},Ticks->{{},{}}]

The absolute value and real component of the stator current, shown in **Fig. 6.5**, are obtained by executing

In[13]:=Plot[Abs[U/(1+Z1[R1,X1,f]/Z0[RFe,X0,f]+Z1[R1,X1,f]
/ Z2c[R2c,s,X2c,f])(1/Z0[RFe,X0,f]+1/Z2c[R2c,s,X2c,f])],
{s,-2,1},PlotRange-> {{1,-2},Automatic},DefaultFont ->
{"Times-Italic",9},AxesLabel-> {"slip"," |Is|"},Ticks->
{{},{}}]
In[14]:=Plot[Re[U/(1+Z1[R1,X1,f]/Z0[RFe,X0,f]+Z1[R1,X1,f]/
Z2c[R2c, s,X2c,f])(1/Z0[RFe,X0,f]+1/Z2c[R2c,s,X2c,f])],{s,-
2,1},PlotRange-> {{1,-2},Automatic},DefaultFont ->{"Times-
Italic",9},AxesLabel-> {"slip"," Re{Is}"},Ticks->{{},{}}]

In order to obtain the torque–speed characteristic shown in **Fig. 6.6**, the following statements have to be executed

```
In[15]:=ws=314;m1=3;
In[16]:=Plot[m1(Abs[I2[U,R1,X1,RFe,X0,R2c,s,X2c,f]])^2 R2c/
(-s ws), {s,-2,1},PlotRange->{Automatic,Automatic},Default-
Font ->{"Times-Italic" ,9},AxesLabel->{"slip","T "},Ticks->
{{},{}}]
```

The torque–speed characteristic of a current fed machine, shown in **Fig. 6.10**, is plotted by executing

```
In[17]:=Plot[- m1 X0^2 100/ws (R2c/s)/((R2c/s)^2+
(X0+X2c)^2),{s,-2,1}, PlotRange->{Automatic,Automatic},
DefaultFont ->{"Times-Italic",9},Axes Label->{"slip","T
[Nm]"}, Ticks->{{},{-10,10}}]
```

The relationship between the relative frequency and slip during the capacitor braking of an induction machine, shown in **Fig. 6.16**, is obtained by typing

```
In[18]:=Xc=5.3;X1=6.4;X2c=4.31;R1=2.763;R2c=3.019;
In[19]:=Plot3D[l^4 s(s R1 X2c^2+R2c X1^2)+l^2 R2c(s(R1^2-
2X1 Xc) +R1 R2c)+s R2c Xc^2,{1,0,1},{s,-1,0},Shading->
False,AxesLabel->{"lambda"," s",""},DefaultFont ->{"Times-
Italic",9},PlotRange -> {Automatic , Automatic,
{0,50}},Ticks->{{0.5,1},{-1,-0.5},{0,50}}]
In[20]:=Show[%,ViewPoint->{-1.357, -2.885, 1.134}]
```

The dependence of the ratio between the induced and applied voltage on the rotor slip and relative frequency of the applied voltage, shown in **Fig. 6.37**, is obtained by executing

```
In[21]:=Plot3D[Abs[1/(1+Z1[R1,X1,f]/Z0[RFe,X0,f]+
Z1[R1,X1,f]/Z2c[R2c,  s,X2c,f])],{s,0,1},{f,.001,.1},
PlotRange->{Automatic,Automatic, {0,1}},Default Font->
{"Times-Italic",9},AxesLabel->{"slip","rel. freq.","| Es/V
|"} ,Ticks->{{0.5,1},{0,0.1},{0,1}},Shading->False]
```

Appendix

A 6.5

The performances of a single phase induction machine are evaluated in this section. First, define the rotor positive Z2p and negative Z2n sequence impedances

```
In[1]:=Z2p[X2c_,R2c_,s_,RFe_,Xm_]:=I Xm RFe (R2c/s+I X2c)/
(I Xm RFe+ RFe (R2c/s+I X2c)+I Xm (R2c/s+I X2c))
In[2]:=Z2n[X2c_,R2c_,s_,RFe_,Xm_]:=I Xm RFe (R2c/(2-s)+I
X2c)/(I Xm RFe+RFe (R2c/(2-s)+I X2c)+I Xm (R2c/(2-s)+I
X2c))
```

The motor's positive Zp and negative Zn sequence impedances are then

```
In[3]:=Zp[X2c_,R2c_,s_,RFe_,Xm_,R1_,X1_]:=R1+I X1+Z2p[X2-
c,R2c,s,RFe,Xm]
In[4]:=Zn[X2c_,R2c_,s_,RFe_,Xm_,R1_,X1_]:=R1+I X1+Z2n[X2-
c,R2c,s,RFe,Xm]
```

The external impedance Zext is defined as

```
In[5]:=Zext[w_,Cap_,a_,Ra_,R1_,Xa_,X1_]:=-I/(a^2 w Cap)+Ra/
a^2-R1+ I(Xa/a^2-X1)
```

The coefficient γ, introduced in Eq. (6.107), is equal to

```
In[6]:=gamma [X2c_,R2c_,s_,RFe_,Xm_,R1_,X1_,w_,Cap_,a_,Ra_,
Xa_]:= (Zext[w,Cap,a,Ra,R1,Xa,X1] + Zp [X2c,R2c,s,RFe,Xm,
R1,X1](1+I/a))/ (Zext[w,Cap,a,Ra,R1,Xa,X1]+Zn[X2c,R2c,s,
RFe,Xm,R1,X1](1-I/a))
```

The positive Imp and negative Imn sequence components of the main phase current are

```
In[7]:=Imp[V_,X2c_,R2c_,s_,RFe_,Xm_,R1_,X1_,w_,Cap_,a_,Ra_,
Xa_]:=V/(Zp[X2c,R2c,s,RFe,Xm,R1,X1]+gamma[X2c,R2c,s,RFe,
Xm,R1,X1,w,Cap,a,Ra, Xa]Zn[X2c,R2c,s,RFe,Xm,R1,X1])
In[8]:=Imn[V_,X2c_,R2c_,s_,RFe_,Xm_,R1_,X1_,w_,Cap_,a_,Ra_,
Xa_]: = gamma[X2c,R2c,s,RFe,Xm,R1,X1,w,Cap,a,Ra,Xa]Imp[
V,X2c,R2c,s,RFe,Xm,R1,X1,w,Cap,a,Ra,Xa]
```

The total main phase current Imain is equal to the sum of its positive and negative sequence components

```
In[9]:= Imain[V_,X2c_,R2c_,s_,RFe_,Xm_,R1_,X1_,w_,Cap_,a_,
Ra_,Xa_]:= Imp[V,X2c,R2c,s,RFe,Xm,R1,X1,w,Cap,a,Ra,Xa]+
Imn[ V,X2c,R2c,s,RFe,Xm, R1,X1,w,Cap,a,Ra,Xa]
```

The positive Iap and negative Ian sequence components of the auxiliary phase current are

```
In[10]:=Iap[V_,X2c_,R2c_,s_,RFe_,Xm_,R1_,X1_,w_,Cap_,a_,Ra_
,Xa_]:=I Imp[V,X2c,R2c,s,RFe,Xm,R1,X1,w,Cap,a,Ra,Xa]/a

In[11]:=Ian[V_,X2c_,R2c_,s_,RFe_,Xm_,R1_,X1_,w_,Cap_,a_,Ra_
,Xa_]:=-I Imn[V,X2c,R2c,s,RFe,Xm,R1,X1,w,Cap,a,Ra,Xa]/a
```

The total auxiliary phase current is then

```
In[12]:= Iaux[V_,X2c_,R2c_,s_,RFe_,Xm_,R1_,X1_,w_,Cap_,a_,
Ra_,Xa_]:= Iap[V,X2c,R2c,s,RFe,Xm,R1,X1,w,Cap,a,Ra,Xa]+
Ian[V,X2c,R2c,s,RFe,Xm,R1,X1,w,Cap,a,Ra,Xa]
```

The reflected rotor phase current also has two components: positive I2pc, and negative sequence rotor current I2nc

```
In[13]:= I2pc[V_,X2c_,R2c_,s_,RFe_,Xm_,R1_,X1_,w_,Cap_,a_,
Ra_,Xa_]:= Z2p[X2c,R2c,s,RFe,Xm]Imp[V,X2c,R2c,s,RFe,Xm,
R1,X1,w,Cap,a,Ra,Xa]/(R2c/s+I X2c)

In[14]:= I2nc[V_,X2c_,R2c_,s_,RFe_,Xm_,R1_,X1_,w_,Cap_,a_,
Ra_,Xa_]:= gamma[X2c,R2c,s,RFe,Xm,R1,X1,w,Cap,a,Ra,Xa]
Z2n[X2c,R2c,s,RFe,Xm] Imp[V,X2c,R2c,s,RFe,Xm,R1,X1,w,Cap,
a,Ra,Xa]/(R2c/(2-s)+I X2c)
```

The positive sequence component Tp of the torque is equal to

```
In[15]:=Tp[V_,X2c_,R2c_,s_,RFe_,Xm_,R1_,X1_,w_,Cap_,a_,Ra_,
Xa_,p_]:=2 Abs[I2pc[V,X2c,R2c,s,RFe,Xm,R1,X1,w,Cap,a,Ra,
Xa]]^2 R2c p/(s w)
```

and the negative sequence component Tn

Appendix

```
In[16]:=Tn[V_,X2c_,R2c_,s_,RFe_,Xm_,R1_,X1_,w_,Cap_,a_,Ra_,
Xa_,p_]:= 2 Abs[I2nc[V,X2c,R2c,s,RFe,Xm,R1,X1,w,Cap,a,
Ra,Xa]]^2 R2c p/((2-s) w)
```

The voltage drop across the capacitor Ucap is

```
In[17]:= Ucap[V_,X2c_,R2c_,s_,RFe_,Xm_,R1_,X1_,w_,Cap_,a_,
Ra_,Xa_]:=I Iaux[V,X2c,R2c,s,RFe,Xm,R1,X1,w,Cap,a,Ra,Xa]/
(w Cap)
```

and the voltage drop across the auxiliary phase Uaux

```
In[18]:= Uaux[V_,X2c_,R2c_,s_,RFe_,Xm_,R1_,X1_,w_,Cap_,a_,
Ra_,Xa_]:=V+Ucap[V,X2c,R2c,s,RFe,Xm,R1,X1,w,Cap,a,Ra,Xa]
```

The single phase induction machine, the performance of which is calculated in this example, has the following parameters

```
In[19]:= V=220;X1=6.4;X2c=4.31;R1=2.763;R2c=3.019;RFe=600;
Xm=55;w=314; Cap=.00025;a=.7;Ra=3.6;Xa=8.9;p=1;
```

The torque–speed characteristic of this machine with run capacitor of 25 µF, shown in **Fig.6.28 (a)**, is obtained by executing

```
In[20]:= Plot [ Tp[V,X2c,R2c,s,RFe,Xm,R1,X1,w,Cap,a,Ra,Xa,
p] - Tn[V,X2c,R2c,s,RFe,Xm,R1,X1,w,Cap,a,Ra,Xa,p],
{s,0.001,1}]

In[21]:=Cap=0.000025

In[22]:=Plot[Tp[V,X2c,R2c,s,RFe,Xm,R1,X1,w,Cap,a,Ra,Xa,p]-
Tn[V,X2c,R2c,s,RFe,Xm,R1,X1,w,Cap,a,Ra,Xa,p],{s,0.001,1}]

In[23]:=Show[Out[20],Out[22],Ticks->None]
```

The currents Imain and Iaux, shown in **Fig. 6.28 (b)**, are plotted by typing

```
In[24]:=Plot[Abs[Imain[V,X2c,R2c,s,RFe,Xm,R1,X1,w,Cap,a,Ra,
Xa]], {s,0.001,0.2},PlotRange->{0,15}]

In[25]:=Plot[Abs[Iaux[V,X2c,R2c,s,RFe,Xm,R1,X1,w,Cap,a,Ra,X
a]],{s,0.001,0.2}]
```

In[26]:=Show[Out[24],Out[25],Ticks->None]

The voltage drops in the circuit are evaluated and plotted by utilizing the following statements

In[27]:= Plot[Abs[Ucap[V,X2c,R2c,s,RFe,Xm,R1,X1,w,Cap,a,
Ra,Xa]], {s,0.001,1},PlotRange->{0,300}]

In[28]:= Plot[Abs[Uaux[V,X2c,R2c,s,RFe,Xm,R1,X1,w,Cap,a,
Ra,Xa]],{s,0.001,1},PlotRange->{0,300}]

In[29]:=Plot[V,{s,0.001,1},PlotRange->{0,300}]

In[30]:=Show[Out[27],Out[28],Out[29],Ticks->None]

The natural frequencies of a single phase induction machine with a capacitor in the auxiliary phase are found from the condition that the imaginary part of the auxiliary phase impedance is equal to zero. The following statements define the imaginary part of the auxiliary phase impedance.

First import the package ReIm, and declare real all machine's parameters which appear in computation

In[1]:= <<Packages/Algebra/ReIm.m

In[2]:= TagReal[x_]:=(x /: Re[x]=x; x /: Im[x]=0)

In[3]:= TagReal[Ra]

In[4]:= TagReal[w]

In[5]:= TagReal[La]

In[6]:= TagReal[L2c]

In[7]:= TagReal[R2c]

In[8]:= TagReal[s]

In[9]:= TagReal[RFe]

In[10]:= TagReal[L0]

In[11]:= TagReal[Cap]

In[12]:= TagReal[Rbc]

In[13]:= TagReal[Lbc]

In[14]:= TagReal[kab]

Define the external impedance Zpc, as well as its real and imaginary part

Appendix

```
In[15]:= Zpc[Cap_,kab_,w_,Rbc_,Ra_,Lbc_,La_]:=1/(I w Cap
kab^2)+Rbc-Ra+I w (Lbc-La)
In[16]:= ImZpc[Cap_,kab_,w_,Rbc_,Ra_,Lbc_,La_]:=
Im[Zpc[Cap,kab,w,Rbc,Ra,Lbc,La]]
In[17]:= ReZpc[Cap_,kab_,w_,Rbc_,Ra_,Lbc_,La_]:=
Re[Zpc[Cap,kab,w,Rbc,Ra,Lbc,La]]
```

Define the reflected rotor forward Z2f and backward Z2b impedances, and their real and imaginary parts

```
In[20]:= Z2f[w_,L0_,RFe_,R2c_,s_,L2c_]:=(I w L0) RFe (R2c/
s+I w L2c)/((I w L0) RFe+ RFe (R2c/s+I w L2c)+I w L0 (R2c/
s+I w L2c))
In[21]:= ReZ2f[w_,L0_,RFe_,R2c_,s_,L2c_]:=Re[Z2f[w,L0,RFe,
R2c,s,L2c]]
In[23]:= ImZ2f[w_,L0_,RFe_,R2c_,s_,L2c_]:=Im[Z2f[w,L0,RFe,
R2c,s,L2c]]
In[25]:= Z2b[w_,L0_,RFe_,R2c_,s_,L2c_]:=(I w L0) RFe (R2c/
(2-s)+I w L2c)/((I w L0) RFe+RFe (R2c/(2-s)+I w L2c)+I w L0
(R2c/(2-s)+I w L2c))
In[26]:= ReZ2b[w_,L0_,RFe_,R2c_,s_,L2c_]:=Re[Z2b[w,L0,RFe,
R2c,s,L2c]]
In[28]:= ImZ2b[w_,L0_,RFe_,R2c_,s_,L2c_]:=Im[Z2b[w,L0,RFe,
R2c,s,L2c]]
```

The forward Zf and backward Zb components of the auxiliary phase impedance are

```
In[30]:= Zf[w_,L0_,RFe_,R2c_,s_,L2c_,Ra_,La_]:=Z2f[w,L0,RFe
,R2c,s,L2c] +Ra+I w La
In[31]:= Zb[w_,L0_,RFe_,R2c_,s_,L2c_,Ra_,La_]:=Z2b[w,L0,RFe
,R2c,s,L2c] +Ra+I w La
```

The coefficient γ, defined in Eq. (6.107), can be written as

```
In[32]:= gamma[Cap_,kab_,w_,Rbc_,Ra_,Lbc_,La_,L0_,RFe_,
R2c_,s_,L2c_]:= (Zpc[Cap,kab,w,Rbc,Ra,Lbc,La]+Zf[w,L0,RFe,
R2c,s,L2c,Ra,La](1+I/kab))/(Zpc[Cap,kab,w,Rbc,Ra,Lbc,La]+
```

```
Zb[w,L0,RFe,R2c,s,L2c,Ra,La](1-I/kab))
```

The auxiliary impedance Zaux is represented in terms of its two components Zaux1 and Zaux2, the imaginary parts ImZaux1 and ImZaux2 of which generate the plot in **Fig.6.30**

```
In[35]:= Zaux1[Cap_,kab_,w_,Rbc_,Ra_,Lbc_,La_,L0_,RFe_,
R2c_,s_,L2c_]:= (1-gamma[Cap,kab,w,Rbc,Ra,Lbc,La,L0,RFe,
R2c,s,L2c])(Ra+I w La)

In[36]:= Zaux2[Cap_,kab_,w_,Rbc_,Ra_,Lbc_,La_,L0_,RFe_,
R2c_,s_,L2c_]:= (gamma[Cap,kab,w,Rbc,Ra,Lbc,La,L0,RFe,R2c,
s,L2c] Z2b[w,L0,RFe,R2c, s,L2c])

In[37]:= ImZaux1[Cap_,kab_,w_,Rbc_,Ra_,Lbc_,La_,L0_,RFe_,
R2c_,s_,L2c_] :=Im[Zaux1[Cap,kab,w,Rbc,Ra,Lbc,La,L0,RFe,
R2c,s,L2c]]

In[38]:= ImZaux2[Cap_,kab_,w_,Rbc_,Ra_,Lbc_,La_,L0_,RFe_,
R2c_, s_,L2c_]:=Im[Zaux2[Cap,kab,w,Rbc,Ra,Lbc,La,L0,RFe,
R2c,s,L2c]]
```

A 8.3

The torque–angle curve in **Fig. 8.8** is generated by executing

```
In[1]:=Curve1=Plot[1.25 Sin[x+e]-Sin[e],{x,0,Pi}, Axes
Label->{"delta", "T / Tmax"},DefaultFont ->{"Times-Italic",
9}, PlotStyle->{GrayLevel [0.5]}]

In[2]:=Curve2=Plot[1.25 Sin[x+e/1000]-Sin[e/1000],{x,0,Pi},
AxesLabel-> {"delta", "T / Tmax"},DefaultFont ->{"Times-
Italic",9}]

In[3]:=Tot=Show[Curve1,Curve2]
```

The constant power factor curves, shown in **Fig. 8.11 (a)** for a machine with rated voltage 220 V and synchronous reactance 10 Ω, are obtained by executing

```
In[4]:= ContourPlot[(E^2-220^2-(Ist 10)^2)/(2 220 Ist 10),
{E,0.1,450}, { Ist,0.1,22}, ContourShading->False, Con-
tourSmoothing->True,PlotPoints-> 50,PlotRange->{-1,1},Con-
tours->{-1,-.866,-.5,0,.5,.866,1},AxesLabel-> {"E", "I"},
```

Appendix

```
DefaultFont ->{"Times-Italic",9}]
```

The constant power curves for the same machine, shown in **Fig. 8.11 (b)**, are obtained by typing

```
In[5]:=ContourPlot[If[Ist>Abs[(Ein-220)/10],220 Ist Sqrt[
Abs[1((Ein^2-220^2-(Ist 10)^2)/(2 220 Ist 10))^2]],0],
{Ein,0.1,450}, {Ist,0.1,22}, ContourShading->False, Con-
tourSmoothing->True,PlotPoints-> 50,PlotRange->{0,3000},
Contours->{50,500,1000,1500,2000,2500,3000}, AxesLabel->
{"E", "I"},DefaultFont ->{"Times-Italic",9}]
```

Besides the parametric plot in **Fig. 8.11 (b)**, the active power can be represented in the form of three-dimensional dependence on the stator current and induced voltage. The plot of the active power in the same synchronous machine as a function of the induced voltage and stator current, shown in Fig. 8.12, is the result of the following statement

```
In[6]:=Plot3D[If[Ist>Abs[(Ein-220)/10],220 Ist Sqrt[Abs[1-
((Ein^2-220^2-(Ist 10)^2)/(2 220 Ist 10))^2]],0], {Ein,0.1,
450},{Ist,0.1,25}, Shading->False,PlotPoints->15, Ticks->
{Automatic,Automatic,{0,5000}}, PlotRange->Automatic,
AxesLabel->{"E [V]", "I [A]","P [W] "},Default Font ->
{"Times-Italic",9}]
```

The torque–speed characteristic in **Fig. 8.17** which illustrates the action of the permanent magnets as exciters of a synchronous machine is obtained by executing

```
In[7]:=Plot[{Sin[x]-.6Sin[2x]},{x,-Pi,Pi},AxesLabel->
{"delta", "T / Tmax"},DefaultFont ->{"Times-Italic",9}]
```

A 9.2

First define the open circuit characteristic EICurve of the motor, i.e. the dependence of the armature induced voltage on the field current

```
In[1]:=EICurve={{0,0},{.25,71},{.5,133},{.75,170},{1,195},{
1.5,220},{2,232},{2.5,240},{3,246},{4,250},{5,253},{6,255}}
In[2]:=EofI=Interpolation[EICurve]
```

The `EofI` curve, shown in Fig. 9.4, is plotted by typing

```
In[3]:=Plot[EofI[x],{x,0,6},AxesLabel->{FontForm["i [A]",
{"Times- Italic",9}],FontForm["E [V] ",{"Times-
Italic",9}]}, PlotRange->{0,260}]
```

The dependence of the main field coil inductance on the field current, shown in **Fig. 9.4**, is generated by executing

```
In[4]:=LICurve=Table[{EICurve[[i,1]], wf EICurve[[i,2]]/(ce
n (EICurve [[i,1]]+.001))},{i,12}]
In[5]:=LofI=Interpolation[LICurve]
```

The motor parameters are

```
In[6]:=p=1;z=640;a=2;n=1000;ce=p z/(30 a);cm=30 ce/Pi;
wf=250;
```

and the plot of inductance on the field current, shown in **Fig. 9.4**, is obtained by executing

```
In[7]:=Plot[LofI[x],{x,0,6},AxesLabel->{FontForm["i [A]",
{"Times-Italic",9}], FontForm["L ",{"Times-Italic",9}]},
PlotRange->{0,8}]
```

The dependence of the main flux on the field current is given as

```
In[8]:=FICurve=Table[{EICurve[[i,1]],EICurve[[i,2]]/(ce n)}
,{i,12}]
In[9]:=FofI=Interpolation[FICurve]
```

Define the armature (a) and field (f) electric parameters of the motor, its inertia J, electrical `Tel` and electromechanical `Tem` time constant and load torque `Tl`

```
In[10]:= La=0.01;Va=220;Ra=1.2;Lfmax=LICurve[[1,2]];Vf=220;
Rf=150; J=0.08;Tel=La/Ra;Tem=N[(4 J Ra)/(ce cm 38.2 FICurve
[[2,2]])];Tl=25; if=Vf/Rf;
```

Appendix

The current and speed transient curves, shown in **Fig. 9.1**, are obtained by executing

```
In[11]:=LinFluxCon=NDSolve[{ia'[t]==(Va-Ra ia[t]-ce Lfmax
if/wf n[t])/La,n'[t]==30(cm Lfmax if/wf ia[t]-Tl)/(J
Pi),ia[0]==n[0]==0},{ia,n}, {t,0.1}]
In[12]:=Plot[Evaluate[n[t] /. LinFluxCon],{t,0,.1},Default-
Font -> {"Times-Italic",9},AxesLabel->{"time [s]","n [rpm]"
},PlotRange-> {0,700}]
In[13]:=Plot[Evaluate[ia[t] /. LinFluxCon],{t,0,.1},De-
faultFont -> {"Times-Italic",9},AxesLabel->{"time [s]","I
[A]"}]
```

The same transients, but with a ten times larger inertia on the shaft, are obtained by executing

```
In[14]:=J=.8;
In[15]:=LinFluxConInert=NDSolve[{ia'[t]==(Va-Ra ia[t]-ce
Lfmax if/wf n[t])/La,n'[t]==30(cm Lfmax if/wf ia[t]-Tl)/(J
Pi),ia[0]==n[0]==0},{ia ,n},{t,.25}]
In[16]:=Plot[Evaluate[n[t] /. LinFluxConInert],{t,0,.25},
DefaultFont-> {"Times-Italic",9},AxesLabel->{"time [s]","n
[rpm]"}]
```

and shown in **Fig. 9.2**.

The transients in shunt motors are evaluated by executing the following statements

```
In[17]:=if=.
In[18]:=LinFluxSim=NDSolve[{ia'[t]==(Va-Ra ia[t]-ce Lfmax
if[t]/wf n[t])/La,if'[t] == (Vf-if[t] Rf)/Lfmax,-
n'[t]==30(cm Lfmax if[t]/wf ia[t]-Tl)/(J Pi),ia[0]==n[0]==
if[0]==0},{ia,n,if},{t,.25}]
In[19]:=Plot[Evaluate[n[t] /. LinFluxSim],{t,0,.25},De-
faultFont -> {"Times-Italic",9},AxesLabel->{"time [s]","n
[rpm]"}]
In[20]:=Plot[Evaluate[ia[t] /. LinFluxSim],{t,0,.25},De-
faultFont -> {"Times-Italic",9},AxesLabel->{"time [s]","I
[A]"}]
```

The results of these computations are shown in **Fig. 9.3**. The transients in the same shunt motor, the main magnetic circuit of which is nonlinear, are calculated by utilizing the following equations

```
In[21]:=NonLinFluxSim=NDSolve[{ia'[t]==(Va-Ra ia[t]-ce
LofI[if[t]]if[t]/wf n[t])/La,if'[t] == (Vf-if[t] Rf)/LofI[
if[t]],n'[t]==30(cm LofI[if[t]]if[t]/wf ia[t]-Tl)/(J Pi),
ia[0]==n[0]==if[0]==0},{ia,n, if},{t,.25}]
```

The plot in **Fig. 9.5** is obtained by typing

```
In[22]:=Plot[Evaluate[n[t] /. NonLinFluxSim],{t,0,.25},De-
faultFont -> {"Times-Italic",9},AxesLabel->{"time [s]","n
[rpm]"}]
In[23]:=Plot[Evaluate[ia[t] /. NonLinFluxSim],{t,0,.25},De-
faultFont -> {"Times-Italic",9},AxesLabel->{"time [s]","I
[A]"}]
```

The transients in a series D.C. machine are calculated by utilizing the following statements

```
In[24]:=EISCurve={{0.1,0},{8,114},{12,164},{16,205},{20,237
},{24,259},{28,278},{32,294},{36,308},{40,320},{44,330},{48
,338},{52,345},{56,351},{60,356},{64,360},{68,363},{72,365}
}
In[25]:=EofIS=Interpolation[EISCurve]
```

Define the machine's parameters

```
In[26]:=p=1;z=640;a=2;n=1000;ce=p z/(30 a);cm=30 ce/Pi;wfs=
25;La=0.01; Va=220;Ra=1.2;Lfsmax=LISCurve[[1,2]];Rf=1.5;
J=0.08;Tl=10.
```

Define the dependence of the main flux and the main inductance on the armature current

```
In[27]:=FISCurve=Table[{EISCurve[[i,1]],EISCurve[[i,2]]/
(ce n)}, {i,18}]
```

Appendix

```
In[28]:=FofIS=Interpolation[FISCurve]
In[29]:=LISCurve=Table[{EISCurve[[i,1]], wfs EISCurve[[i,2
]]/(ce n (EISCurve[[i,1]]+.001))},{i,18}]
In[30]:=LofIS=Interpolation[LISCurve]
In[31]:=n=.
```

The system of differential equations which describes transient and steady state behaviour of a series motor is

```
In[32]:=NonLinFluxSer=NDSolve[{ia'[t]==(Va-(Ra+Rf) ia[t]-ce
LofIS[ ia[t]] ia[t]/wfs n[t])/(La+LofIS[ia[t]]),n'[t]==
30(cm LofIS[ia[t]] ia[t]/wfs ia[t]-Tl)/(J Pi),
ia[0]==n[0]==0},{ia,n},{t,2.5},MaxSteps-> 10000]
```

The following two steps give the curves in **Fig. 9.6**

```
In[33]:=Plot[Evaluate[ia[t] /. NonLinFluxSer],{t,0,2.5},De-
faultFont -> {"Times-Italic",9},AxesLabel->{"time [s]","I
[A]"}]
In[34]:=Plot[Evaluate[n[t] /. NonLinFluxSer],{t,0,2.5},De-
faultFont -> {"Times-Italic",9},AxesLabel->{"time [s]","n
[rpm]"}]
```

A 9.3

First define the equivalent circuit parameters of the induction machine, the transients in which are analyzed

```
In[1]:=Lm=1;Lr=Ls=1.05;sigma=1-Lm^2/(Lr Ls);omegas=314;
Rs=Rr=100; Um=220;J=0.0015;p=2;
```

The transient lasts 0.5 sec

```
In[2]:=tFinal=.5
```

Define the system of the machine's differential equations

```
In[3]:= Run1=NDSolve[{Psi1'[t]==Um-Rs/(sigma Ls)Psi1[t]+
omegas Psi2[t]+Rs Lm/(sigma Lr Ls) Psi3[t], Psi2'[t] ==-
omegas Psi1[t]-Rs/(sigma Ls)Psi2[t]+ Rs Lm/(sigma Lr Ls)
Psi4[t],Psi3'[t]==Rr Lm/(sigma Lr Ls)Psi1[t]-Rr/(sigma
Lr)Psi3[t]+Psi5[t]omegas Psi4[t],Psi4'[t]==Rr Lm/(sigma Lr
Ls)Psi2[t]-Psi5[t]omegas Psi3[t]-Rr/(sigma Lr)Psi4[t],
Psi5'[t]==3 Lm p^2/(sigma Lr Ls)/(2 J omegas)(Psi1[t]
Psi4[t]-Psi2[t] Psi3[t]), Psi1[0]==0, Psi2[0]==0,
Psi3[0]==0, Psi4[0]==0, Psi5[0]==1}, {Psi1,Psi2,Psi3,Psi4,
Psi5},{t,0,tFinal},MaxSteps->Infinity]
```

Plot the torque–angular speed curve shown in **Fig. 9.7**

```
In[4]:=ParametricPlot[Evaluate[ {omegas(1-Psi5[t]),-
Psi5'[t]} /. Run1], {t,0,tFinal},PlotRange->All]
```

Plot the currents in **Fig. 9.8**: first, the d-axis current

```
In[5]:=Plot[Evaluate[ {Psi1[t]/(sigma Ls)-Psi3[t] Lm/(sigma
Lr Ls)} /. Run1],{t,0,.5}]
```

then, the D-axis current

```
In[6]:=Plot[Evaluate[ {-Lm Psi1[t]/(sigma Ls Lr)+Psi3[t]/
(sigma Lr)} /. Run1],{t,0,.5}]
```

phase a current

```
In[7]:=Plot[Evaluate[ {Cos[omegas t](Psi1[t]/(sigma Ls)-
Psi3[t] Lm/(sigma Lr Ls))-Sin[omegas t](Psi2[t]/(sigma Ls)-
Psi4[t] Lm/(sigma Lr Ls))} /. Run1],{t,0,.5},PlotPoints->
500]
```

and phase A current

```
In[8]:=Plot[Evaluate[ {Cos[omegas t-(1-Psi5[t])omegas t](-
Lm Psi1[t]/(sigma Ls Lr)+Psi3[t]/(sigma Lr))-Sin[omegas t-
(1-Psi5[t])omegas t] (- Lm Psi2[t]/(sigma Ls Lr)+Psi4[t]/
(sigma Lr))} /. Run1],{t,0,.5},Plot Points->500]
```

Repeat the computation for a ten times smaller inertia

```
In[9]:=J=0.00015;tFinal=.2;
```

```
In[10]:=Run2=NDSolve[{Psi1'[t]==Um-Rs/(sigma Ls)Psi1[t]
+omegas Psi2[t]+Rs Lm/(sigma Lr Ls) Psi3[t], Psi2'[t]==-
omegas Psi1[t]-Rs/(sigma Ls)Psi2[t]+Rs Lm/(sigma Lr Ls)
Psi4[t],Psi3'[t]==Rr Lm/(sigma Lr Ls)Psi1[t]-Rr/(sigma
Lr)Psi3[t]+Psi5[t]omegas Psi4[t],Psi4'[t]==Rr Lm/(sigma Lr
Ls)Psi2[t]-Psi5[t]omegas Psi3[t]-Rr/(sigma Lr)Psi4[t],
Psi5'[t]==3 Lm p^2/(sigma Lr Ls)/(2 J omegas)(Psi1[t]
Psi4[t]-Psi2[t] Psi3[t]),Psi1[0]==0,Psi2[0]==0,Psi3[0]==0,
Psi4[0]==0,Psi5[0]==1}, {Psi1,Psi2,Psi3,Psi4,Psi5},{t,0,
tFinal},MaxSteps->Infinity]
```

The torque–angular speed curve, shown in **Fig. 9.9** is obtained by executing

```
In[11]:=ParametricPlot[Evaluate[ {omegas(1-Psi5[t]),-
Psi5'[t]} /. Run2],{t,0,tFinal},PlotRange->All]
```

The currents in **Fig. 9.10** will be plotted next. First, the d-axis current

```
In[12]:=Plot[Evaluate[ {Psi1[t]/(sigma Ls)-Psi3[t] Lm/
(sigma Lr Ls)} /. Run2],{t,0,.2},PlotPoints->500,
PlotRange-> {0,1.5}]
```

then, the D-axis current

```
In[13]:=Plot[Evaluate[ {-Lm Psi1[t]/(sigma Ls Lr)+Psi3[t]/
(sigma Lr)} /. Run1],{t,0,.2},PlotPoints->500,PlotRange->
{-1.,.2}]
```

next, phase a current

```
In[14]:=Plot[Evaluate[ {Cos[omegas t](Psi1[t]/(sigma Ls)-
Psi3[t] Lm/(sigma Lr Ls))-Sin[omegas t](Psi2[t]/(sigma Ls)-
Psi4[t] Lm/(sigma Lr Ls))} /. Run1],{t,0,.2},PlotPoints->
500]
```

and, finally, phase A current

```
In[15]:=Plot[Evaluate[ {Cos[omegas t-(1-Psi5[t])omegas t]
```

```
(-Lm Psi1[t]/(sigma Ls Lr)+Psi3[t]/(sigma Lr))-Sin[omegas
t-(1-Psi5[t])omegas t](- Lm Psi2[t]/(sigma Ls Lr)+Psi4[t]/
(sigma Lr))} /. Run1],{t,0,.2},Plot Points->500,PlotRange->
{-1.,.7}]
```

References

1. M. G. Say: *Alternating Current Machines*, John Wiley, New York, 1983
2. M. G. Say, E. O. Taylor: *Direct Current Machines*, Pitman, 1983
3. S. Wolfram: *Mathematica–A System for Doing Mathematics by Computer*, Second Edition, Addison-Wesley, 1991
4. R. Wolf: *Uvod u Teoriju Elektricnih Strojeva*, Skolska Knjiga, Zagreb, 1975
5. C. S. Siskind: *Direct Current Machinery*, McGraw-Hill, New York, 1952
6. V. Ostovic: *Dynamics of Saturated Electric Machines*, Springer Verlag, New York, 1989
7. B. J. Chalmers: *Electromagnetic Problems of Electric Machines*, Chapman and Hall, London, 1965
8. R. Richter: *Elektrische Maschinen I*, Birkhauser Verlag, Basel und Stuttgart, 1967
9. R. Richter: *Elektrische Maschinen II*, Springer Verlag, Berlin, 1930
10. D.C. Jiles and D.L. Atherton: Theory of Ferromagnetic Hysteresis, *Journal of Magnetism and Magnetic Materials*, Vol. 61 (1986), pp. 48–60, North-Holland, Amsterdam
11. G.R. Slemon: *Magnetoelectric Devices*, John Wiley, New York, London, Sydney, 1966
12. E.F. Fuchs, A.J. Vandenput, J. Hoell and J.C. White: Design Analysis of Capacitor-Start, Capacitor-Run Single-Phase Induction Motors, *IEEE Transactions on Energy Conversion*, Vol. 5, No. 2, June 1990, pp. 327–336
13. V. Ostovic: *Capacitor Braking Method of Induction Machines*, M.S. Thesis, University of Zagreb, 1978
14. B. Jurkovic and V. Ostovic: Capacitor Braking Method of Induction Machines, part one: Definition of Torque-Speed Curve, *Elektrotehnika, Zagreb*, 2/1980, pp. 261–266
15. V. Ostovic and B. Jurkovic : Capacitor Braking Method of Induction Machines, part two: Energy Flow and Definition of Parameters , *Elektrotehnika, Zagreb*, 2/1981, pp. 77–83
16. B. Jurkovic and V. Ostovic: Capacitor Braking Method of Induction Machines, part three: Single Phase Braking, *Elektrotehnika, Zagreb*, 1–2/1982, pp. 3–12

Index

A.C. commutator motors, 268
active parts of an electric machine, 52, 74
air gap, 9, 12, 18, 19, 25, 26, 32, 35, 36, 51,
    power, 285
    reluctance, 38
    specific permeance, 190, 224
    variable width, 115
Ampère's law, 8, 11, 14, 40, 51, 55, 58, 60, 80, 85, 154
ampereconductors, 80
ampereturns, 9, 13, 15, 24, 28, 32, 35, 40, 51, 55, 60, 69, 80, 86, 129
angle
    electrical, 78
    geometrical, 78
Animate, 454, 457, 460, 466, 476, 477
armature reaction, 156, 181, 249, 354
auxiliary phase, 318

B–H curve: *see* Magnetization curve
Biot–Savart's law, 196
boundary conditions, 9, 60, 69
brushes, 131, 346, 354, 356, 365, 368

capacitor motor, 323

Carter factor, 151, 161, 314
charge, 8
chopper, 369
chorded coil, 86
circular
    currents, 74
    diagram, 291
coercive force, 33
coil pitch, 67
commutation, 356
    linear, 357
    overcommutation, 357
    strong overcommutation, 357
    undercommutation, 356
    strong undercommutation, 356
commutating pole, 156
commutator, 271
    pitch, 133
    segment, 132
consequent poles, 343
constant
    active power curve, 388
    power factor curve, 389
copper loss, 58, 72, 187, 288, 398
core, 13, 18, 25, 29, 37, 44
    loss, 25, 28

current
    bar, 138
    density, 2, 8, 19, 57, 61, 64, 65, 70
    discontinuous, 373
    fed induction machine, 299
    magnetizing, 284
    no load field, 378
    rectangular, 126, 256
    ring, 139
    sheet, 79, 84, 277
    source, 23, 35

Dahlander, 344
d-axis, 146, 166
D.C. generators,
    compoundly excited, 363
    separately excited, 358
    series excited, 362
    shunt excited, 360
D.C. motors
    compoundly excited, 367
    separately excited, 365
    series excited, 366
    shunt excited, 365
deep bars, 69
diamagnetic materials, 3
dielectric constant, 1, 2
differential equation, 6
    ordinary, 59
    partial, 58
    voltage, 44, 47, 321
distribution factor, 76
domains, 4, 5, 28, 250
duty cycle, 371

eddy currents, 28, 57, 65, 72

loss, 30
effective number of turns, 96
efficiency, 288
electric
    current, 2, 7, 14, 19
    field strength, 19, 58
    time constant, 44, 47, 402
electrical conductivity, 1, 8, 19, 30, 63
electromechanical time constant, 403
electromotive force, 9, 19
elementary currents, 4
end winding, 51, 73
    leakage flux, 51, 146
equal area criterion, 385
equivalent
    circuit, 185, 221, 228, 282, 321
    resistance, 68
external characteristic, 288

Faraday's law, 15, 18, 57
ferromagnetic materials, 3
FindMinimum, 445, 486
FindRoot, 434, 438, 489
finite element method, 191, 199, 354
flux
    density, 3, 9, 13, 19, 28, 33, 50, 63
    differential, 181
    electric, 7
    fringing, 16, 149
    lines, 11, 38, 51, 60, 67, 146
    linkage, 17, 35, 42, 52, 68, 187
    magnetic, 7, 9, 14, 16, 30, 51
    main, 147
    source, 23
    total, 185
    tube, 16, 19, 22, 37

Index 517

zig-zag, 146
force, 180, 236, 249
Fourier, 68, 186, 198, 201, 212, 231
full pitch, 88
fundamental harmonic, 69

Gramme, 346

harmonics
    of main inductance, 199
    odd, 95
    spatial, 68, 89, 140
    third, 126
    time, 43, 86
heat
    conduction, 376
    convection, 376
    radiation, 376
heteropolar machine, 14
homopolar machine, 14, 363
hydrogenerator, 167, 375
hysteresis, 6, 33, 50
    loss, 28

induced voltage, 28, 35, 230, 349
inductance, 17, 24, 28, 42, 45, 56, 66
    leakage, 51, 186
    main, 187, 190
    mutual, 188, 204, 209
infinite bus, 302
inrush current, 47
intensity of magnetization, 4, 250
`Interpolation`, 437, 445, 488, 505, 508
inverse air gap width, 198, 203
inverter, mechanical, 131

Jacobi, 346

kinetic energy, 238
Kirchhoff's voltage law, 9, 32
Kloss equation, 299
Kraemer drive, 339

linear current density, 79
Langevin's equation, 5
linear medium, 2

magnetic circuit, 9, 14, 18, 22, 25, 27, 32, 35, 40, 46, 50, 187
magnetic equivalent circuit method, 19, 36, 191
magnetic field
    coenergy, 25, 235
    energy, 19, 24, 28, 188, 235, 238
    rotating, 109
    strength, 1, 4, 8, 11, 14, 19, 22, 28, 33, 52, 54, 63
magnetic
    neutral, 156, 279
    permeability, 1, 8, 12, 20, 63
    scalar potentials, 40
magnetization curve
    anhysteretic, 5, 31
    hysteretic, 6
    linearized, 3, 44
    nonlinear, 4, 16, 17, 36, 64, 191
magnetomotive force (mmf), 8, 21, 74, 217
    distribution, 86, 103, 147, 153
    drop, 22, 26
    elliptical, 221
    pulsating, 98, 222

rotating, 222
source, 28
wave, 113
main
   flux, 51
   reactance, 224
Maxwell's equations, 1, 14, 58, 230
mechanical power, 226
motional voltage, 17, 368

NDSolve, 440, 441, 507, 509, 511
negative sequence, 113, 119, 124
   voltage, 320
node potential method, 40
nonlinear
   medium, 2
   circuit, 38
normalized conductor height, 67

Ohm's law, 2, 19
orthogonality, 145

parallel paths, 132
paramagnetic materials, 3
per unit representation, 390
permanent magnets, 3, 33, 395
permeance, 35, 53
pitch
   diametral, 189
   factor, 76
phase, 61
pole
   pair, 78
   pitch, 78, 83
positive sequence, 113, 119, 220
   voltage, 253

power chart, 389
pulsating mmf, 117
pulsation factor, 348

q-axis, 146

R–L circuit, 23
rectifier, 369
reference frame, 128
   rotating, 128
ReIm.m, 496, 502
relative permeability, 3, 12, 20,
   of permanent magnets, 35
reluctance, 19, 28, 30, 35, 38, 187, 191
   of permanent magnets, 35
residual flux density, 33
resistance, 19, 51
resonance, 243
Roebel bar, 74
rotating field, 109
RungeKutta.m, 435, 436

salient poles, 198, 312
saturation of the iron, 5, 12, 42, 63, 115
Scherbius drive, 263
self excitation, 308
short circuit ratio, 392
silicon, 31
Simplify, 493–494
Simpson's rule, 21
single phase induction machine, 318
skewing, 208
   factor, 209
skin
   depth, 59
   effect, 55, 58, 296

Index

slip, 139
    pull out, 298
slot
    ampereturns, 51
    harmonics, 203
    leakage flux, 51, 55, 185
    leakage inductance, 52, 56, 208
    opening, 147
    pitch, 78
    specific permeance, 53
sparking, 279
squirrel cage, 69, 77, 137, 280
    bars, 77
    rings, 77
stability
    of an induction machine, 303
    of a synchronous machine, 386
starting torque, 69, 232
state variables, 236, 402
    electromagnetic, 236
    mechanical, 236
stranded conductors, 73
subsynchronous drive, 341
subtransient
    period, 424
    reactance, 426
    time constant, 428
superposition law, 143, 188
synchronism, 178
synchronous machine
    overexcited, 375, 378
    underexcited, 375, 379
synchronous
    impedance, 382
    reactance, 377, 384

tensors, 2
Thevenin, 41
tooth, 18, 21, 54
torque, 187
    angle, 378
    asynchronous, 254, 282, 304
    electromagnetic, 238, 351
    equation, 237
    function, 256
    maximum, 297
    pulsating, 242, 257
    reluctance, 242
    synchronous, 254, 305
transform matrix, 129, 332
transformer voltage, 17, 368
transient
    period, 424
    reactance, 426
    time constant, 428
transients, 397
`Transpose`, 441
transposed conductors, 74
`Trigonometry.m`, 493
`TrigReduce`, 493–494
turbogenerator, 167, 375

universal motor, 135, 368

V–I characteristic, 26
vector star, 120
voltage
    fed induction machine, 288
    source, 22, 35

waves
    rotating, 113

standing, 113
travelling, 117
Weiss's theory, 4
winding
    armature, 131, 153
    commutating pole, 153
    compensating, 153
    compound field, 154
    duplex, 132
    factor, 96
    function, 95, 188, 199, 204
    main field, 156
    multiplicity, 104
    series field, 153
    shunt field, 153
    simplex, 132
windings, 18
    commutator, 102
    concentrated, 76
    concentric, 76, 89, 169
    distributed, 76, 93
    double layer, 76
    lap, 76, 93
    single layer, 76
    wave, 76, 93
wound rotor induction motor, 281
    speed control of, 295, 337

yoke, 18